3rd International Conference on Mechanical Engineering and Automation Science (ICMEAS 2017)

IOP Conference Series: Materials Science and Engineering
Volume 280

Birmingham, United Kingdom
13 – 15 October 2017

ISBN: 978-1-5108-5498-7
ISSN: 1757-8981

Printed from e-media with permission by:

Curran Associates, Inc.
57 Morehouse Lane
Red Hook, NY 12571

CURRAN ASSOCIATES INC. proceedings.com

Some format issues inherent in the e-media version may also appear in this print version.

Printed by Curran Associates, Inc. (2018)

Institute of Physics
Dirac House, Temple Back
Bristol BS1 6BE UK

Phone: 44 1 17 929 7481
Fax: 44 1 17 920 0979

techtracking@iop.org

Additional copies of this publication are available from:

Curran Associates, Inc.
57 Morehouse Lane
Red Hook, NY 12571 USA
Phone: 845-758-0400
Fax: 845-758-2633
Email: curran@proceedings.com
Web: www.proceedings.com

Table of contents

Volume 280

3rd International Conference on Mechanical Engineering and Automation Science (ICMEAS 2017)

13–15 October 2017, Birmingham, UK

Accepted papers received: 4 December 2017
Published online: 20 December 2017

Preface

Papers

Mechanical

Automation

IOP Conf. Series: Materials Science and Engineering **280** (2017) 011001 doi:10.1088/1757-899X/280/1/011001

PREFACE

ICMEAS 2017 was held in University of Birmingham, UK, from 13-15 October, 2017. ICMEAS 2017 was Co-organized by University of Birmingham and Hong Kong Society of Mechanical Engineers. The conference provides a useful and wide platform both for display the latest research and for exchange of research results and thoughts in Mechanical Engineering and Automation Science. The participants of the conference were from almost every part of the world, with background of either academia or industry, even well-known enterprise. The success and prosperity of the conference is reflected high level of the papers received.

The proceedings are a compilation of the accepted papers and represent an interesting outcome of the conference. This book cover 3 chapters: Mechanical, Automation and Design & Manufacturing.

We would like to acknowledge all of those who supported ICMEAS 2017. Each individual and institutional help were very important for the success of this conference. Especially we would like to thank the organizing committee for their valuable advices in the organization and helpful peer review of the papers.

We sincerely hope that ICMEAS 2017 will be a forum for excellent discussions that will put forward new ideas and promote collaborative researches. We are sure that the proceedings will serve as an important research source of references and the knowledge, which will lead to not only scientific and engineering progress but also other new products and processes.

Prof. Kyle Jiang, University of Birmingham, UK

IOP Conf. Series: Materials Science and Engineering **280** (2017) 011002 doi:10.1088/1757-899X/280/1/011002

COMMITTEES

Conference Chairman

Prof. Kyle Jiang, University of Birmingham, UK

Conference Committee Co-Chairs

Prof. Du Ruxu,The Chinese University of Hong Kong, Hong Kong

Prof. Ian McAndrew, Embry Riddle Aeronautical University, UK

Prof. Kai Cheng, Brunel University, UK

Program Committee Chair

Prof. Stefan Dimo, University of Birmingham, UK

Prof. Yong Zhao, Nazarbayev University, Kazakhstan

Local Chair

Dr. Raya Al-Dadah, University of Birmingham, UK

Advisory Committee

Prof. Hui Ding, Harbin Institute of Technology, China

Prof. Enrico Vezzetti, Politecnico di Torino, Italy

Prof. Egon Müller, Technische Universität Chemnitz, Germany

International Technical Committee

Prof. Josua Meyer, University of Pretoria, South Africa

Prof. Sandy To, Hong Kong Polytechnic University, Hong Kong

Prof. Antonis I.G. Vardulakis, Aristotle University of Thessaloniki, Greece

Prof. David Barton, University of Leedsm, UK

Prof. Shinn-Liang Chang, National Formosa University, Taiwan

Prof. Tang Sai Hong, Universiti Putra Malaysia, Malaysia

IOP Conf. Series: Materials Science and Engineering **280** (2017) 011003 doi:10.1088/1757-899X/280/1/011003

Peer review statement

All papers published in this volume of *IOP Conference Series: Materials Science and Engineering* have been peer reviewed through processes administered by the proceedings Editors. Reviews were conducted by expert referees to the professional and scientific standards expected of a proceedings journal published by IOP Publishing.

IOP Conf. Series: Materials Science and Engineering **280** (2017) 012001 doi:10.1088/1757-899X/280/1/012001

Numerical and Experimental Investigations on the Hydrodynamic Performance of a Tidal Current Turbine

Xiaohui SU[1,a], Jiantao Zhang[1], Yong Zhao[2,b] , Huiying ZHANG[3], Guang ZHAO[3] and Yao CAO[3]

[1]School of Hydraulic Engineering, Dalian University of Technology, China, 116023
[2]School of Engineering, Nazarbayev University, Astana, Republic of Kazakhstan
[3]School of Energy and Power Engineering, Dalian University of Technology, China, 116023

E-mail:[a] sxh@dlut.edu.cn; [b]yong.zhao@nu,edu,kz

Abstract. In this paper, numerical and experimental investigations are presented on the hydrodynamic performance of a horizontal tidal current turbine (TCT) designed and made by our Dalian University of Technology (DUT) research group. Thus it is given the acronym: DUTTCT. An open source CFD solver, called PimpleDyMFoam, is employed to perform numerical simulations for design analysis, while experimental tests are conducted in a DUT towing tank. The important factors, including self-starting velocity, tip speed ratio (TSR) and yaw angle, which play important roles in the turbine output power, are studied in the investigations. Results obtained show that the maximum power efficiency of the newly developed turbine (DUTTCT) could reach up to 47.6% and all its power efficiency is over 40% in the TSR range from 3.5 to 6; the self-starting velocity of DUTTCT is about 0.745m/s; the yaw angle has negligible influence on its efficiency as it is less than 10°.

1. Introduction

Tidal current energy has been receiving more and more attention since the late 1990's, mainly due to four reasons: the increasing demand for electricity from all human activities; the depletion of fossil energy; the increasingly stringent regulations on CO2 emission due to climate change considerations and the last, also the most important one, the significant improvement of energy conversion technology [1-7]. The technology of energy conversion in the specialized field of tidal current energy is different from that of the tidal barrage, which is used to exploit the resources of the tidal range. The tidal current energy is harnessed by an underwater turbine located in a tidal current. Compared with the tidal barrage technology, the newly developed turbine technology has at least two advantages: 1) there is no need to build a barrage which requires a large amount of capital investment; 2) there is little negative impact on the environment and ecology in the neighborhood of the turbine.

The DUTTCT is a horizontal TCT which is designed and made by the DUT research group [8]. The DUTTCT is designed for use in the Bohai sea and the Huanghai sea in China, where there is low tidal current velocity ranging from 1m/s to 2m/s. At the current stage, the turbine is only designed for the purpose of theoretical and experimental studies. Numerical and experimental investigations have been conducted for examining the turbine hydrodynamic performance. Here the methods and results of the investigations are detailed in this paper.

IOP Conf. Series: Materials Science and Engineering **280** (2017) 012001 doi:10.1088/1757-899X/280/1/012001

2. Experimental Tests

The sketch of the DUTTCT impeller is shown in Figure 1, and its detailed parameters are shown in Table 1.

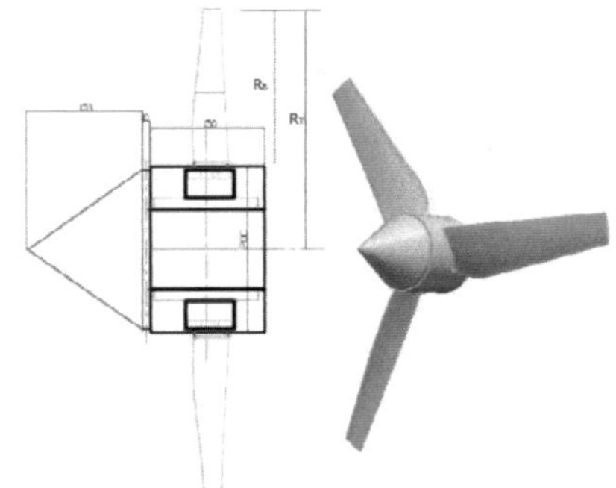

Figure 1. The sketch of DUTTCT impeller
(left: geometry design; right: 3D model)

Table 1. The blade parameters of DUTTCT

item	parameters
Radius of impeller	R_T=0.57m
Length of blade	R_B=0.47m
Number of blades	Z_T=3
Swept area	S_T=1.02m^2

According to the experiments planned, in each TSR and yaw experiment, a predetermined flow speed is prescribed, and the torque of magnetic powder brake is then adjusted in order to obtain the required turbine rotating speed, thus the predetermined TSR. The signals of rotational speeds and torques during the whole process are recorded. The time averaged torque and rotating speed are used in experimental data analysis. The characteristic curves of energy conversion efficiency vs. tip speed ratio are shown in Figure 2.

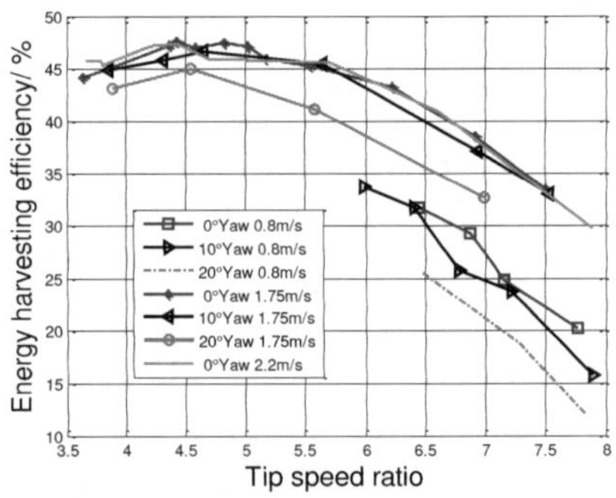

Figure 2. Efficiencies versus TSR for the DUTTCT.

The results show that energy conversion efficiency increases as TSR increases from 3.5 to 4.5 and then decreases as TSR further increases from 4.5 to 7.5. The optimum tip speed ratio is about 5 for the proposed turbine and the corresponding efficiency is 47.6% in the experiment. The lowest tip speed ratio is limited to 3.64 since the corresponding rotation speed, 110.6 rpm, is beyond the measurement range of the system.

Yaw effects are also investigated in this section. There are two sets of data for yaw effects obtained by experiments corresponding to incoming velocities at 0.8m/s and 1.75m/s, respectively. The yaw angle for each incoming flow is set to 0° (no yaw angle), 10° and 20°, respectively. Figure 2 shows that energy conversion efficiency with 10° yaw, decreases slightly, by less than 1%, compared with the corresponding result of the zero-yaw case for an incoming velocity at 0.8 m/s. As for the case of incoming velocity at 1.75 m/s, the efficiency value does not decrease too much, by only about 2-3%, compared with the value of zero-yaw angle. There is a maximum value, at about 45% with TSR = 4.5 for all cases. When the yaw angle increases to 20°, the situations for the two incoming velocities are very different and their efficiency difference is large by comparison with the results of lower yaw angles. For example, for the case of incoming velocity at 0.8 m/s, the profile of efficiency at 20° yaw angle is almost parallel to that of zero yaw angle and the efficiency of the former is lower than the latter by about 5-6%. With incoming velocity at 1.75 m/s and a yaw angle of 20°, different from the smooth characteristic curves for zero yaw angle or 10° yaw angle as TSR varies from 3.5 to 6, a sharp profile of efficiency is obtained for the same TSR range. These indicate that larger yaw angles will have greater negative impacts on the conversion of energy, which should be avoided during operations.

3. Numerical Simulations and Investigations

Numerical simulations and investigations are conducted after the experiment for the purpose of mutual validation with experimental data and supplementing incomplete experimental data, as well as for the in-depth analysis of flow patterns. For example, there is a lack of data for working characteristic curves as TSR is less than 3.64 in the experiment due to the mismatch of magnetic powder brake.

In this paper, the PimpleDyMFoam, which is an ALE solver based on the hybrid PISO-SIMPLE (PIMPLE) numerical solution algorithm and dynamic meshes (DYM) with the AMI from the open source software OpenFOAM4.x [9] and its predecessor called TurboDyMFoam, is used to perform the numerical simulations. In the following subsections, the details of the governing equations for fluid flows used in the software and numerical simulations could be found in Refs. [10-14].

3.1. Computational domain of the DUTTCT

Figure 3 shows the rotation direction of the blades and the arbitrary mesh interface (AMI) boundary in the computational domain. Pink cylinder denotes the whole computational domain, the surface of yellow green cylinder the AMI boundary, and the red part the impeller. The dimensions of the computational domain are shown in Table 2.

Table 2 The sizes of computational domain

item	sizes
Diameter of outer cylinder	D_O=5m
Height of outer cylinder	H_o=8.5m
Diameter of inner cylinder	D_i=3.2m
Height of inner cylinder	H_i=4.6m
Diameter of AMI cylinder	D_{AMI}=2m
Height of AMI cylinder	H_{AMI}=0.5327m

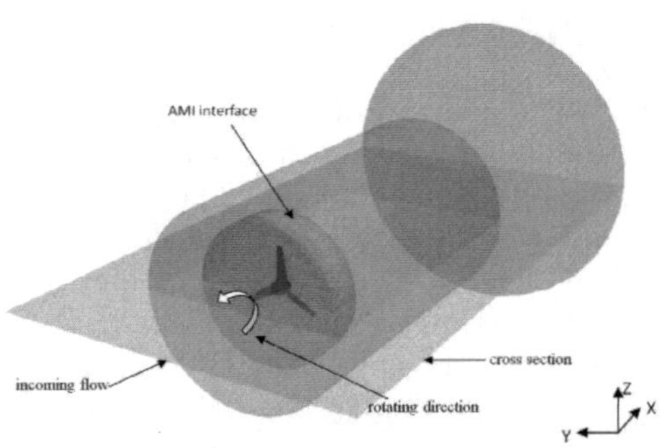

Figure 3 The sketch of AMI for DUTTCT in computational domain.

(The pink cylinder denotes the computational domain; the yellow green cylinder surface represents the AMI; the green cylinder indicates the refinement domain around the turbine and wake region)

3.2. The simulation cases

In the simulations performed to obtain its characteristic curves, both incoming flow velocity and rotational angular velocity are fixed in each simulation and the TSR is varied from 2 to 7 which are summarized in Table 3.

Table 3 Simulation cases for working characteristic curves

Case number	Current speed ($m \cdot s^{-1}$)	Tip Speed Ratio (t)
Case-2-c-t* 1-255	1,1.25,1.5,1.75,2.0	2,2.5,3,3.5,4, 4.5,5,5.5,6,6.5,7

* Note: the affiliated definitions of case-2-c-t: '2' denotes working a characteristic simulation case; 'c' represents the current speed and 't' denotes the tip speed ratio. For example, case-2-1.25-2.5 means the working characteristic simulation case with a current speed of 1.25 *m/s* and a tip speed ratio of 2.5. With those current speeds and TSRs shown in Table 3, there are totally 55 cases for working characteristic simulations.

3.3. Numerical results

The working characteristic curves show three most important factors of the turbine: the highest efficiency of the turbine; the optimal TSR; and the high efficiency range of TSR. After the completion of all simulation cases, the characteristic curves of the DUTTCT are summarized in Figure 4. It shows that the highest efficiency is about 45% at an optimal TSR of 5.5, and for TSR ranging from 3.5 to 7, the efficiency of the DUTTCT is above 40%. The output efficiency does not show significant variations with various incoming velocities. For the purpose of comparison, those experimental data given in the towing tank experiment are also plotted in Figure 4. A good agreement between the simulation and experimental results is obtained for larger TSRs from TSR=5. However, the efficiencies obtained from experiment shown in Figure 4 are slightly larger than the numerical results in the range from TSR=3 to 5. Nevertheless, based on the curves obtained in Figure 4, the DUTTCT has indeed good working performance with various low incoming velocities in a wide range of TSRs.

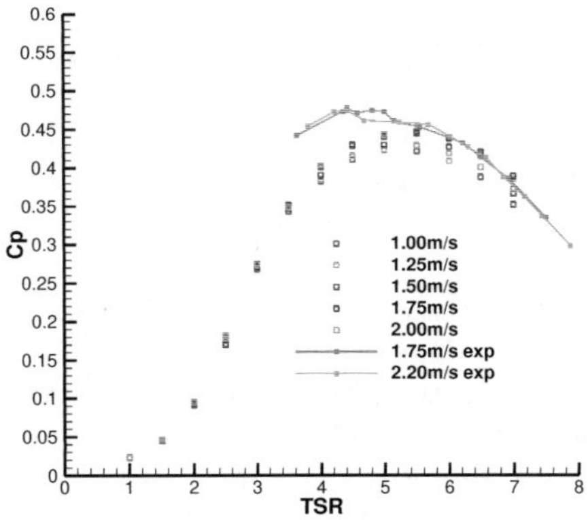

Figure 4. The working characteristic curves of the DUTTCT.

4 Conclusions and Discussions

In this paper, the hydrodynamic performance of the DUTTCT has been investigated experimentally and numerically. The findings obtained in the numerical and experimental investigations are very valuable for further improvement of the turbine performance and also provide valuable experience for future research work on tidal current turbines. The major conclusions about the DUTTCT are as follows:

1) The results show that the maximum power conversion efficiency of the DUTTCT could reach up to 47.6% and all the power efficiencies are over 40% for TSRs ranging from 3.5 to 6. Its self-starting velocity is about 0.745m/s, while the self-starting torque is about 2 Nm. The yaw angle does not affect its efficiency significantly when it is less than 10°. The DUTTCT has good working performance with low free-stream velocities in a wide range of TSRs.

2) Numerical simulations and experiment tests should be conducted simultaneously to complement and validate each other. For example, the power characteristic curves obtained from experiments lack lower-range TSR values because of the mismatched magnetic powder brake. However, numerical simulation results provide valuable low range TSR data.

Acknowledgement

DUTTCT project is financially Supported by the Natural Science Foundation of Liaoning Province, China (Project No: 2015020629). The financial support is gratefully acknowledged.

Reference

[1] Uihlein A. and Magagna D. Wave and tidal current energy - A review of the current state of research beyond technology. Renewable and Sustainable Energy Reviews, 2016, Vol.58, 1070-1081.

[2] Liu H.W., Ma S., Li W., Gu H.G., Lin Y.G., Sun X.J. A review on the development of tidal current energy in China. Renewable and Sustainable Energy Reviews, 2011, Vol.15(2), 1141-1146.

[3] Li D., Wang S.J.,Yuan P. An overview of development of tidal current in China: Energy resource, conversion technology and opportunities. Renewable and Sustainable Energy Reviews, 2010, 14, 2896-2905.

[4] Li, W., Zhou H.B., Liu H.W., Lin Y.G., Xu Q.K. Review on the blade design technologies of tidal current turbine. Renewable and Sustainable Energy Reviews, 2016, 63, 414(9)

[5] Li Y. On the definition of the power coefficient of tidal current turbines and efficiency of tidal current turbine farms. Renewable Energy, Volume 68, August 2014, 868-875.

IOP Conf. Series: Materials Science and Engineering **280** (2017) 012001 doi:10.1088/1757-899X/280/1/012001

[6] Zhao Y. and Su X.H. Tidal Energy: Technologies and Recent Developments. IEEE International Energy Conference& Exhibition, Bahrain, 18-22 December 2010, 618-623.
[7] Zhou Z.B., Benbouzid M., Charpentier J.F., Sculler F., Tang T.H. Developments in large marine current turbine technologies - A review. Renewable and Sustainable Energy Reviews, 2017, 71, 852-858.
[8] Zhang K.L. Numerical Study of Horizontal Tidal Current Turbine by using PimpleDyMFoam. Dalian University of Technology, 2013, Master thesis.
[9] OpenFOAM. OpenFOAM Programmer's Guide. 2011.
[10] Mathematics, Numerics, Derivations and OpenFOAM®, Tobias Holzmann, Book, December 2016, DOI: 10.13140/RG.2.2.27193.36960
[11] Jasak, H. (1996). Error analysis and estimation for the finite volume method with applications to fluid flows. PhD thesis in Imperial College: University of London.
[12] Jasak H. and Gosman A.D. Element residual error estimate for the finite volume method. Computers and Fluids, 2003, 32(2), 223-248.
[13] Jasak H., Weller H.G., Gosman A. D. High resolution NVD differencing scheme for arbitrarily unstructured meshes. International Journal for Numerical Methods in Fluids, 30 September 1999, 31(2), 431-449.
[14] Weller H.G., Tabor G., Jasak H., Fureby C. A tensorial approach to computational continuum mechanics using object-oriented techniques. Computers in Physics, November 1998, 12(6), 620-631.

IOP Conf. Series: Materials Science and Engineering **280** (2017) 012002 doi:10.1088/1757-899X/280/1/012002

Numerical Study on Biaxial Seismic Performance of Reinforced Concrete Box Piers Based on OpenSees

Chuncheng Liu[1], Chunlei Lv[1], Chongyang Wang[2], Geng Yang[1], Dapeng Huang[1] and Yiling Zhang[1]

[1]Northeast Electric Power University Institute of Civil Engineering, Jilin, China
[2]Beijing Electric Power Transmission and Transformation Company, Beijing, China

1978968364@qq.com

Abstract. In order to study the seismic performance of reinforced concrete box piers under biaxial loading, the hysteretic model of reinforced concrete box piers was established by selecting appropriate material model, constitutive relation, structural element type, and the corresponding boundary conditions and loading method based on OpenSees. The numerical simulation of the ductility of reinforced concrete box piers with different axial compression ratio, stirrup ratio and slenderness ratio under biaxial horizontal loading was conducted. The results of biaxial quasi-static tests and that of numerical simulation were compared. The analysis showed that the hysteretic behaviour of reinforced concrete box piers under biaxial loading could be well simulated and the degradation of strength and stiffness of box piers in the process of bidirectional cyclic loading could be reflected by OpenSees on the basis of the appropriate material constitutive model and structural element type. Furthermore, the influence of reinforcement ratio and concrete strength on the ductility of reinforced concrete box pier was studied by OpenSees.

1. Introduction

With the rapid development of the large bridge engineering in the high seismic intensity area of Western China, the box pier has been widely used for its high stiffness mass ratio, strength mass ratio and good ductility. However, the research of the box pier mostly focuses on the pseudo static test under horizontal seismic action, and the seismic performance of the box pier under biaxial loading is still relatively weak [1] [2]. In fact, whether the movement of surface seismic wave or the seismic response of bridge structures are multidimensional while the earthquake occur, and the results of bridge damage show that the damage degree of bridge under biaxial loading was increased. It is mainly because of the mutual coupling of the load in different directions reduces the bearing capacity and deformation capacity of structure, and a pronounced gap exists as compared with the seismic performance of pier under horizontal uniaxial loading [3].

The push-over analysis, modal analysis and nonlinear dynamic analysis of structure can be realized with OpenSees. In addition, the bridge engineering, geotechnical engineering and other fields can be well simulated by OpenSees, which has attracted the attention of many scholars at home and abroad [4] [5]. Therefore, the numerical simulation of reinforced concrete box piers under bidirectional cyclic loading was conducted by selecting the appropriate material model, constitutive relation and structure unit type with OpenSees, and then the results of biaxial quasi-static tests and that of numerical simulation were compared. Finally, the influence of reinforcement ratio and concrete strength on the ductility of reinforced concrete box pier was studied.

2. Establishment and verification of model

2.1. General situation about biaxial quasi-static test

In this paper, the numerical simulation of the specimens B3, B4 and B5 in the biaxial quasi-static test of reinforced concrete box piers was conducted based on OpenSees analysis program. The detailed design parameters of the piers can be found in reference [6].

2.2. Establishment of model

In this paper, concrete02 which considers the tension of concrete is used to simulate concrete material, and the simulation of reinforcement adopts Steel02 in the model of reinforced concrete box pier based on OpenSees. And the three parameters that control the transition from elastic stage to hardening phase of Steel02 are 18.5, 0.925 and 0.15 according to the suggestions given by OpenSees program.

The bottom of reinforced concrete box pier was set as fixed end, and the top was selected as free end in the case of establishing box pier model, because the specimen was fixed by the base and the top loading device could slide freely when the biaxial quasi-static test was carried out. At the same time, a constant axial load is applied at the top of the pier, and the axial load remains unchanged in the simulation. In the terms of biaxial horizontal loading, the simplification of loading method was taken into account, and the displacement control was adopted during the whole process of loading. According to the test results, the yield displacement of each specimen was between 4 and 16 mm, so 2mm was taken as the displacement increment. And after yielding, the increment of displacement was 3mm. Each load cycle was once. The displacement amplitude ratio of the strong and weak axis of the section was 1:1. Figure 1 shows the loading system of pier.

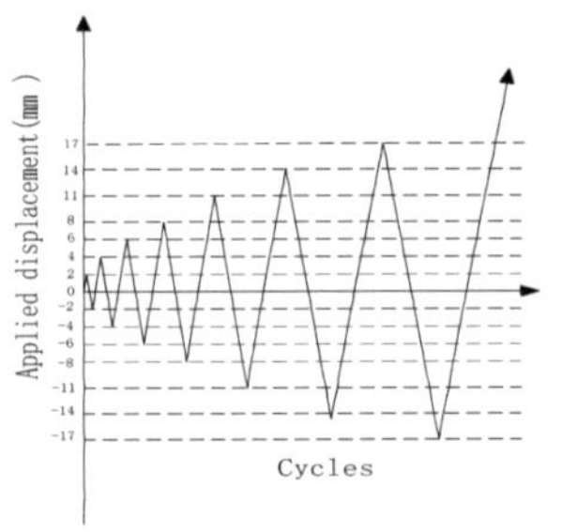

Figure 1. Loading protocol of bridge piers

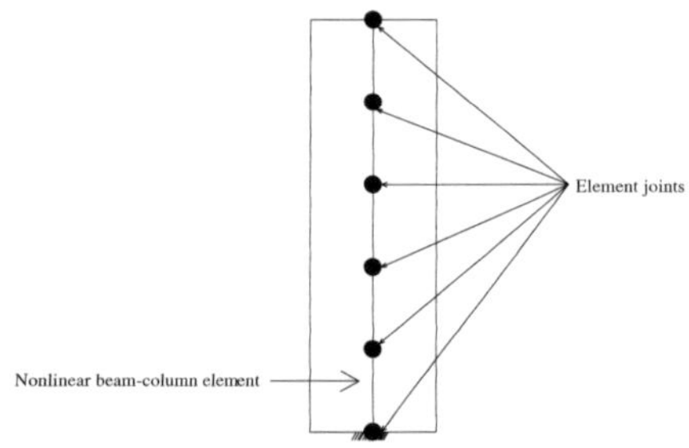

Figure 2. Model of reinforced concrete box piers

According to mechanical properties and geometrical features of pier, the steel reinforcement and concrete of the cross section were divided into fiber of concrete protective layer, fiber of core concrete and reinforcement fiber; the encrypted area and non-encrypted area of stirrup in the bottom of pier were distinguished. The pier was modular in structure, and the element type selected nonlinear beam-column. Moreover, a number of integration points were set along the axial direction of pier. And then the hysteretic analysis model of reinforced concrete box pier was established, as shown in Figure 2.

2.3. Calculated values compared with test values

The comparison of the hysteresis curves obtained by numerical simulation with that obtained with the test measurements of B3 specimen is presented in Figure 3 to Figure 5.

Results of the comparison show that the hysteretic analysis model of reinforced concrete box pier has fine precision with the simulation of specimens. The hysteresis curves obtained by the numerical simulation are in good agreement with the experimental results of reinforced concrete box piers, and the rising segment, the strength degradation and the pinch effect of curves can be simulated well. Moreover, the difference of bearing capacity and deformation in different loading directions of box pier under bidirectional cyclic loading is reflected, and the simulation error of y direction in cross section is larger. The error may result from the coupling effects of different loading directions have a greater influence on the bearing capacity and deformation of y direction in cross section, but overall, the agreement between the simulation and test demonstrates that simulation program for reinforced concrete box piers is reliable.

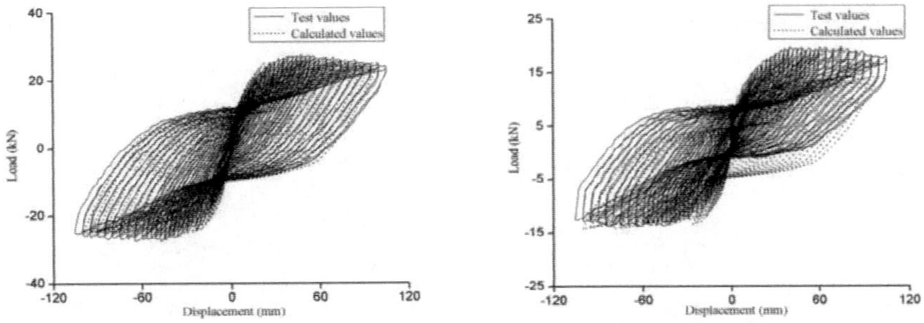

(a) x-direction (b) y-direction

Figure 3. Comparisons of the experimental and theoretical hysteretic curves of the bridge pier of B3

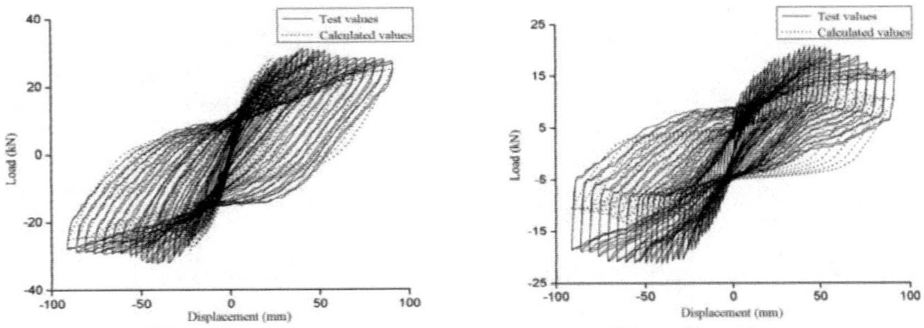

(a) x-direction (b) y-direction

Figure 4. Comparisons of the experimental and theoretical hysteretic curves of the bridge pier of B4

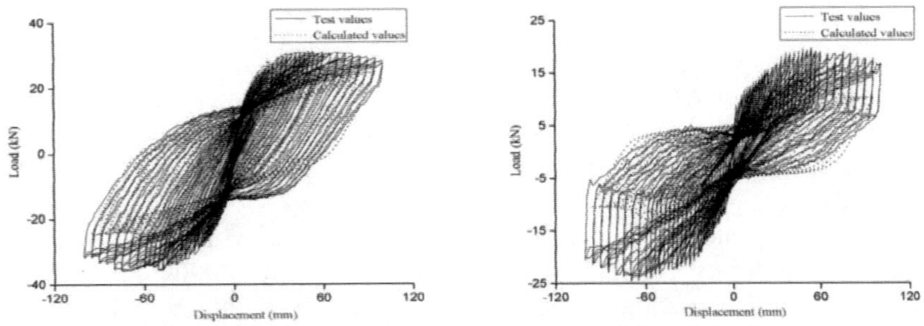

(a) x-direction (b) y-direction

Figure 5. Comparisons of the experimental and theoretical hysteretic curves of the bridge pier of B5

By comparing skeleton curve of simulation and test of B3 specimen, it could be found that the calculating results have accordance well with those experimental observations and the calculated curve can reflect the yield segment and descent stage of specimen accurately, as shown in Figure 6. For the destruction phase of skeleton curve, there are some differences between the calculated values and the test values because the bond slip between steel and concrete is not considered in the numerical simulation.

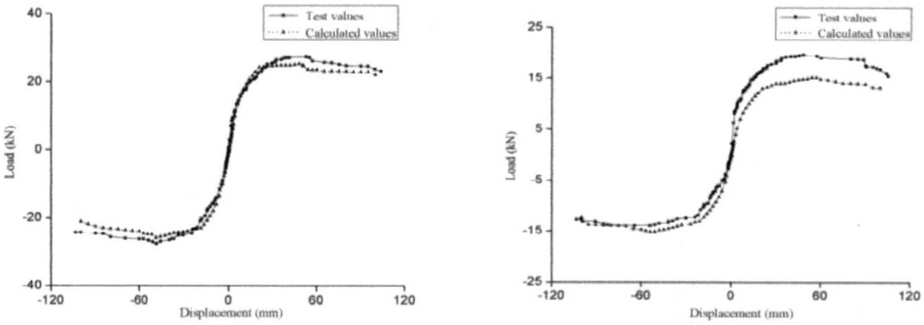

(a) x-direction (b) y-direction

Figure 6. Comparisons of the experimental and theoretical skeleton curves of the bridge pier of B3

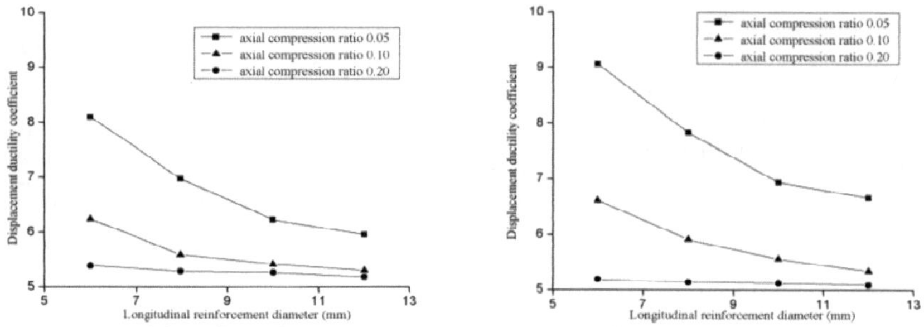

(a) x-direction (b) y-direction

Figure 7. Variation of displacement ductility coefficient with longitudinal reinforcement diameter

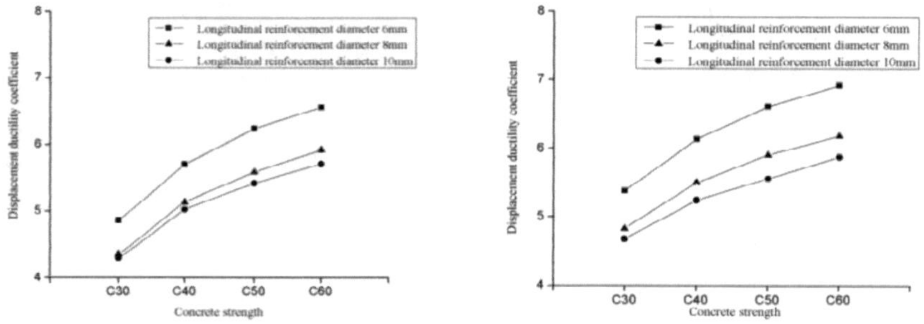

(a) x-direction (b) y-direction

Figure 8. Variation of displacement ductility coefficient with concrete strength

3. Parameter analysis

Based on the established hysteretic analysis model of reinforced concrete box pier, the effect of reinforcement ratio and concrete strength on the ductility of box pier was conducted for the further analysis of the ductility of reinforced concrete box piers under biaxial horizontal loading.

The ductility of structure under cyclic loading refers to the inelastic deformation capacity of the structure which can still maintain a certain bearing capacity. Ductility index is an important index to evaluate the seismic performance of structures, and ductility coefficient is usually used to reflect the deformation capacity of the specimens. The displacement ductility factor is

$$\mu = \Delta_u / \Delta_y \tag{1}$$

Where Δ_y = yield displacement; Δ_u = ultimate displacement.

3.1. Diameter of longitudinal reinforcement

The diameter of longitudinal reinforcement is one of the most important factors which affect the ductility of reinforced concrete box pier. Figure 7 shows the variation of displacement ductility coefficient in different loading direction with longitudinal reinforcement diameter when axial compression ratio is 0.05, 0.10 and 0.20, respectively. It can be seen that the displacement ductility coefficient in different loading direction with the increase of longitudinal reinforcement diameter decreases significantly, when the axial compression ratio is less than 0.20. This is because the damage of box pier is the failure of reinforcement when the axial compression ratio is relatively small. Although the yield displacement of box pier raises with the increase of diameter of longitudinal reinforcement, the change of the ultimate displacement of pier is smaller, this makes the ductility of piers decreased; The displacement ductility coefficient in different loading direction does not change obviously with the increase of longitudinal reinforcement diameter when the axial compression ratio is greater than 0.2, and the specimen showed brittleness characteristic.

3.2. Concrete strength

The concrete strength has important influence on the ductility of reinforced concrete box piers. In this paper, the concrete strength of C30, C40, C50 and C60 were selected to study. Figure 8 shows the variation of displacement ductility coefficient in different loading direction with concrete strength when longitudinal reinforcement diameter is different. It can be found that the displacement ductility coefficient in different loading direction with the increase of concrete strength rises significantly when the strength of concrete is C30 to C60, which is due to the greater ultimate displacement with the increase of concrete strength, but the growth of displacement ductility coefficient is slower.

4. Conclusions

This paper conducted the numerical simulation of biaxial quasi-static test and parametric analysis of reinforced concrete box piers on the basis of the existing material model and structural element in OpenSees. The following conclusions can be obtained:

(1)Based on the appropriate material model, the constitutive relation and the fiber beam-column element, the established fiber element model of reinforced concrete box pier can simulate the hysteretic curve and skeleton curve of box pier under biaxial loading well, and it can reflect the degradation of strength and stiffness of the bridge pier specimen during biaxial loading.

(2)The calculated curves of numerical simulation of the piers are in good agreement with the test results on the whole, and the calculated values can reflect the bearing capacity and deformation characteristics of the pier specimens under bidirectional cyclic loading accurately.

(3)The displacement ductility coefficient in different loading direction decreases with the increase of longitudinal reinforcement diameter, and rises with the increase of concrete strength. But the growth of displacement ductility coefficient with the increase of concrete strength is slower.

References

[1] Yeh Y K, Mo Y L and Yang C Y. Seismic performance of rectangular hollow bridge columns [J]. Struet. Engrg. 2002, 128(1): 60–80.

[2] Won K L and Sarah L B. Modeling residual displacements of concrete bridge columns under earthquake loads using fiber elements [J]. Journal of Bridge Engineering. 2010, 240–249.

[3] Zhenhua Dong, Xiuli Du and Qiang Han. Study on hysteretic model of RC bridge pier with rectangular hollow section under lateral-birectional cyclic load [J]. Engineering Mechanics. 2016, 33(4): 24–33(In Chinese).

[4] Kent D C and Park R. Flexural members with confined-concrete [J]. Journal of the Structural Division. 1971, 97(7): 1969–1990.

[5] Scott B D, Park R and Priestley M J N. Stress-strain behavior of concrete confined by overlapping hoops at low and high strain rates [J]. Aci Journal. 1982, 79(1): 13–27.

[6] Zhanghua Xia, Zhouhong Zong and Rumian Zhong. Seismic performance of reinforced concrete thin-walled piers with rectangular hollow cross-sections based on bi-axial quasi-static testing [J]. Journal of Southeast University (Natural Science Edition). 2013, 43(1): 180-187(In Chinese).

IOP Conf. Series: Materials Science and Engineering **280** (2017) 012003 doi:10.1088/1757-899X/280/1/012003

Study on Vibration of Heavy-Precision Robot Cantilever Based on Time-varying Glowworm Swarm Optimization Algorithm

T H Luo[1], S Liang[1] and C B Miao[1]

[1]School of Mechanotronics and Vehicle Engineering ,Chongqing Jiaotong University,Chongqing 400074,China

137377539@qq.com

Abstract. A method of terminal vibration analysis based on Time-varying Glowworm Swarm Optimization algorithm is proposed in order to solve the problem that terminal vibration of the large flexible robot cantilever under heavy load precision.The robot cantilever of the ballastless track is used as the research target and the natural parameters of the flexible cantilever such as the natural frequency, the load impact and the axial deformation is considered. Taking into account the change of the minimum distance between the glowworm individuals, the terminal vibration response and adaptability could meet. According to the Boltzmann selection mechanism, the dynamic parameters in the motion simulation process are determined, while the influence of the natural frequency and the load impact as well as the axial deformation on the terminal vibration is studied. The method is effective and stable, which is of great theoretical basis for the study of vibration control of flexible cantilever terminal.

1. Introduction

The design and control of flexible manipulator are widely studied and applied in the fields of robotics and aerospace technology. The traditional robots can guarantee the high speed and high precision performance with the rigid manipulator, while the solution of the arm inverse will increase the computational complexity and reduce the accuracy of the robot with the increase of the number of joints. In particular, due to the heavy load conditions, its long arm and bearing large, the robot cantilever prone to deformation in the process of rotation or amplitude, such as track robot cantilever components. Therefore, a new optimization algorithm is needed to solve the high-precision inverse of the redundant robot arm. As a new intelligent algorithms, the Glowworm Swarm Optimization (GSO) algorithm has been successfully applied to various engineering fields and multi-modal function optimization.

Song[1] Considered the dynamic modeling and simulation of flexible manipulator in complex load conditions. Leung et al[2] focused on the analysis of nonlinear dynamic multi-body flexible systems. Yang et al[3] on the hydraulic excavator mechanical cantilever for the determination of the natural frequency and sensitivity calibration. Xu et al[4] simulated the natural frequency of the crane boom and analyzed the frequency change. Viktorova et al[5] explored the causes of hand-arm vibration and frequency response analysis. Sayahkarajy et al[6] proposed a two-link flexible manipulator control method for mixed vibration and no-load heavy loads. The above content in the robot dynamics modeling, trajectory planning and control made a lot of research results. However, the nonlinear dynamic effects of rigid and the effect of flexible deformation of the whole system in the space-time field are neglected. In addition, Wang et al[7] use the GSO algorithm to optimize the BP neural network initial weight and threshold. Yu et al[8] used the randomness of chaos system to initialize GSO to improve the algorithm's global search ability. Gong et al[9] solve the multi-dimensional problem effectively based on hybrid artificial GSO algorithm. Wu et al[10] improve the artificial bee colony algorithm and particle swarm optimization algorithm, and through experiments to verify the GSO algorithm is better. Shen et al[11] proposed a new multi-objective optimization genetic algorithm to optimize the robot joint angle. At present, there is no scholar using GSO algorithm and robot arm inverse into combination.

In this paper, a new method based on Glowworm Swarm Optimization of Time-varying（TVGSO）is proposed to solve the vibration characteristics of robot cantilever under heavy load precision condition, which is due to automatic control, deformation of bar, joint vibration and inertial mass. Based on the golden ratio segmentation theory, the Boltzmann selection mechanism is used to analyze the time-varying step, while the time-varying process is based on the minimum distance between the glowworm and the individual glowworm choose the direction and position in the search process. Finally, compared with the original GSO algorithm, the accuracy and stability of the TVGSO algorithm are better, which is of great theoretical basis for the study of the vibration control of the flexible cantilever terminal.

2. Dynamic model of heavy-precision cantilever for flexible robot

2.1 Establishment of Cantilever Generalized Coordinate System for Flexible Robot

Cantilever model of ballastless tracked robot as shown in Figure 1; the Euler-Bernoulli beam model with simplified flexible cantilever as shown in Figure 2. The basic parameters of the cantilever are as follows: concentrated mass M_0, unit length mass m, initial length L, flexibility δ, in addition, H_0, H_1 for the hydraulic drive cylinder position. XOY is the inertia coordinate, and the angle θ between the coordinate axes O_{X1} and O_X is a rigid angle. Hydraulic flexible mechanical arm shows worse accuracy and stability, due to servo hydraulic cylinder vibration, hydraulic cylinder and the arm joint card and mechanical arm rod elastic vibration and other factors, while affecting its dynamic characteristics and control characteristics.

2.2 Modeling of Flexible Manipulator Dynamics Equation

According to the theory of dynamic manipulator of flexible manipulator, the dynamic equation is established[12]. The kinetic equation of the flexible arm is obtained by the hypothetical modal method. As shown in Figure 2, the flexural displacement of the flexible arm is set according to $w(l,t)$ in the ideal state. The length of the hydraulic cylinder l, the actual output force F:

Figure 1. Cantilever model of ballastless tracked robot

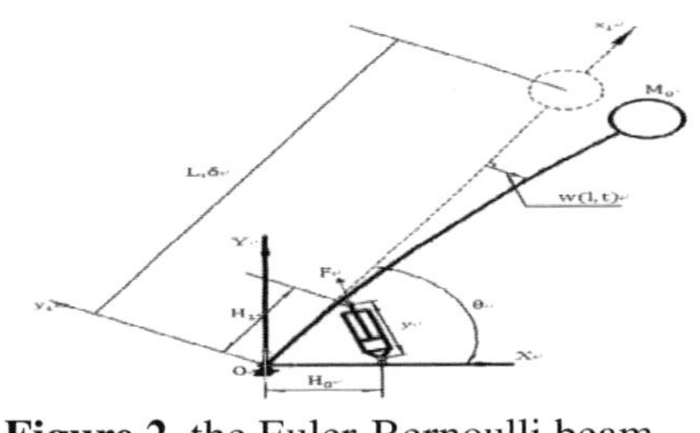

Figure 2. the Euler-Bernoulli beam model with simplified flexible cantilever

$$w(l,t)=\sum_{i-1}^{n}\Phi_i(l)q_i(t) \tag{1}$$

Where $\Phi_i(l)$ denotes the modal function of the hinge-free end model, $q_i(t)$ denotes the generalized modal coordinates of the articulated-free end model, and n is the modal order. The kinetic energy and potential energy of the flexible arm system are denoted by C and G respectively, and the potential energy G is the sum of the potential energy potential and the elastic potential energy. The kinetic equation of the flexible manipulator is based on the Lagrangian equation as follows:

$$H(q)\ddot{q} = C(q,\breve{q})\breve{q} + G(q) = \tau \tag{2}$$

$h_{ij} = \left(\frac{\partial^0 A_k}{\partial p_i} J_k \frac{\partial ({}^0 A_k)^T}{\partial p_j} \right)$ is the inertial force h_{ij} proportional to the generalized acceleration; $H(q)\ddot{q}$ is the quadratic form of the generalized velocity, where the term containing $\dot{q}_i^2$ is the centrifugal force and the term of $\dot{q}_i\dot{q}_j(i \neq j)$;G (q) is the gravitational term; τ is the driving force vector of the hydraulic drive system actuator on the flexible manipulator.

According to the equation of the velocity of the flexible manipulator, the elements of the i-th row of the Jacobian matrix J are:

$$J_{ij}(q) = \frac{\partial s_i(q)}{\partial q_i} \tag{3}$$

2.3 Space-time operator fractal transport equation of flexible robot cantilever

Ignore the damping C, according to the operator splitting mechanism constraint equation

$$M_f \ddot{q}_f + K_f q_f = Q_f + Q_{vf} \tag{4}$$

$$Q_{vf} = -M_f \ddot{q}_f + \frac{1}{2}\frac{\partial}{\partial q_f}(\dot{q}_f^T M_f \dot{q}_f) \tag{5}$$

Where Q_{vf} is the generalized force elastic component corresponding to the velocity quadratic term; Q_f is the elastic component corresponding to the generalized external force; M_f is the mechanism mass matrix; K_f is the stiffness matrix.

$$M_f = M_f^1 + M_f^2 \tag{6}$$

$$M_f^1 = M_{f1} + M_{f2} \tag{7}$$

M_f^1 is the single-order time matrix of the generalized elasticity of the cantilever of the robot; M_f^2 is the generalized elastic multilevel time matrix with the end concentrated mass; M_{f1} is the elastic single-order time matrix which does not consider the coupling deformation; M_{f2} Elastic Multi - order Time Matrix Contributing to Coupled Deformation.

The single-order time matrix of the flexible mechanical cantilever system is:

$$M = M_{ff}^1 + M_{ff}^1 = (m + m_1) \begin{bmatrix} \frac{1}{3} & 0 & 0 \\ 0 & \frac{26}{35} & -\frac{221}{210} \\ 0 & -\frac{221}{210} & \frac{21^2}{105} \end{bmatrix} \tag{8}$$

The elastic stiffness matrix of the manipulator can be obtained by the elastic potential of the relative moving coordinate system:

$$U = \frac{1}{2}\int_0^1 [EI(u_{fy}^{''})^2]dx = \frac{1}{2}\int_0^1 [u_{fx}^{'} \cdot u_{fy}^{''}] \begin{bmatrix} EA & 0 \\ 0 & EI \end{bmatrix} \begin{bmatrix} u_{fx}^{'} \\ u_{fy}^{''} \end{bmatrix} dx \tag{9}$$

Where u_x is the lateral displacement of the neutral axis of the manipulator, and u_y is the axial displacement.

2.4 Optimization of Cantilever Inverse Target for Flexible Robot

In solving the problem of plane redundant robot arm, if there is no condition constraint, then the solution will be uncertain. First, $\theta^* = [\theta_1^*, \theta_2^*, \cdots, \theta_{N-1}^*]$ satisfy the robot arm solution equation for the target position; Secondly, when the robot arm moves from the initial position to the target position, the joint angle of the robot arm should be as small as possible, that is, to satisfy $f_1(\theta)$ take the minimum value, so the robot arm inverse solution involves multi-objective function optimization.

In this paper, the sum of the absolute values of the difference between the solution value and the theoretical value of the manipulator is less than the given minimum value, and $f_1(\theta)$ is the objective function. The optimal objective function satisfies the redundancy Robot arm inverse solution accuracy. The optimization model as follows:

$$\min f_1(\theta) = (\theta_0 - \theta_0^{ini})^2 + \sum_{k=1}^{N-1}\left[(\theta_k - \theta_k^{ini}) - \theta_{k-1} - \theta_{k-1}^{ini}\right]^2 \quad (10)$$

3. Time-varying glowworm swarm optimization algorithm

3.1 Basic Glowworm Swarm Optimization Algorithm and Its Defects

The GSO algorithm has the advantages of low parameters and good stability [8], while its disadvantage of slow convergence and low accuracy of the optimal value affects the overall search performance of the algorithm. The choice of the direction of the glowworm algorithm is in the form of roulette, which leads to the decline of the population diversity, which makes the algorithm prematurely converged and is not conducive to searching the optimal solution.

3.2 Time-varying step strategy and Boltzmann selection mechanism

In order to compensate for the shortcomings of the step in the GSO algorithm, the dynamic time-varying step search mechanism is used in the algorithm iteration process. In this paper, the TVGSO algorithm quantifies the minimum distance between the glowworm i and all the glowworm individuals in the field and changes the steps according to the change of the minimum distance. If the minimum distance is less than s_0, if the distance is less than s_0, if the minimum distance is less than s_0, if the minimum distance is less than s_0, then take the minimum step size $s_i(t) = 0.618 s_0$.

$$d_{\min} = \min\{\|x_i - x_j\|, j \in N_i(t)\} \quad (11)$$

$$0.618 < \frac{d_{\min}}{s_i(t-1)} < 1, s_i(t) = .0618 s_0 \quad (12)$$

$$0.618^2 < \frac{d_{\min}}{s_i(t-1)} < 0.618, s_i(t) = .0618^2 s_0 \quad (13)$$

Instead of the original roulette way of the glowworm, the Boltzmann selection mechanism dynamically adjusts the direction and position of the search process and the selection pressure during the solution. The probability that the glowworm i moves in the direction of the j-only glowworm in the adjacent area:

$$p_{ij}(t) = \frac{\exp(\frac{l_j(t) - l_i(t)}{T})}{\sum \exp(\frac{l_k(t) - l_i(t)}{T})} \quad T = T_0(\alpha^{t-1}) \quad (14)$$

Where T is the temperature of the simulated annealing algorithm, T_0 is the initial temperature of simulated annealing, α is the attenuation factor, and t is the number of TVGSO iterations.

3.3 Time-varying glowworm swarm optimization algorithm description

The overall flow of the TVGSO algorithm is described below:

Step 1 Initialize $\rho, \gamma, l_i(0)$ and other constant parameters;

Step 2 Initialize glowworm position randomly, calculate the fitness of each glowworm;

Step 3 The fluorescein value of each glowworm is updated;

Step 4 Enters the motion phase, and the domain set $N_i(t)$ of the glowworm individual i is calculated;

Step 5 Calculate the glowworm i field according to. All glowworm are selected as the probability $p_{ij}(t)$ of the moving object. The minimum distance between glowworm i and glowworm j in the field is calculated. If the distance is less than the initial step, the steps are updated in real time

Step 6 In order to prevent the dynamic decision radius of the glowworm individual from being reduced to 0, the radius of the field of the glowworm i is updated .

Step 7 $t = t+1$; if there is no iteration to a given number of times, then step 3, otherwise stop the calculation, the output.

4. Numerical calculation of cantilever model of flexible robot

Figure 3 Boltzmann selection mechanism object model

As shown in Figure 3, the Boltzmann selection mechanism model for the glowworm is modeled, and ω_1 corresponds to the $j-th$ transverse bending natural frequency of the rotating cantilever beam[13]. The elastic modulus is $E = 6.875\times10^{10} N/m^2$, the poisson's ratio is $\mu = 0.33$, the shear correction coefficient $k = \frac{10(1+\mu)}{(12+11\mu)} = 0.85$, the length of the beam is l, the width is b, the density $\rho = 2.767kg/m^3$, the center rigid body radius $R = 1m$, and the angular velocity θ is variable.

$$\begin{bmatrix} \omega_1^2 m_{y11} & \cdots & \omega_{Ny}^2 m_{y1Ny} \\ \vdots & \ddots & \vdots \\ \omega_1^2 m_{yNy1} & \cdots & \omega_{Ny}^2 m_{yNyNy} \end{bmatrix} \bullet \begin{bmatrix} \ddot{B}_1 \\ \vdots \\ \ddot{B}_{Ny} \end{bmatrix} = \begin{bmatrix} \dot{\theta}^2 H_{11} - \dot{\theta}^2 m_{y11} + V_{y11} & \cdots & \dot{\theta}^2 H_{1Ny} - \dot{\theta}^2 m_{y1Ny} + V_{y1Ny} \\ \vdots & \ddots & \vdots \\ \dot{\theta}^2 H_{Ny1} - \dot{\theta}^2 m_{yNy1} + V_{yNy1} & \cdots & \dot{\theta}^2 H_{NyNy} - \dot{\theta}^2 m_{yNyNy} + V_{yNyNy} \end{bmatrix} \bullet \begin{bmatrix} \ddot{B}_1 \\ \vdots \\ \ddot{B}_2 \end{bmatrix} \quad (15)$$

The height ratio $\lambda = a/l$ is defined. The variation of the natural frequency p(rad/s) of the first six orders is compared with the different angular velocity Ω. The numerical results are shown in the following Tables 1 and 2.

Table 1. Value of basis parameters

Parameter	Ω/rad	l/m	b/m	λ
Value	5	8	0.05	0.00625

Table 2. The numerical results of modele

Order	A	B	C	D
1	5.4426	6.8738	6.8906	5.5836
2	28.3889	33.7303	33.7787	28.5290
3	73.8918	80.2785	80.2996	73.8938
4	142.2326	149.4281	149.4080	142.1568
5	260.4303	267.2562	267.1883	260.3188
6	415.6248	422.3192	422.2321	415.4292

The simulation results show that the dynamic response is mainly affected by the low-order frequency, so for the Timoshenko beam, after adding the deformation mode, although the beam has a greater range of deformation, and thus mixed space-time coupling produced a certain "softening" But the magnitude of the frequency is reduced. It is shown that in the generalized gyroscope G and the generalized stiffness matrix K, the coupling term $\overline{W}_{21}, \overline{W}_{22}$ of the bending deformation is weak, and the coupling term in the longitudinal displacement and its D, C term in the generalized stiffness matrix K Caused by the "power rigid" phenomenon is still more significant.

5.Application of Boltzmann Selection Mechanism in Cantilever Model of Flexible

In this paper, the model parameters of the plane redundant robot arm are as followed: ,the length of the manipulator is $L_1 = 1.2m, L_2 = 1m, L_3 = 0.6m, L_4 = 0.6m, L_5 = 0.4m$; the initial angle $\theta = [\frac{\pi}{8}, \frac{11\pi}{36}, \frac{5\pi}{9}, \frac{7\pi}{9}, -\frac{17\pi}{18}]$the initial position of the robot arm terminal is $(0.6898m, 2.3824m)^T$, the robot arm terminal moves to $(2.3m,2m)^T$, $(2m,2m)^T$, $(1m,2.8m)^T$, $(1.5m,2m)^T$ four target location of the inverse solution. The objective function is optimized by the TVGSO algorithm, and the minimum value of the rotation angle f_1 is found in the 100 optimization results, and the optimal value of the rotation angle is obtained. Set the total exercise time to 1s after simulation. In the end of the arm to add the MARK point, the simulation can be measured from the initial position $(0.6898m, 2.3824m)^T$ moved to the target position of the trajectory curve, as shown in Figure 4 ~9.

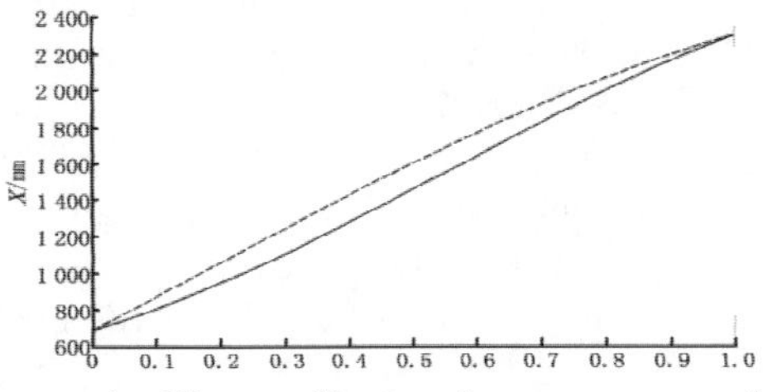

Figure 4 X coordinate change curve when the target position is (2.3m, 2m)

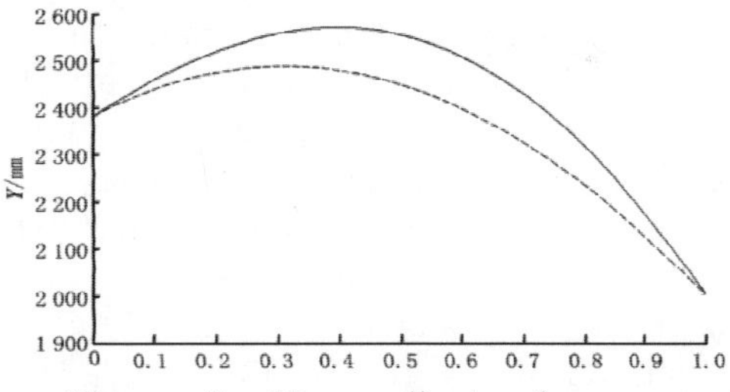

Figure 5 Y coordinate change curve when the target position is (2.3m, 2m)

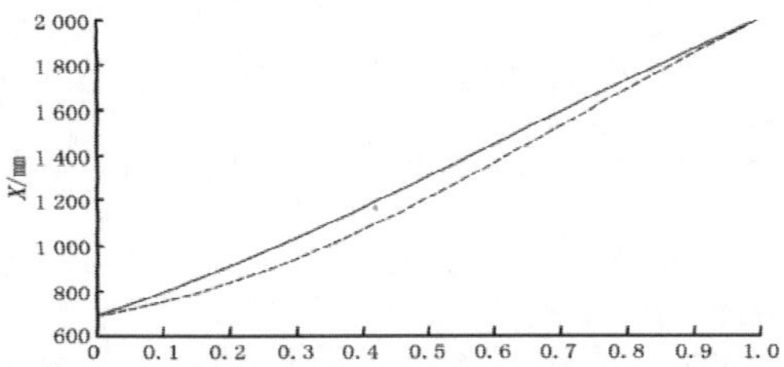

Figure 6 X coordinate change curve when the target position is (2m, 2m)

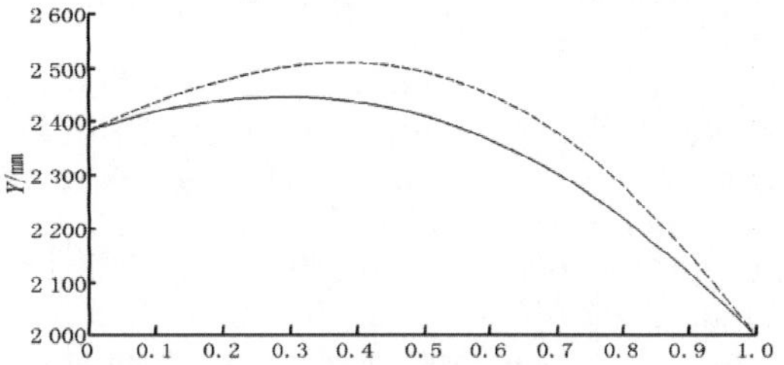

Figure 7 Y coordinate change curve when the target position is (2m, 2m)

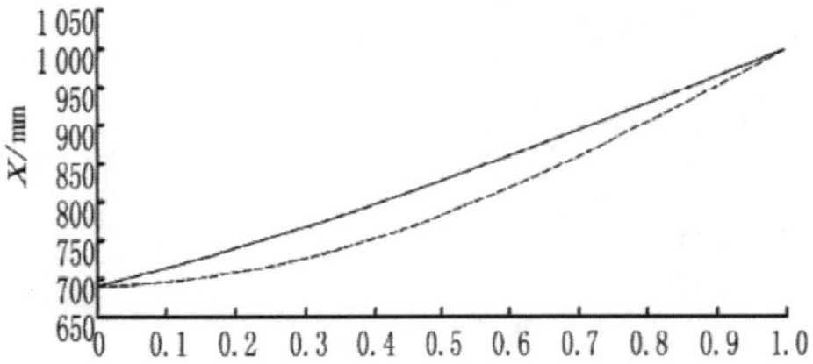

Figure 8 X coordinate change curve when the target position is (1m, 2.8m)

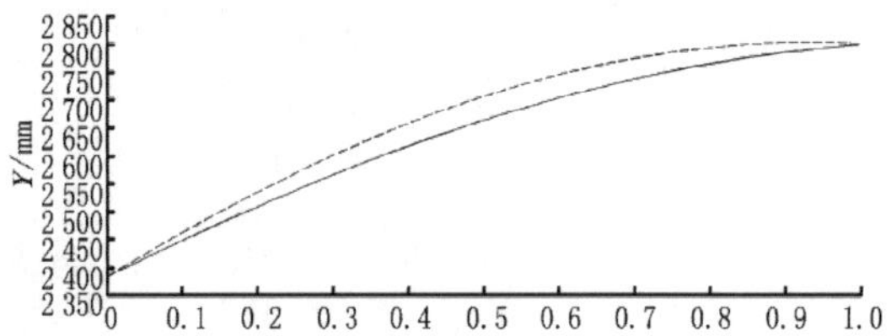

Figure 9 Y coordinate change curve when the target position is (1m, 2.8m)

By using TVGSO algorithm to optimize the inverse of the arm of the plane redundant robot, the optimal rotation angle value of the end position and the expected position error is less than the given minimum value in the inverse of the arm of a plurality of planar redundant robots. Because this paper focuses on the inverse of the robot arm, so when the end position and the desired position error take the minimum value of the rotation angle f_1 value is relatively large. The simulation curves of the robot arm in ADAMS (Figure 4 ~ 9) verify the correctness and accuracy of the inverse solution. At the same time, the target position error optimized by comparing the TVGSO algorithm with the optimized GSO algorithm Target position error, it can also be seen that the TVGSO algorithm is superior to the GSO algorithm. Therefore, based on TVGSO algorithm to solve the redundant robot arm inverse solution is very effective. The error obtained by MATLAB simulation is somewhat different from the error Δ (ADAMS) and Δ (MATLAB) obtained by ADAMS simulation, which is caused by the limitation of π (3.14159) and the number of effective digits of rotation angle.

6. Summary

In this paper, the flexible manipulator is regarded as the Euler-Bernoulli equal-section beam model with the existing cantilever of the ballastless trackless orbiter. The space independent parameters such as the natural frequency, the load impact and the axial deformation are the characteristic, A method of terminal vibration mechanism analysis based on time-varying glowworm swarm optimization algorithm is proposed.

Based on the nonlinear dynamic model, this paper proposes a TVGSO algorithm to apply to the split equation of the flexible cantilever operator. According to the boltzmann selection mechanism, the dynamic parameter in the motion simulation process is determined. Based on golden ratio segmentation ,the algorithm gives a time-varying strategy. The method is based on the change of the minimum distance between the glowworm individuals in the field, and the algorithm can satisfy the dynamic response and adaptability of the terminal vibration. The simulation results show that the method is effective and stable, and it has a great theoretical basis for the study of vibration control of flexible cantilever terminals.

References

[1] Song Jianlong. Study on dynamic modeling and simulation of flexible manipulator[D].Tianjin: Tianjin University,2011

[2] Leung A Y T, Wu, Zheng W F. Nonlinear dynamic analysis of flexible multibody system [J]. Acute Mechanical Solidi Sonica, 2004 17 (4):330-336

[3] Yang Wei, Wang Jiaxu, Qin Datong. Experimental sensitivity of natural frequency of hydraulic excavator working device.[J]Journal of Agricultural Mechanization, 2016, 37 (2): 21-24

[4] Xu Yan, Liu Jie, Sun Guangfu, et al. Effect of Rigid-flexible Coupling on Natural Frequency of Crane Crane Vibration [J] .Journal of Mechanical Engineering, 2016, 42 (8): 142-145

[5] Irina Viktorova , Matthew Fleck, Muhammed Kose .HAVS and HAV-nots: Investigating Resonance in the Human Arm Caused by Contact with Machinery[J].Materials Science & Engineering Journal, Clemson University,2016,06:26-31

[6] Mostafa Sayahkarajy , Z Mohamed , A.A.M. Faudzi , E. Supriyanto .Hybrid vibration and rest-to-rest control of a two-link flexible robotic arm using H∞ loop-shaping control

IOP Conf. Series: Materials Science and Engineering **280** (2017) 012003 doi:10.1088/1757-899X/280/1/012003

design, Engineering Computations, 2016. 33:395 – 409
[7] Wang Gaige, Guo Lihong, Duan Hong, Liu Luo, Wang Heqi. Estimation of target threat based on firefly algorithm for BP neural network optimization[J]. Journal of Jilin University (Engineering Science), 2013,04: 1064-1069
[8] Yu Shuhao, Su Shoubao. Study and Application of Optimization Algorithm for Chaotic Firefly [J] .Computer Science & Exploration, 2014, 03: 352-358
[9] Gong Qiaoqiao, ZHOU Yongquan, LUO Qifang. Hybrid artificial glowworm swarm optimization algorithm for solving multi-dimensional knapsack problem[J].Procedia Engineering,2011(10): 2880-2884
[10] Wu Bin, Qian Cunhua, Ni Weihong, et al. The improvement of glowworm swarm optimization for continuous optimization problems[J]. Expert Systems with Applications, 2012, 39(7):6335-6342
[11] Shen Xiaoning, Li Sheng, Guo Yu, Chen Qingwei, Hu Weili. An inverse algorithm for redundant manipulators based on multi-objective genetic algorithm [J]. Journal of Systems Simulation, 2008,02: 399-403
[12] Basher H A. Modeling and simulation of flexible robot manipulator with a prismatic; joint [C]. Proceedings, IEEE. Southeast Con,2007
[13] Mehram Sadri, Davood Youneian , Ebrahim Esmailzadeh. Nonlinear harmonic vibration and stability analysis of a cantilever beam carrying an intermediate lumped mass, Nonlinear Dyn .2016. 84:1667–1682

Study on the Riveting Forming of Aluminum Ball Joint

H B Tian [1,a], X H Liu [1], X L Li [1], W J Yuan [1] and J Feng [2]

[1] Department of Mechanical Engineering, Shanghai Second Polytechnic University, Shanghai 201209, P. R. China;
[2] Shanghai Huizhong Automotive Manufacturing Corporation, Shanghai 200122, P. R. China

[a] hbtian@sspu.edu.cn

Abstract. Aluminum ball joint is an advanced structure using to connect steering parts with the merits of simple and lighting. Usually the ball sealed in the Aluminum housing through riveting forming method which has an impact on the push out strength and ball joint torques directly. So how to control the riveting process and achieve the best properties match is very important. In the paper, the relations in the riveting processes, push out strength and torque are studied using experimental method. The parameters affected the push out strength and torque are studied and the influence rules are achieved. The analytical model to present the push out strength is established.

1. Introduction

Automobile steering system is mostly used Aluminum and connected by ball joint to reduce weight. Figure 1 shows a typical type of ball joint structure which is assembled by ball stud, housing, bearing, housing cover, etc. Generally the housing is a part of control arm and made by Aluminum material, the ball stud is made by steel, the bearing is made by high polymer material and the housing cover is made by soft steel or Aluminum. During the assembly process, the ball stud, bearing, housing cover are put in the housing firstly, then the sloped edge on the housing is turned down (shown in the Figure 1-a)), then the housing cover and ball stud were sealed. References [1-2] show push out strength is the most major parameters to value the assembly quality and small push out strength can result in disaster. How to obtain the analytical model to present the push out strength is important to the application.

Riveting forming method is usually used to connect two parts in the automobile factory and aerospace factory [3]. But for the structure shows in Figure 1, it is difficult formed by the general riveted method, the die must be spun in the riveting processes to achieve the seal results [4]. References [4-5] show that housing size and shape have an impact on the push out strength. Thus, it is impossible to obtain the analytical model to present the push out strength. In this paper, the processes of ball joint sealing of Aluminum are studied, and the impacts of housing shape on the sealing quality are researched, which will provide necessary technical support for the ball joint sealing of control arms.

IOP Conf. Series: Materials Science and Engineering **280** (2017) 012004 doi:10.1088/1757-899X/280/1/012004

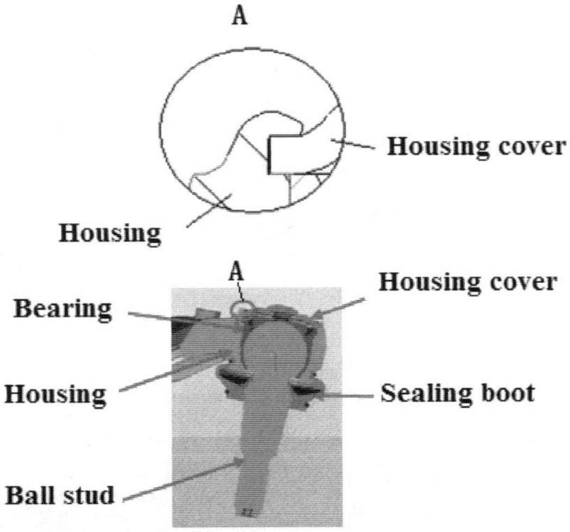

Figure 1. The structure diagram of ball joint

2. Die design and equipment

In the riveting process, the equipment and die show as Figure 2. In the forming, the die spinning and stepped down to make the part deform. In order to keep the force balance, two rollers are used. The part shows as Figure 3. In order to obtain the rule between part shape and push out strength, the parameters of parts show as table1. In the experiment, 800r/min and 2mm/s are used.

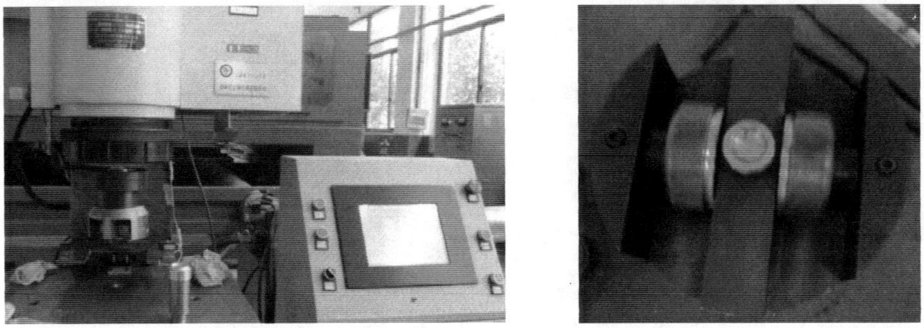

a) Riveting equipment b) die

Figure 2. Riveting equipment and die

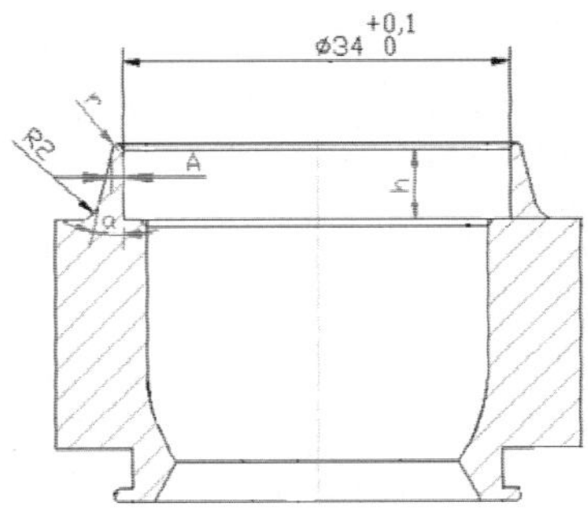

Figure 3. The shape of part

Table 1. Dimension of parts.

	a (°)	h (mm)	A (mm)	r (mm)
1#	13	5.09	1.24	0.62
2#	15	5.09	1.24	0.62
3#	18	5.09	1.24	0.62
4#	13	5.7	1.6	∞
5#	10	5.7	1.77	∞

3. Experiments and results

After the riveting processes, the following parts can be achieved (Figure 4).

a) 1#

b) 2#

c) 3#

d) 4#

e) 5#

Figure 4. Parts after deformation

It can be seen that all of the parts have a good surface quality after riveting forming. Figure 5 show the push out strength-displacement curves and parts after push out test.

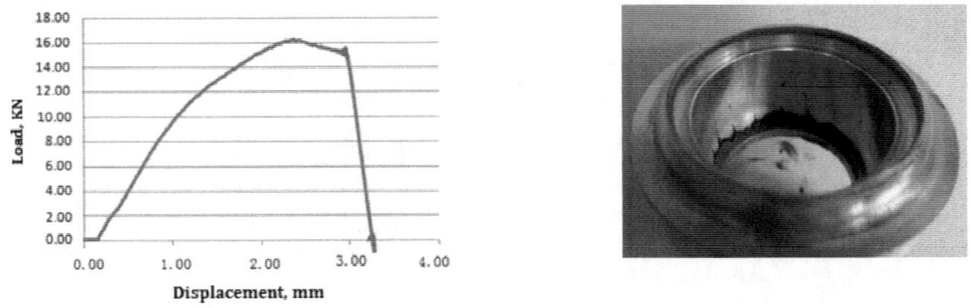
18.00
16.00
14.00
12.00
10.00
8.00
6.00
4.00
2.00
0.00
Load, KN
0.00
1.00
2.00
3.00
4.00
Displacement, mm

a) 1#

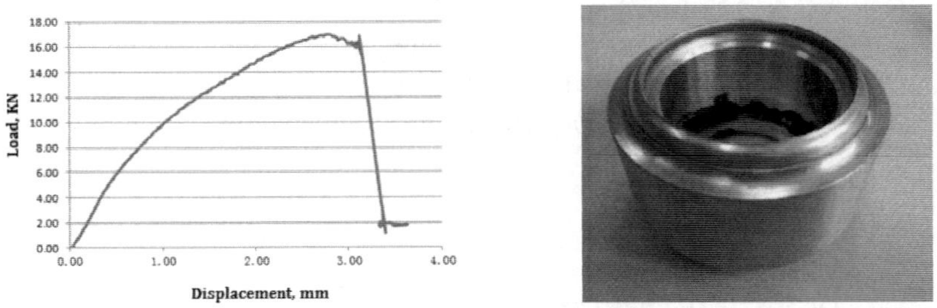
18.00
16.00
14.00
12.00
10.00
8.00
6.00
4.00
2.00
0.00
Load, KN
0.00
1.00
2.00
3.00
4.00
Displacement, mm

b) 2#

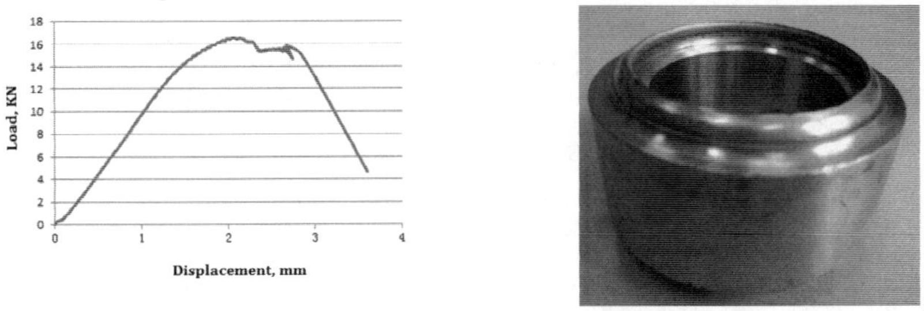
18
16
14
12
10
8
6
4
2
0
Load, KN
0
1
2
3
4
Displacement, mm

c) 3#

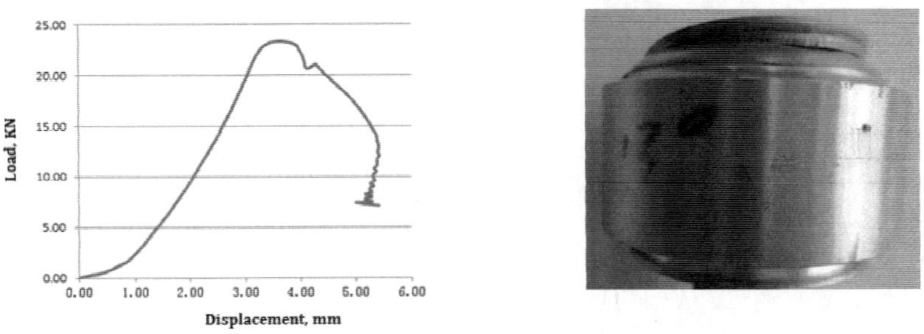
25.00
20.00
15.00
10.00
5.00
0.00
Load, KN
0.00
1.00
2.00
3.00
4.00
5.00
6.00
Displacement, mm

d) 4#

IOP Conf. Series: Materials Science and Engineering **280** (2017) 012004 doi:10.1088/1757-899X/280/1/012004

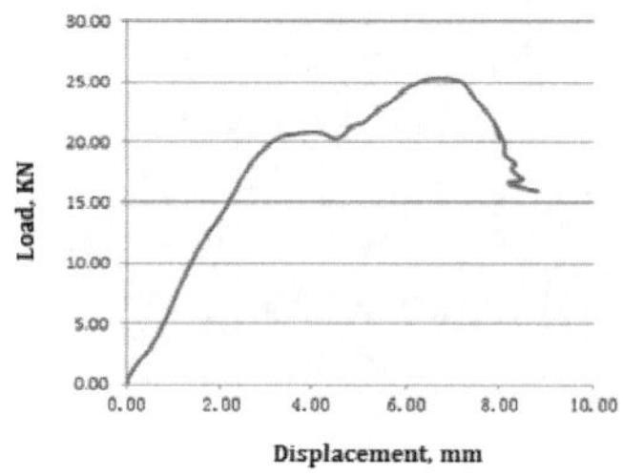

e) 5#

Figure 5. Push out force-displacement curves and parts after push out test

It can be seen from Figure 5 and Table 2 that push out strength is related to the shape of housing, and the push out strengths come from part 4# and 5# are larger than that come from part 1# to 3#. According to shearing theory, the push out strength can be calculated according the Equation (1).

$$\text{Pos} = \sigma_t S \quad (1)$$

Where, Pos is present the push out strength, MPa;
σ_t is present the shear strength of housing material;
S is present the shear area in the push out test.

For the housing material, its strain-stress curve shows as Figure 6 and the shear strength is mostly 350MPa. The shearing section area can be calculated from the shearing section thickness and list in the Table 2.

Table 2. Push out strength and shearing section area.

	Shearing section thickness,mm	shearing section area,mm^2
1#	0.39	36.738
2#	0.4	37.68
3#	0.38	35.796
4#	0.56	52.75
5#	0.592	55.76

According Equation (1) , The push out strengths can be calculated and showed in the Figure 7.

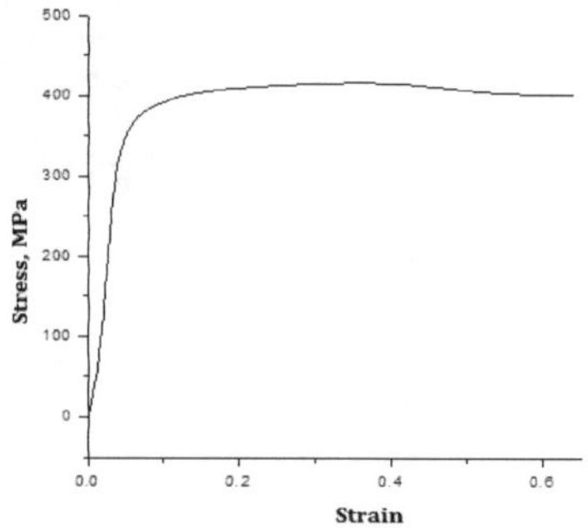

Figure 6. Strain-stress curve

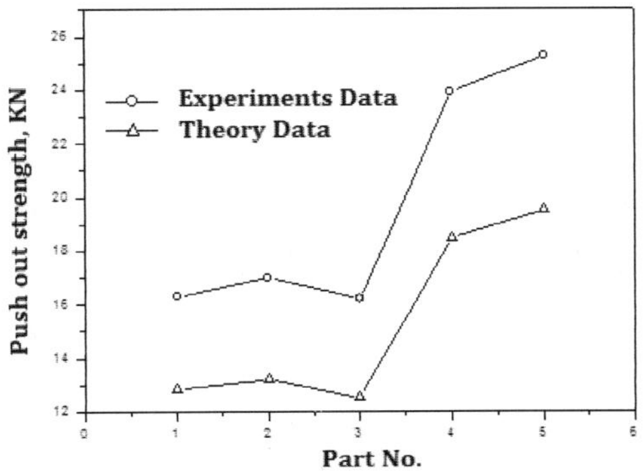

Figure 7. The push out strength

It can be seen from Figure 7 that push out strength obtained from experiment has a same trend with that come from calculation from equation (1) and higher than that obtained from calculation, this is because in the forming the material strength can be improved by the deformation.

4. Conclusions

1) The shape of housing has a major impact on the push out strength, and more material deformed more push out strength;
2) The push out strength is depended on the thickness of shearing section, the thicker thickness the bigger push out strength can be obtained;
3) The strength can be improved through riveting forming and the push out strength can be predicted by the thickness of shearing thickness.

Acknowledgments

This research was supported by the National Natural Science Foundation of China (Grant No. 51405284), and Leading Academic Discipline Project of Shanghai Polytechnic University (Grant No. XXKZD1601).

References

[1] Burcham M N, Escobar R, Yenusah C O, Stone T W and Berry G N 2017 *J Fail. Anal. and Preven* **17** 262–74
[2] Baynal K, Makara M and Gulbudak K 2010 *Int J Auto Tech* **11** 701-10
[3] Chen N, Thonnerieux M, Ducloux R, Wan M and Chenot J L 2014 *Int J Mater Form* **7** 65–79
[4] Jang B, Lee KH 2014 *Part C: J Mech Eng Sci* **228** 146-51
[5] Tian H B, Liu X H and Yu Z T 2012 *Mater Sci Forum* **704-705** 758-61

IOP Conf. Series: Materials Science and Engineering **280** (2017) 012005 doi:10.1088/1757-899X/280/1/012005

Numerical Modelling and Analysis of Hydrostatic Thrust Air Bearings for High Loading Capacities and Low Air Consumption

Yunluo Yu, Guang Pu and Kyle Jiang*

University of Birmingham, Birmingham B15 2TT, United Kingdom

*K.Jiang@bham.ac.uk

Abstract. The paper presents a numerical simulation study on hydrostatic thrust air bearings to assess the load capacity, compressed air consumptions, and the dynamic response. Finite Difference Method (FDM) and Finite Volume Method (FVM) are combined to solve the non-linear Reynolds equation to find the pressure distribution of the air bearing gas film and the total loading capacity of the bearing. The influence of design parameters on air film gap characteristics, including the air film thickness, supplied pressure, depth of the groove and external load, are investigated based on the proposed FDM model. The simulation results show that the thrust air bearings with a groove have a higher load capacity and air consumption than without a groove, and the load capacity and air consumption both increase with the depth of the groove. Bearings without the groove are better damped than those with the grooves, and the stability of thrust bearing decreases when the groove depth increases. The stability of the thrust bearings is also affected by their loading.

1. Introduction

Hydrostatic air film thrust bearings are very important machine elements to support lightly loaded, high speed rotating machinery such as spindles, gas expanders, compressors, and micro gas turbines. Previous studies on the design of thrust air bearings mainly focused on static characteristics [1-3]. However, in a high speed rotational system, such as a spindle, pressure fluctuations and bearing air film variations in thrust bearings caused by various factors, such as different loads and rotating speeds, induce excitations of the bearing dynamic response, which can lead to bearing instability. Therefore, developing an effective tool to analysis the dynamic responses as well as the static characteristics of hydrostatic thrust air bearings is of great importance for designing high quality and high speed rotational machinery.

Numerical analysis is the main methodology used in the investigation of the modern thrust air bearings. Developing accurate numerical models for the design of thrust air bearings is one of the major tasks in the field. Fundamentally the work is to solve the Reynolds Equation numerically. Three discrete schemes can be used together with Newton's method to solve the second order partial differential equation: Finite Difference Method (FDM), Finite Element Method (FEM) and the Finite Volume Method (FVM). Florin suggested that by applying Newton's method, it can reduce computational time significantly [4]. Newton's method is also proven to have a good stability in getting converged solutions [5]. Nenzi Wang carried out a series of studies on numerical methods which can be applied to solve a compressible Reynolds Equation [6-8]. The work includes comparisons of different methods with respect to computational time and stability. He also suggested different stopping criterion in an iterative process and analysed the truncation error of numerical approaches [6-8]. The above mentioned approaches are all based on FDM and they have good agreement with the experiment results of hydrostatic air bearings. FVM is also suggested by some researchers. It is based on the continuity equation and Green's theorem. This approach shows advantages in dealing with film discontinuities and some pocket restrictor configuration [9].

In this paper, a numerical model of hydrostatic thrust air bearings is proposed combining the advantages of FDM and FVM for solving the Reynolds Equation. Then motion equations are added into the model to simulate the dynamic response of the thrust air bearing. Finally, the effect of several design and working parameters on air film characteristics, including the air film thickness, supplied pressure, depth of the groove and external load, are investigated based on the proposed model.

2. Modeling

2.1. Reynolds equations and the finite difference scheme

In this section, a numerical model based on the finite different method and finite volume method is developed for hydro-static air bearings in order to study the effect of several design parameters on air film gap characteristics, including the air film thickness, supplied pressure, depth of the groove and external load.

After introducing the dimensionless variables:

$$R = \frac{r}{r_2}, P = \frac{p}{p_a}, H = \frac{h}{h_0}, u = \frac{\omega r_2}{2}, \tau = \omega t$$

where r_2 is the outer radius of the bearing, p_a the atmospheric pressure, h_0 the reference film thickness, h the film thickness, t time. With the isentropic gas, the Reynolds equation of compressible flow can be written as Eq.(1):

$$\frac{\partial}{\partial R}\left(RPH^3\frac{\partial P}{\partial R}\right) + \frac{1}{R}\frac{\partial}{\partial \theta}\left(PH^3\frac{\partial P}{\partial \theta}\right) = \Lambda R\frac{\partial(PH)}{\partial \theta} + 2\Lambda R\frac{\partial(PH)}{\partial \tau} \tag{1}$$

$$\Lambda = \frac{6\mu\omega r_2}{p_a {h_0}^2} \tag{2}$$

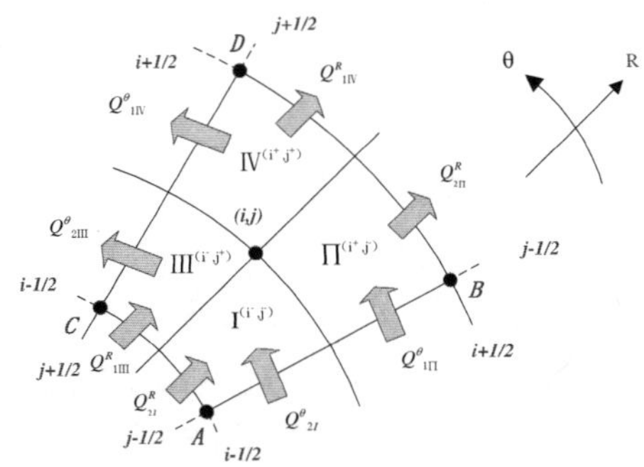

Figure 1. The defined control volume

where Λ is the compressibility number or bearing number and μ the viscosity of air. To solve the Reynolds Equation, a control volume is defined in Fig.1, and the mass flow rate Q^R across the boundary of $R = constant$ and the mass flow rate Q^θ across the boundary of $\theta = constant$ are calculated respectively as follows:

$$Q^R = \int_{R_1}^{R_2}(-RPH^3\frac{\partial P}{\partial R} + \Lambda RPH)dR \tag{3}$$

$$Q^\theta = \int_{\theta_1}^{\theta_2}(-PH^3\frac{\partial P}{\partial \theta})d\theta \tag{4}$$

The mass flow rate due to squeezing motion inside the control volume is expressed as follows:

$$Q^{\tau} = \int_{\theta_1}^{\theta_2} \int_{R_1}^{R_2} (2\Lambda R \frac{\partial PH}{\partial \tau}) dR d\theta \tag{5}$$

The following relationship can be found from the mass flow rate balance on the control volume:

$$Q^{\theta}{}_{2I} + Q^{\theta}{}_{1III} - Q^{\theta}{}_{2II} - Q^{\theta}{}_{1IV} + Q^{R}{}_{2I} + Q^{R}{}_{1II} - Q^{R}{}_{2III} - Q^{R}{}_{1IV} = Q^{\tau} + Q^{orifice} \tag{6}$$

where $Q^{orifice}$ is the mass flow rate of the orifice. The dimensionless mass flow rate of the orifice can be calculated according to the following equations under isentropic assumption:

$$Q^{orifice} = \begin{cases} HC_s \left\{ \frac{2}{\gamma-1} \left[\left(\frac{P_r}{P_s} \right)^{2/\gamma} - \left(\frac{P_r}{P_s} \right)^{(\gamma+1)/\gamma} \right] \right\}^{1/2}, \ if \ \frac{P_r}{P_s} \geq \left(\frac{2}{\gamma+1} \right)^{(\gamma-1)/\gamma} \\ HC_s \left(\frac{2}{\gamma+1} \right)^{\left(\frac{(\gamma+1)}{2(\gamma-1)} \right)}, \ if \ \frac{P_r}{P_s} < \left(\frac{2}{\gamma+1} \right)^{(\gamma-1)/\gamma} \end{cases} \tag{7}$$

$$C_s = \frac{12 \mu \pi d C_d P_s}{h_0^2 p_{at}^2} \sqrt{\gamma \Re T} \tag{8}$$

where T is the atmospheric temperature at supply conditions, C_d the discharge coefficient, d the orifice diameter, μ the dynamic viscosity of the fluid, γ the isentropic index and $\Re$ the gas constant. Powel [10] regarded C_d as a function of the pressure ratio $\frac{P_r}{P_s}$. However, many researchers take the coefficient as 0.8 for all flow conditions [11]. In this study, C_d is taken as 0.8 in all the cases. By solving Eq.(1), the pressure distribution of the lubricating can be found.

2.2. Calculation of static characteristics

For a thrust bearing experiencing a small vibration in the axial direction, the minimum air film thickness and pressure can be expressed as:

$$h = h_0 + \varepsilon e^{j\omega_f t}$$

$$p = p_0 + \varepsilon p_t e^{j\omega_f t} \tag{9}$$

where ε represents the function of air film thickness and the bound of ε is approximately within ±10 percent of the steady state air film thickness. Subscripts 0 and t mean the physical quantities under the steady and unsteady operation conditions, respectively.

Based on the equilibrium on the control volume, the following relationships can be obtained:

$$F_0(P_0) = Q^{\theta}{}_{2I0} + Q^{\theta}{}_{1III0} - Q^{\theta}{}_{2II0} - Q^{\theta}{}_{1IV0} + Q^{R}{}_{2I0} + Q^{R}{}_{1II0} - Q^{R}{}_{2III0} - Q^{R}{}_{1IV0} - Q^{orifice} = 0 \tag{10}$$

$$F_t(P_t, P_0) = Q^{\theta}{}_{2Is} + Q^{\theta}{}_{1IIIs} - Q^{\theta}{}_{2IIs} - Q^{\theta}{}_{1IVs} + Q^{R}{}_{2Is} + Q^{R}{}_{1IIs} - Q^{R}{}_{2IIIs} - Q^{R}{}_{1IVs} - Q^{\tau}{}_{s} - Q^{orifice} = 0 \tag{11}$$

Eq. (10) is a nonlinear equation which can be solved by applying the Newton-Raphson iteration algorithm. Eq.(11) is a linear equation which can be solved based on the steady state pressure P_0 obtained through Eq.(10).

In this work, maximization of the bearing load capacity and minimization of the consumption of compressed air are considered as the way to the optimal design of the hydrostatic thrust bearings. The consumption of compressed air is the air flow rate of all the orifice restrictors. The load capacity can be calculated using Eq. (12):

$$w = \int_{r_1}^{r_2} \int_{0}^{2\pi} (p_0 - p_a) r dr d\theta \tag{12}$$

2.3. Calculation of dynamic responses

The calculation requires to solve the Reynolds equation with the equation of motion [12-13] simultaneously. The Reynolds equations have been introduced in Section 2.2. The equation of motion can be written as:

$$\Delta w(t) = m\ddot{h} \tag{13}$$

where m is the mass supported by the bearing, Δw is the variation of the resultant pressure force. The dynamic response of thrust air bearing can be calculated using the following scheme:

1. Choose an equilibrium position of the simulated thrust bearing for dynamic calculation.
2. Input initial step displacements and velocities for step by step method.
3. Dynamic calculation of the movement (acceleration, velocity, displacement, pressure and load capacity, etc.)

3. Numerical simulation

3.1. Thrust bearings for simulation

In this section, 4 types of hydrostatic thrust air bearings, as shown in Fig. 2 and Tab.1, are simulated using the FDM and FVM. The influences of several design parameters on air film characteristics, including the air film thickness, the pressure of supplied air, the depth of the groove, and the external loading, are investigated to provide information in the future design of hydrostatic thrust air bearing.

Table 1. Description of bearings simulated

Geometric parameters	Thrust bearing without a groove	Thrust bearings with a groove		
	Type 1	Type 2	Type 3	Type 4
r_2(mm)	15	15	15	15
r_1(mm)	6	6	6	6
h_g(um)	-	10	20	30

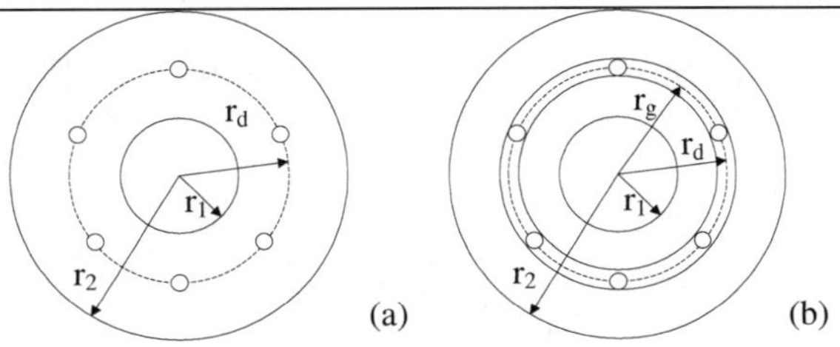

Figure 2. (a) Hydrostatic thrust air bearing without grooves and (b) Hydrostatic thrust air bearing with a groove.

Other fixed parameters are listed in Table 2.

Table 2 Fixed parameters of bearings studied

Bearing parameters	Value	Bearing parameters	Value
Reference gap h_0(um)	20	Coefficient of discharge C_d	0.8
Radius of orifices r_d	10.5×10^{-3}	Isentropic index γ	1.4
Radius of the groove r_g	10.5×10^{-3}	Gas constant $\Re$ (J kg^{-1} K^{-1})	287
Diameter of orifices d_0	250×10^{-6}	Dynamic viscosity u (Pa s)	18.38×10^{-6}
Supply pressure p_s (Pa)	6×10^{-5}	Temperature at supply conditions T (°K)	293

3.2. Analysis of static characteristics

If air consumption needs to be reduced as much as possible while maintaining a reasonable load capacity, the influence of air film thickness, supplied pressure, depth of the groove and the effect of supplied pressure on load capacity and mass flow rate can be investigated.

The load capacity and mass flow rate of all the orifice restrictors calculated versus the air film thickness are respectively presented in Figs.3 and 4. The results are for $p_s = 3\times10^5$ Pa, $n_{ori} = 6$ and $\omega = 0$ rpm. From Figs.3 and 4, it can be seen that the load capacity is inversely proportional to the air film thickness while the mass flow rate is proportional to the air film thickness, which means a lower air film thickness can lead to a higher load capacity and lower compressed air consumption. In comparison of the load capacity and the mass flow rate of the 4 thrust air bearings, it can be seen that the bearings with groove, namely Type 2, Type 3 and Type 4, have a higher load capacity but larger air consumption than those without grooves. In addition, a deeper groove leads to a higher load capacity and larger air consumption, such as Type 4.

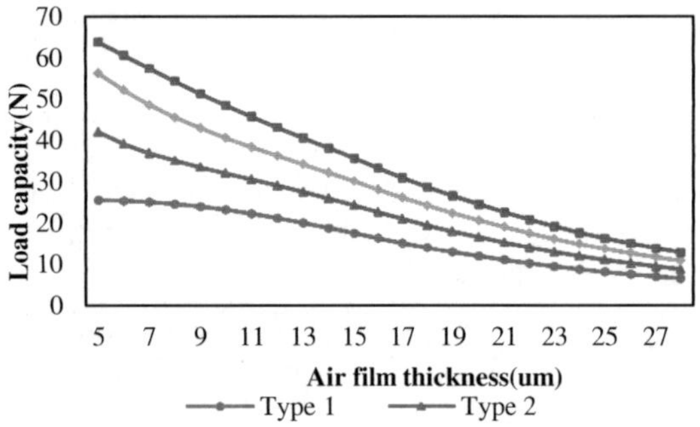

Figure 3. Static load capacity versus air film thickness for the 4 types of thrust bearings

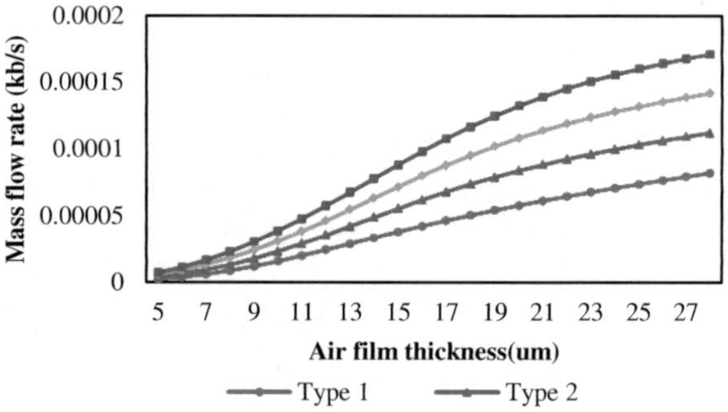

Figure 4. Mass flow rate versus air film thickness for the 4 types of thrust bearings

The load capacity and mass flow rate of all the orifice restrictors calculated versus the supplied pressure are respectively presented in Fig.5 and Fig.6. Results are obtained for $h_0 = 20\times10^6$ um, $n_{ori} = 6$ and $\omega = 0$ rpm.

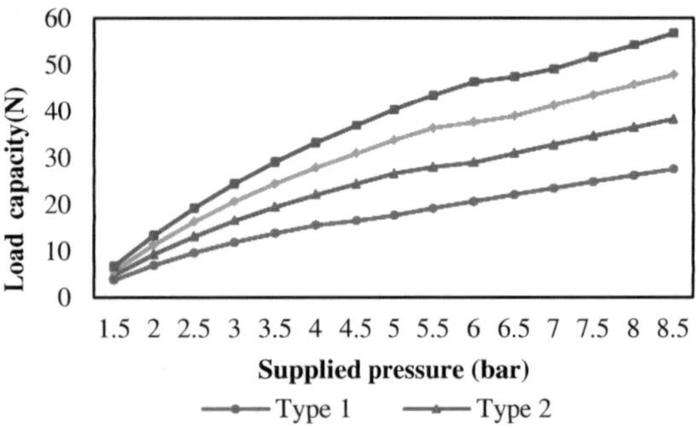

Figure 5. Static load capacity versus supplied pressure for the 4 thrust bearings

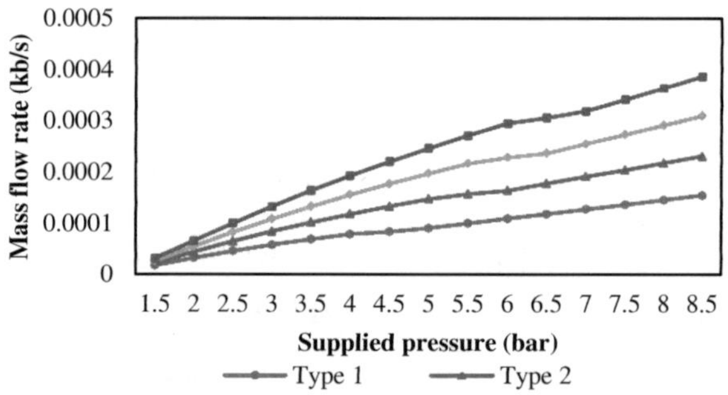

Figure 6. Mass flow rate versus supplied pressure for the 4 thrust bearings

Figs.5 and 6 show that the load capacity and the mass flow rate are both proportional to supplied pressure for the 4 bearings. Comparing their load capacities and the mass flow rates, it can be seen that the bearings 2, 3 and 4 with a groove achieve a higher load capacity and mass flow rate than the bearing without a groove, i.e. bearing 1. Once again, it proves that the deeper the grooves are, the higher the load capacity and air consumption.

3.3. Analysis of dynamic response

Figs.7 and 8 depict the dynamic step responses of the 4 thrust bearings. The step responses are calculated with the scheme described in Section 2.3. The same initial conditions used in all the cases are $h_0 = 20\mu m, \dot{h_o} = 0, P_s = 6\ bar, \omega = 10000rpm,$ and $dt = 10^{-5}s$. The dynamic simulation took 2 external loadings into consideration: $w_1 = 5N$ and $w_2 = 20N$.

With an external load of 5N, the step responses of the bearings are undamped for Type 2, Type 3 and Type 4; however, the step response of Type 1 bearing is damped, which means thrust bearings without a groove have a better stability compared with those with a groove. The same conclusion can be found in Reference [14], which tested the nature frequency of the thrust bearings with and without a groove experimentally, and proved that the bearings without a groove is better damped than the bearing with a groove. Nevertheless, it can be concluded from Fig 8 that the stability decreases when the groove depth increases.

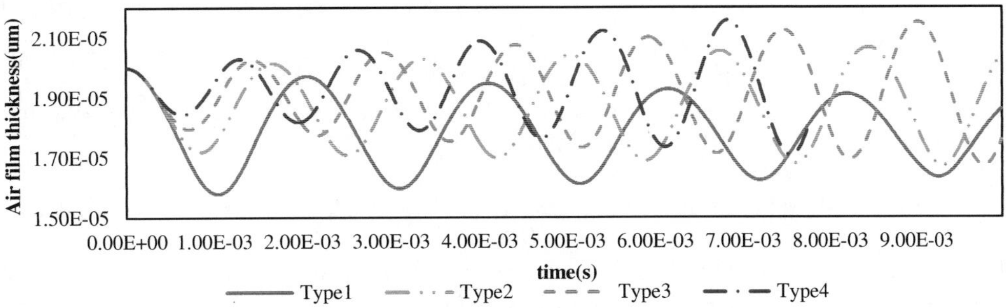

Figure 7. The dynamic step responses of the 4 types thrust air bearings for $w_1 = 5N$.

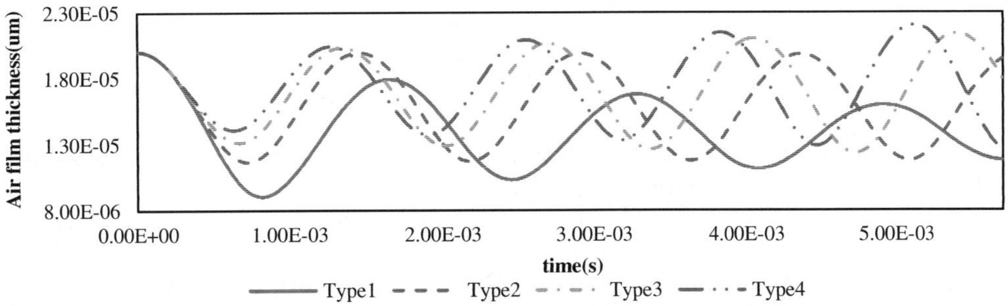

Figure 8. The dynamic step responses of the 4 types thrust air bearings for $w_2 = 20N$

With an external loading of 20N, the dynamic step responses of Types 3 and 4 are still not damped, but damped for Type 1 and Type 2 thrust bearings, and the Type 1 thrust bearing still performs the best stability. It needs to be noticed that for the external load $w_1 = 5N$, Type 2 cannot get a damped response; but for the external load $w_2 = 20N$, the dynamic response for Type 2 thrust bearing got damped. This means the stability of thrust bearing can be greatly affected by the external load. Same conclusion can be found in Reference [14].

4. Conclusions

In this paper, a numerical model of thrust air bearings, based on FDM and FVM, is presented. The load capacity, compressed air consumptions, and the dynamic response of hydro-static thrust air bearings are investigated based on the proposed model. The numerical results show that the thrust bearings with a circular groove can always get a higher load capacity and higher air consumption than the thrust bearing without the groove, and the load capacity and the air consumption is proportional to the groove depth. The bearing without the groove is better damped than the bearings with the groove, which means the stability of thrust bearing decreases when the groove depth increases. In addition, hydrostatic thrust bearings with a circular groove would lose the stability with a high external load. These findings can be taken into consideration in the design of a thrust air bearing.

Acknowledgments

The research is supported, in part, by European Horizon 2020 project 644971, Department for Transport T-TRIG 2017 project, and the scholarship from China Scholarship Council (CSC).

References

[1] Lipschitz A, Basu P, Johnson R P. A bi-directional gas thrust bearing[J]. Tribology transactions, 1991, 34(1): 9-16.

[2] Huitric J, Tournerie B. Finite element analysis of grooved gas thrust bearings and grooved gas face seals[J]. Journal of tribology, 1993, 115: 349.

[3] Artiles A, Shapiro W, Jones H F. Design analysis of Rayleigh-step floating-ring seals[J]. ASLE transactions, 1984, 27(4): 321-331.

[4]Radu F A, Pop I S, Knabner P. Newton—Type Methods for the Mixed Finite Element Discretization of Some Degenerate Parabolic Equations[M]. Numerical mathematics and advanced applications. Springer Berlin Heidelberg, 2006: 1192-1200.

[5] Waumans T, Peirs J, Reynaerts D, et al. On the dynamic of high-speed gas bearings: stability study and experimental validation[C]. Proceedings of the 3rd Conference on Sustainable Construction and Design. 2011: 342-351.

[6] Wang N, Chang C. An application of Newton's method to the lubrication analysis of air-lubricated bearings[J]. Tribology transactions, 1999, 42(2): 419-424.

[7] Wang N, Chang S H, Huang H C. Comparison of Iterative Methods for the Solution of Compressible-Fluid Reynolds Equation[J]. Journal of Tribology, 2011, 133(2): 021702.

[8] Wang N, Chang S H, Huang H C. Stopping Criterion in Iterative Solution Methods for Reynolds Equations[J]. Tribology Transactions, 2010, 53(5): 739-747.

[9] Arghir M, Alsayed A, Nicolas D. The finite volume solution of the Reynolds equation of lubrication with film discontinuities[J]. International Journal of Mechanical Sciences, 2002, 44(10): 2119-2132.

[10] Powell J W. Design of aerostatic bearings[J]. The Machinery Published Co. Ltd.1970.

[11] Neves MT, Schwarz VA, Menon GJ. Discharge coefficient influence on the performance of aerostatic journal bearings. Tribol Int 2010;43:746–51.

[12] Charki A, Bigaud D, Guérin F. Behavior analysis of machines and system air hemispherical spindles using finite element modeling. Ind Lubr Tribol 2013;65(4).

[13] Bonis M, Charki A. Modélisation des caractéristiques statiques et de la stabilité des paliers de butée aérostatiques par la méthode des éléments Finis[J]. Revue Européenne des Eléments, 2001, 10(6-7): 755-767.

[14] Charki A, Diop K, Champmartin S, et al. Numerical simulation and experimental study of thrust air bearings with multiple orifices[J]. International Journal of Mechanical Sciences, 2013, 72: 28-38.

IOP Conf. Series: Materials Science and Engineering **280** (2017) 012006 doi:10.1088/1757-899X/280/1/012006

Modeling and Analysis of the Static Characteristics and Dynamic Responses of Herringbone-grooved Thrust Bearings

Yunluo Yu, Guang Pu and Kyle Jiang*

University of Birmingham, Birmingham B15 2TT, United Kingdom

*K.Jiang@bham.ac.uk

Abstract. This paper describes a theoretical investigation of static and dynamic characteristics of herringbone-grooved air thrust bearings. Firstly, Finite Difference Method (FDM) and Finite Volume Method (FVM) are used in combination to solve the non-linear Reynolds equation and to find the pressure distribution of the film and the total loading capacity of the bearing. The influence of design parameters on air film gap characteristics, including the air film thickness, depth of the groove and rotating speed, are analyzed based on the FDM model. The simulation results show that hydrostatic thrust bearings can achieve a better load capacity with less air consumption than herringbone grooved thrust bearings at low compressibility number; herringbone grooved thrust bearings can achieve a higher load capacity but with more air consumption than hydrostatic thrust bearing at high compressibility number; herringbone grooved thrust bearings would lose stability at high rotating speeds, and the stability increases with the depth of the grooves.

1. Introduction

Herringbone grooved thrust air bearings have been widely employed in high speed and light weight rotational machinery, such as small precision motors, gas expanders, spindles because of its excellent load capacity and stability compared with other types of bearings. Herringbone grooves inscribed on either a rotating or stationary machine component pump lubricating air inward to raise air film stiffness and the threshold speed of the whirling stability. The grooves can also be engraved fully or partially on a surface with a certain angle inclined to the axial direction. The large variety and complex geometries of the grooves pose challenges for the designs of herringbone grooved thrust bearings. Nevertheless, the references dealing with the effective design methodology of herringbone grooved thrust bearings are very limited when compared with the references on the bearing characteristic analysis [1]. Therefore, even though herringbone grooved thrust bearings have been applied for a long time, there is still a need in theoretical studies to predict their performances accurately.

Fundamentally, numerical analysis of herringbone grooved thrust bearings is to study the bearing performance by solving the Reynolds Equation numerically. As a second order partial differential equation, there are three discrete schemes which can be used: Finite Difference Method (FDM), Finite Element Method (FEM) and the Finite Volume Method (FVM). Florin suggested that by applying Newton's method, it can reduce computational time significantly [2]. Newton's method is also proven to have better stability in getting converged solutions [3]. Nenzi Wang carried out a series of studies on numerical methods which can be applied to solve a compressible Reynolds Equation [4-6]. The work includes comparisons of different methods with respect to computational time and stability. He also proposed various criteria to stop an iterative process and analyzed the truncation errors of numerical approaches [4-6]. The above mentioned approaches are all based on FDM and they have good agreement with experiment results for static air bearings. The FVM is also adopted by some researchers. It is based on the continuity equation and Green's theorem. This approach shows advantages in dealing with film discontinuities and some pocket restrictor configurations [7]. In reference to the advantages of the FDM and FVM, these two methods are combined to solve the Reynolds Equation in this study.

In this paper, a numerical model for herringbone grooved thrust bearings is proposed based on the FDM and FVM; and then, the motion equations are added into the model to get the dynamic response of herringbone grooved thrust bearings; finally, the effects of several designs and working parameters on air film characteristics of herringbone grooved thrust bearings, including the air film thickness, rotating speed and groove depth, are investigated and compared with hydrostatic thrust air bearings.

2. Modeling

2.1. Reynolds equations and the finite difference scheme

In this section, a numerical model based on the FDM and FVM is developed for herringbone grooved thrust air bearings in order to study the effect of design parameters on air film characteristics, including the air film thickness, groove depth and rotating speed.

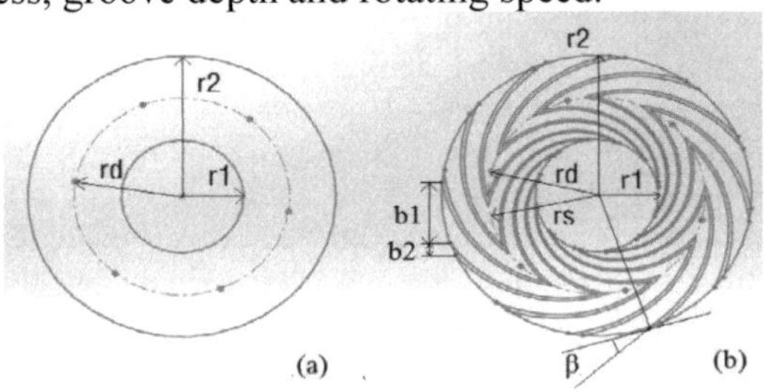

Figure 1. The geometries of (a) the hydrostatic thrust bearing and (b) the herringbone grooved thrust air bearing

The geometries of the herringbone grooved thrust air bearing studied in this paper and the hydrostatic thrust bearing which used for comparison are shown in Fig.1. In order to model arbitrarily shaped herringbone grooves of thrust bearings effectively, the coordinate system of the bearing surface can be transformed in reference to the boundaries of the grooves as shown in Fig.2.

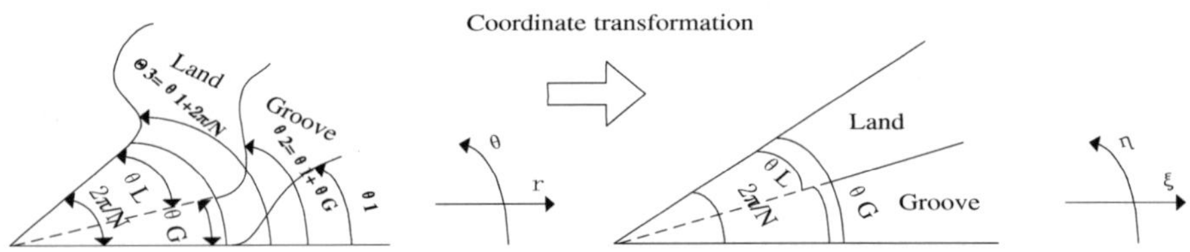

Figure 2. Bearing geometry transformation based on the boundary fitted coordinate system

The transformation functions of the coordinate system are listed as Eq.(1) and Eq.(2):

$$\xi = r \tag{1}$$

$$\begin{cases} Inner: \eta = \theta - In(r/r_s)/tan\beta \\ Outer: \eta = \theta - \left\{In\left(\frac{r_s}{r_2}\right) - In\left(\frac{r}{r_2}\right)\right\}/tan\beta \end{cases} \tag{2}$$

After introducing the dimensionless variables:

$$\zeta = \frac{\xi}{r_2}, P = \frac{p}{p_a}, H = \frac{h}{h_0}, u = \frac{\omega r_2}{2}, \tau = \omega t \tag{3}$$

where r_2 is the outer radius of the bearing, p_a the atmospheric pressure, h_0 the reference film thickness, h the film thickness, t the time. With the assumption that the isothermal perfect gas is used, the Reynolds equation of compressible flow can be written as:

$$\frac{\partial}{\partial\zeta}\left(\zeta PH^3\frac{\partial P}{\partial\zeta}\right)+\frac{1}{\zeta}\frac{\partial}{\partial\eta}\left(PH^3\frac{\partial P}{\partial\eta}\right)=\Lambda\zeta\frac{\partial(PH)}{\partial\eta}+2\Lambda\zeta\frac{\partial(PH)}{\partial\tau} \tag{4}$$

$$\Lambda=\frac{6\mu\omega r_2}{p_a{h_0}^2} \tag{5}$$

where Λ is known as the compressibility number or bearing number, μ the viscosity of the air film. To solve the Reynolds Equation, a control volume is defined as Fig.3. The mass flow rate Q^ζ, which across the boundary of $\zeta = constant$ and the mass flow rate Q^η, which across the boundary of $\eta = constant$ are, respectively, calculated as Eq.(6) and Eq.(7).

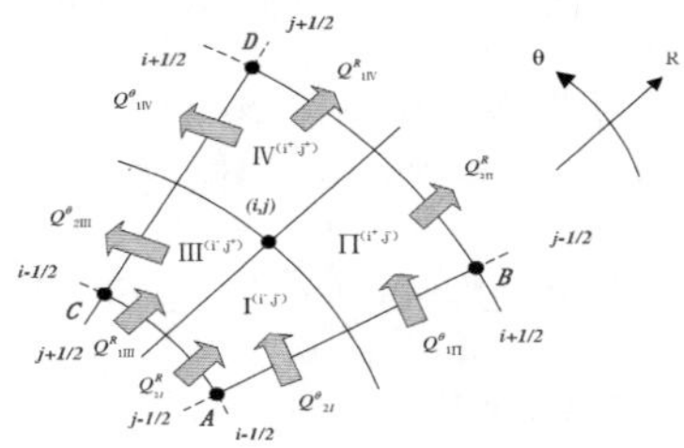

Figure 3. The defined control volume

$$Q^\zeta=\int_{\zeta_1}^{\zeta_2}(-\zeta PH^3\frac{\partial P}{\partial\zeta}+\Lambda\zeta PH)d\zeta \tag{6}$$

$$Q^\eta=\int_{\eta_1}^{\eta_2}(-PH^3\frac{\partial P}{\partial\eta})d\eta \tag{7}$$

The mass flow rate due to squeezing motion inside the control volume is expressed as follows:

$$Q^\tau=\int_{\eta_1}^{\eta_2}\int_{\zeta_1}^{\zeta_2}(2\Lambda\zeta\frac{\partial PH}{\partial\tau})d\zeta d\eta \tag{8}$$

The following relationship can be found from the mass flow rate balance on the control volume:

$$Q^\eta{}_{2I}+Q^\eta{}_{1III}-Q^\eta{}_{2II}-Q^\eta{}_{1IV}+Q^\zeta{}_{2I}+Q^\zeta{}_{1II}-Q^\zeta{}_{2III}-Q^\zeta{}_{1IV}=Q^\tau+Q^{orifice} \tag{9}$$

where $Q^{orifice}$ is the mass flow rate of the orifice. The dimensionless mass flow rate of the orifice can be calculated according to the following equations under isentropic assumption:

$$Q^{orifice}=\begin{cases}HC_s\left\{\frac{2}{\gamma-1}\left[\left(\frac{P_r}{P_s}\right)^{2/\gamma}-\left(\frac{P_r}{P_s}\right)^{(\gamma+1)/\gamma}\right]\right\}^{1/2}, & if\ \frac{P_r}{P_s}\geq\left(\frac{2}{\gamma+1}\right)^{(\gamma-1)/\gamma}\\ HC_s\left(\frac{2}{\gamma+1}\right)^{\left(\frac{(\gamma+1)}{2(\gamma-1)}\right)}, & if\ \frac{P_r}{P_s}<\left(\frac{2}{\gamma+1}\right)^{(\gamma-1)/\gamma}\end{cases} \tag{10}$$

$$C_s=\frac{12\mu\pi dC_dP_s}{h_0^2p_{at}^2}\sqrt{\gamma\Re T} \tag{11}$$

where T is the atmospheric temperature at supply conditions, C_d the discharge coefficient, d the orifice diameter, μ the dynamic viscosity of the fluid, γ the isentropic index and $\Re$ the gas constant. Powel [8] regarded C_d as a function of the pressure ratio $\frac{P_r}{P_s}$. However, many researchers take the coefficient as 0.8 for all flow conditions [9]. In this study, C_d is taken as 0.8 in all the cases. By solving Eq.(1), the pressure distribution of the lubricating can be found.

2.2. Calculation of static characteristics

For a thrust bearing experiencing a small vibration in the axial direction, the minimum air film thickness and pressure can be expressed as:

$$h = h_0 + \varepsilon e^{j\omega_f t}$$

$$p = p_0 + \varepsilon p_t e^{j\omega_f t} \quad (12)$$

where ε represents the function of air film thickness and the bound of ε is approximately within ± 10 percent of the steady state air film thickness. Subscripts 0 and t mean the physical quantities under the steady and unsteady operation conditions, respectively.

Based on the equilibrium on the control volume, the following relationships can be obtained:

$$F_0(P_0) = Q^{\eta}{}_{2I0} + Q^{\eta}{}_{1III0} - Q^{\eta}{}_{2II0} - Q^{\eta}{}_{1IV0} + Q^{\zeta}{}_{2I0} + Q^{\zeta}{}_{1II0} - Q^{\zeta}{}_{2III0} - Q^{\zeta}{}_{1IV0} - Q^{orifice} = 0 \quad (13)$$

$$F_t(P_t, P_0) = Q^{\eta}{}_{2Is} + Q^{\theta}{}_{1IIIs} - Q^{\eta}{}_{2IIs} - Q^{\eta}{}_{1IVs} + Q^{\zeta}{}_{2Is} + Q^{\zeta}{}_{1IIs} - Q^{\zeta}{}_{2IIIs} - Q^{\zeta}{}_{1IVs} - Q^{\tau}{}_{s} - Q^{orifice} = 0 \quad (14)$$

Eq. (13) is a nonlinear equation which can be solved by applying the Newton-Raphson iteration method. Eq.(14) is a linear equation which can be solved based on the steady state pressure P_0 obtained through Eq.(13).

In this work, maximization of the bearing load capacity and minimization of the consumption of compressed air are considered as the way to the optimal design of the herringbone grooved bearings. The consumption of compressed air is the air flow rate of all the orifice restrictors. The load capacity can be calculated using the following equation:

$$w = \int_{\xi_1}^{\xi_2} \int_0^{2\pi} (p_0 - p_a)\xi d\xi d\eta \quad (15)$$

2.3. Calculation of dynamic responses

The calculation requires to solve the Reynolds equation with the equation of motion [10-11] simultaneously. The Reynolds equations have been introduced in Section 2.2. The equation of motion can be written as:

$$\Delta w(t) = m\ddot{h} \quad (16)$$

where m is the mass supported by the bearing, Δw is the variation of the resultant pressure force. The dynamic response of thrust air bearing can be calculated using the following scheme:

1. Choose an equilibrium position of the simulated thrust bearing for dynamic calculation.
2. Input initial step displacements and velocities for step by step method.
3. Dynamic calculation of the movement (acceleration, velocity, displacement, pressure and load capacity, etc.)

3. Numerical simulation

3.1. Thrust bearings for simulation

Table 1. Specification of test bearings

Bearing Type	Specifications
Hydro-static (Type 1)	$r_1 = 6mm, r_2 = 15mm, r_d = 10.5mm,$
Herring bone (Type 2)	$r_1 = 6mm, r_2 = 15mm, r_d = 10.5mm, r_s = 10.5mm, h_g = 10\mu m, \beta = 16deg, \frac{b_1}{b_2} = 1.667N = 24$
Herring bone (Type 3)	$r_1 = 6mm, r_2 = 15mm, r_d = 10.5mm, r_s = 10.5mm, h_g = 20\mu m, \beta = 16deg, \frac{b_1}{b_2} = 1.667, N = 24$
Herring bone (Type 4)	$r_1 = 6mm, r_2 = 15mm, r_d = 10.5mm, r_s = 10.5mm, h_g = 30\mu m, \beta =$

	$16deg, \frac{b_1}{b_2} = 1.667, N = 24$

In this section, the herringbone grooved thrust bearing and the hydrostatic thrust air bearing shown in Fig.1 and Tab.1 are simulated using the proposed FDM and FVM model. Then, the influences of several design parameters of the thrust bearings on its characteristics have been investigated to provide information in the future design of hydrostatic thrust air bearing. The design parameters including the air film thickness, rotating speed, groove depth and external load.
Other fixed parameters are listed in Tab.2.

Table 2. Fixed parameters of bearings studied

Bearing parameters	Value	Bearing parameters	Value
Reference gap h_0(um)	20	Coefficient of discharge C_d	0.8
Radius of orifices r_d	10.5×10^{-3}	Isentropic index γ	1.4
Diameter of orifices d_0	250×10^{-6}	Gas constant $\Re$ (J kg^{-1} K^{-1})	287
Supply pressure p_s (Pa)	6×10^{-5}	Dynamic viscosity u (Pa s)	18.38×10^{-6}
Atmospheric pressure p_{at} (Pa)	1×10^{-5}	Temperature at supply conditions T (°K)	293

3.2. Analysis of static characteristics

In some applications, it is preferable to reduce air consumption while maintaining reasonable load capacity. For this reason, the influence of air film thickness, rotating speed and groove on load capacity and mass flow rate have been investigated .

When the rotor is stationary, the load capacity and mass flow rate of the thrust bearings against the air film thickness are respectively presented in Fig.4, where $p_s = 3\times10^5$ Pa, $n_{ori} = 6$ and $\omega = 0$ rpm. When the rotor speed is at300000 rpm, the load capacity and mass flow rate of all the orifice restrictors against the air film thickness are respectively presented in Fig.5.

Figs.4 (a) and (b) show that when the rotational speed is 0, it is the Type 1 thrust bearing, namely hydro-static thrust bearing, which achieves the best load capacity and consumed least compressed air. Moreover, from Fig.4 (a), it can be seen that with the increase in groove depth, the load capacity decrease and the air consumption increase. This is because that with the grooves, the average air film thickness of herringbone grooved thrust bearings is larger than the one without the grooves. Thus leads to a lower load capacity and higher air consumption.

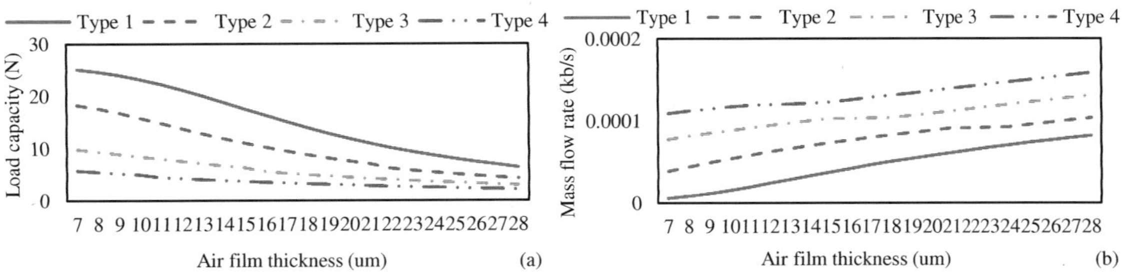

Figure 4. (a)The load capacity and (b) the mass flow rate of the 4 types of thrust air bearings versus the air film thickness when $\omega = 0$

From Fig.5 (a), it can be seen that at the speed of 300000 rpm and with a low air film thickness, Types 2 and 3 herringbone grooved bearings achieve a much higher load capacity than Type1, but Type 1 still achieve a higher load capacity than Type 4; with a high air film thickness, it is always Type 1 that achieves highest load capacity, but the difference with other 3 types of herringbone grooved thrust bearings are not obvious. This is because that the dynamic effect of the herringbone grooved thrust bearings is proportional to the compressibility number Λ. With a low air film thickness and a high rotational speed, the compressibility number Λ can achieve a high value, which means a strong dynamic effect. As illustrated in Fig.5 (a), thrust bearings Types 2 and 3 have a low air film thickness and high rotating speed and the hydrodynamic effect caused by the grooves is strong enough to compensate the loss in load capacity caused by the increase of average air film thickness raised by the grooves; but with a high air film thickness, the compressibility number Λ decrease to a low value, and at this time the hydrodynamic effect caused by the groove is not enough to compensate the loss in load capacity caused by the grooves. Therefore, hydrostatic thrust bearing Type 1 achieves the highest load capacity. However, if the groove depth is very large, the bearing's performance will drop. This can be concluded from the Type 4 bearing, which has the largest air film thickness and thus always get a lower load capacity than the hydrostatic thrust bearing. From Fig.5 (b), it can be seen that the air consumption of the herringbone grooved thrust bearings is higher than that of non-grooved hydrostatic air bearings, and the air consumption increases with the groove depth.

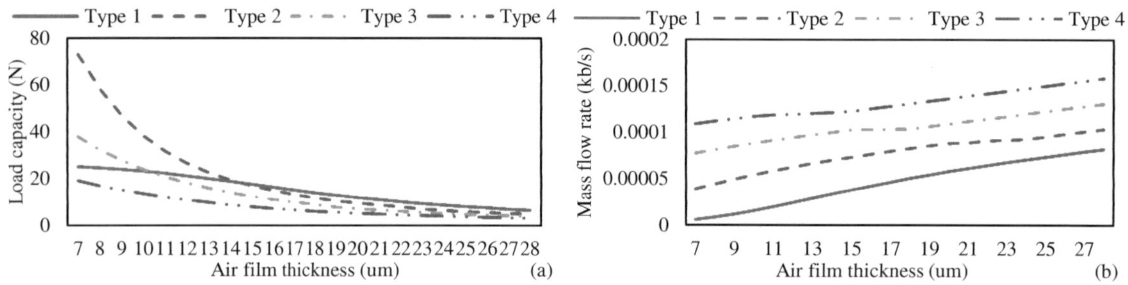

Figure 5.(a) The load capacity and (b) the mass flow rate of the 4 types of thrust air bearings versus the air film thickness when $\omega = 300{,}000$ rpm.

3.3. Analysis of dynamic response

Figs.6-7 give the dynamic step responses of the 4 types of thrust air bearings. The initial conditions used in all the cases are: $h_0 = 20\mu m$, $\dot{h}_o = 0$, $P_s = 6\ bar$, $w = 3N, dt = 10^{-5}$. The dynamic simulations take rotational speeds of $\omega_1 = 100{,}000$ rpm and $\omega_2 = 300{,}000$ rpm respectively.

From Fig.6, it can be seen that at $\omega_1 = 100{,}000$ rpm, the dynamic step responses of all the 4 types of thrust bearings are damped. However, from Fig.7, at speed $\omega_2 = 300{,}000$ rpm, thrust bearings Type 2 and Type 3 become undamped, while Types 1 and 4 remain damped. Therefore, it can be concluded herringbone grooved thrust bearings would lost the stability at high rotating speeds. Nevertheless, the stability of thrust bearings increases with the depth of the grooves; the same conclusion can be found in Reference [12]

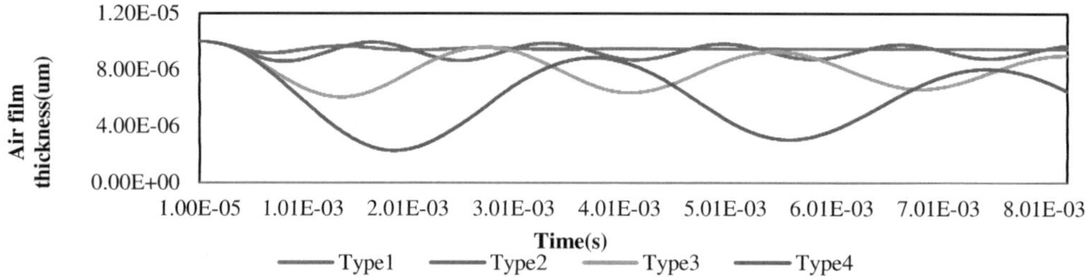

Figure 6. The dynamic step responses of the 4 types of thrust air bearings at a rotational speed of $\omega_1 = 100{,}000$ rpm.

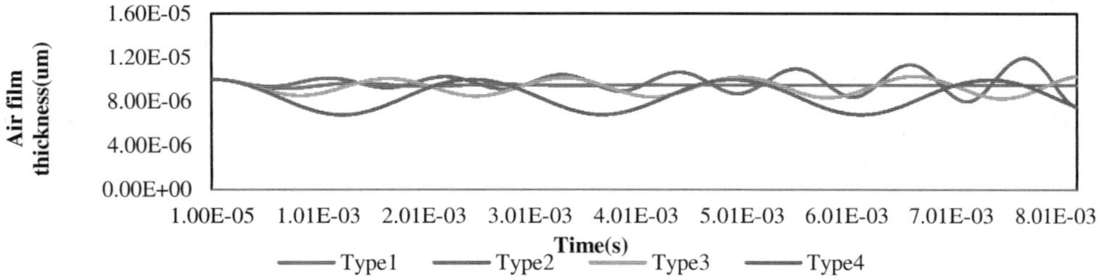

Figure 7. The dynamic step responses of the 4 types of thrust air bearings at a rotational speed of $\omega_2 = 300{,}000$ rpm.

4. Conclusions

In this paper, a numerical model of herringbone grooved thrust bearings, based on FDM and FVM, is presented and the load capacity, consumption of compressed air, and the dynamic response of herringbone grooved thrust bearings are investigated based on the proposed model. The influence of design parameters of the thrust bearings, including the air film thickness, rotating speed, groove depth and external load, on its characteristics have been investigated. Numerical results obtained show that hydrostatic thrust bearing can achieve a better load capacity and less air consumption than herringbone grooved thrust bearings at low compressibility number. Herringbone grooved thrust bearings can achieve a better load capacity but more air consumption than hydro-static thrust bearing at high compressibility number. In addition, herringbone grooved thrust bearings would lose the stability at high rotating speeds, and the stability of thrust bearings increases with the depth of the grooves. Same conclusions can be found in Reference [12].

Acknowledgments

The research is supported, in part, by European Horizon 2020 project 644971, Department for Transport T-TRIG 2017 project, and the scholarship from China Scholarship Council (CSC).

References

[1] Hashimoto, Hiromu, and Masayuki Ochiai. "Theoretical analysis and optimum design of high speed gas film thrust bearings." Journal of Advanced Mechanical Design, Systems, and Manufacturing 1.1 (2007): 102-112.

[2] Radu F A, Pop I S, Knabner P. Newton—Type Methods for the Mixed Finite Element Discretization of Some Degenerate Parabolic Equations[M]. Numerical mathematics and advanced applications. Springer Berlin Heidelberg, 2006: 1192-1200.

[3] Waumans, T., et al. "On the dynamic stability of high-speed gas bearings: stability study and experimental validation." Sustainable Construction and Design 2.2 (2011): 342.

[4] Wang, Nenzi, and Chinyuan Chang. "An application of Newton's method to the lubrication analysis of air-lubricated bearings." Tribology transactions 42.2 (1999): 419-424.

[5] Wang, Nenzi, Shih-Hung Chang, and Hua-Chih Huang. "Comparison of Iterative Methods for the Solution of Compressible-Fluid Reynolds Equation." Journal of Tribology 133.2 (2011): 021702.

[6] Wang, Nenzi, Shih-Hung Chang, and Hua-Chih Huang. "Stopping Criterion in Iterative Solution Methods for Reynolds Equations." Tribology Transactions 53.5 (2010): 739-747.

[7] Arghir, Mihai, Ahmad Alsayed, and Daniel Nicolas. "The finite volume solution of the Reynolds equation of lubrication with film discontinuities." International Journal of Mechanical Sciences 44.10 (2002): 2119-2132.

[8] Powell J W. Design of aerostatic bearings[J]. The Machinery Published Co. Ltd.1970.

[9] Neves MT, Schwarz VA, Menon GJ. Discharge coefficient influence on the performance of aerostatic journal bearings. Tribol Int 2010;43:746–51.

[10] Charki A, Bigaud D, Guérin F. Behavior analysis of machines and system air hemispherical spindles using finite element modeling. Ind Lubr Tribol 2013;65(4).

[11] Bonis M, Charki A. Modélisation des caractéristiques statiques et de la stabilité des paliers de butée aérostatiques par la méthode des éléments Finis[J]. Revue Européenne des Eléments, 2001, 10(6-7): 755-767.
[12] Theoretical analysis and optimum design of high speed air film thrust bearings (Application to optimum design problem).

IOP Conf. Series: Materials Science and Engineering **280** (2017) 012007 doi:10.1088/1757-899X/280/1/012007

Optimum Parameters of a Tuned Liquid Column Damper in a Wind Turbine Subject to Stochastic Load

M H Alkmim[a], M V G de Morais[c] and A T Fabro[b]

Departament of Mechanical Engineering, University of Brasília, Campus Universitário Darcy Ribeiro, Brasília, 70910-900, Brazil

[a]mansour.alk@gmail.com, [b]fabro@unb.br, [c]mvmorais@unb.br

Abstract. Parameter optimization for tuned liquid column dampers (TLCD), a class of passive structural control, have been previously proposed in the literature for reducing vibration in wind turbines, and several other applications. However, most of the available work consider the wind excitation as either a deterministic harmonic load or random load with white noise spectra. In this paper, a global direct search optimization algorithm to reduce vibration of a tuned liquid column damper (TLCD), a class of passive structural control device, is presented. The objective is to find optimized parameters for the TLCD under stochastic load from different wind power spectral density. A verification is made considering the analytical solution of undamped primary system under white noise excitation by comparing with result from the literature. Finally, it is shown that different wind profiles can significantly affect the optimum TLCD parameters.

1. Introduction

Remarkable progress in wind turbine technology has been made over the past years with advances in the field of structural and dynamic analysis, which allow for the creation of larger and more efficient wind turbines [1].

The premise of a vibration control device is that if a secondary system composed of a mass, a damping, and a spring is installed on a primary structure with its natural frequency tuned to a very close dominant mode of the primary structure, reductions of the dynamic response can be achieved.

In particular, among energy dissipation systems, tuned liquid column dampers (TLCDs), first proposed by [2] and [3] are emerging in several specialized publications (e.g. [4]–[7]) and have become a good option due to its relatively low cost and good efficiency.

The TLCD operates based on the movement of the liquid column. In this paper, the TLCD has a "U" shape. The TLCD requires no extra mechanism such as springs or joints, furthermore, its geometry may vary according to design needs, making them very versatile devices

To avoid solving nonlinear simultaneous equations, solutions such as statistical linearization [8] and parameter optimization [9] have been proposed in previous works. Yalla and Kareem [10] derived a closed-form solution for optimized TLCD damping ratio and head loss coefficient. Although the method does not rely on an iterative procedure, in order to solve the minimum variance integrals, a formula is derived indirectly considering some properties of the spectrum of the stationary output of a linear time-invariant system to white noise input. Altay et al. [11] presented an expanded optimization approach, which considers the geometric layout of the damper. A numerical verification is carried out by stochastic inflow turbulence simulator TurbSim and the aero-elastic dynamic horizontal axis wind turbine simulator FAST.

Optimization procedures have also been proposed for structures with mounted TLCD under harmonic load. By maximizing the reduction of peak structural response under harmonic excitation, Gao [12] designed the optimum tuning parameters for a TLCD for a wide frequency range. Shum [13] proposed a close-form optimal solution by optimizing the response of primary structure at two invariant points using the fixed-point method with a perturbation technique.

This paper aims to propose an optimization approach, based on a global direct search optimization algorithm, in order to find the optimum TLCD parameters for reducing vibration levels in slender structures such as wind turbines, when subject to an arbitrary stationary random wind excitation. Four different wind models are investigated given by a power spectral density (PSD) profile and its effects on the optimum parameters are discussed.

This paper is organized as follow. The mathematical description of TLCD and the wind turbine model are presented in Section 2 Section 3 presents the optimization approach as well as a numerical verification. Section 4 shows the stochastic induced wind load models and a FFT based approach for time history simulation. Finally, Section 5 presents some concluding remarks.

2. TLCD and structure modelling

Considering the TLCD rigidly connected on the primary structure as sketched in Figure 1 and the "U" shaped support with negligible mass when compared to fluid and constant cross section, it is possible to model the structure as a one degree of freedom model with equivalent mass, stiffness and damping. The equation describing the motion of the fluid and structure is given by

$$M\ddot{x}(t)+C\dot{x}(t)+Kx(t)=F(t) \tag{1}$$

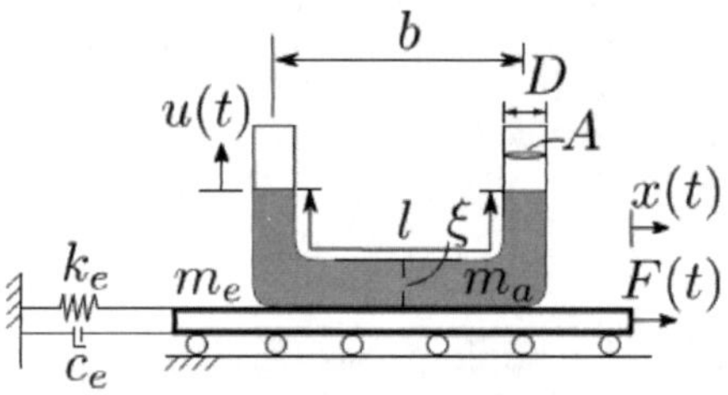

Figure 1: Schematic model of the system.

where M, C, K are the mass, damping and stiffness matrices are given by

$$M=\begin{pmatrix}1+\mu & \alpha\mu\\ \alpha & 1\end{pmatrix},\quad C=\begin{pmatrix}2\omega_e\zeta_e & 0\\ 0 & 2\omega_a\zeta_a\end{pmatrix},\quad K=\begin{pmatrix}\omega_e^{\,2} & 0\\ 0 & \omega_a^{\,2}\end{pmatrix} \tag{2}$$

where the mass matrix is in its dimensionless form, and the system is already linearized. The dimensionless parameters $\mu=m_a/m_e$ and $\alpha=b/l$ are the mass ratio and length ration where m_a and m_e are the damper and structure mass respectively and b and l are the horizontal and total length to the damper's tube. The ω_e and ω_a are the structure and damper natural frequencies and we introduce the dimensionless parameter tuning ratio $\gamma=\omega_a/\omega_e$. Finally, ζ_e and ζ_a are the structure and damper damping ratio respectively.

3. Parameter optimization criteria

From the linearized system, a frequency response function (FRF) is calculated which will be used in the random vibration analysis. The frequency response functions are obtained by assuming the system in Eq. (1) oscillates under harmonic motion. The system response vector is then given

$$\begin{aligned} X(\omega) &= [-\omega^2 M + j\omega C + K]^{-1} F(\omega) \\ &= H(j\omega)F(\omega) \end{aligned} \tag{3}$$

where ω stands for the driving frequency, $H(j\omega)$ is the frequency response function , j is the imaginary unit.

Considering the exciting force $F(t)$ as a stationary random signal with power spectral density (PSD) $S_{ff}(\omega)$, the structural response is also random and with PSD given by $S_{yy} = HS_{ff}H^T$ [14] where the superscript T denotes the matrix transpose.

The desired performance index (cost function) J will be defined by the mean square response. If the exciting force is zero mean then the response also has zero mean, therefore the mean square response equals the variance, thus

$$\begin{aligned} J(\zeta_a, \gamma) &= \int_{-\infty}^{\infty} S_{yy}(\omega) d\omega \\ &= \int_{-\infty}^{\infty} H(j\omega) S_{ff}(\omega) H(j\omega)^T d\omega \end{aligned} \tag{4}$$

where the response PSD $S_{yy}(\omega)$ is real positive and, therefore, the sufficient and necessary conditions for the optimization are met. For the prescribed frequency range $[\omega_l, \omega_u]$ the optimization problem consists of looking for the parameters that minimize the variance of response PSD, i.e.

$$\begin{aligned} &\min_{\zeta_a, \gamma \in \Omega} J_i(\zeta_a, \gamma) \\ &s.t. \qquad \zeta_a, \gamma \geq 0 \\ &I = \{i \mid \omega_l \leq \omega_i \leq \omega_u\} \end{aligned} \tag{5}$$

where Ω is the set of design parameters ζ_a and γ satisfying the constraint. To solve the optimization problem we introduce the optimization algorithm.

The Generalized Patter Search (GPS) is a class of direct search methods, originally proposed for unconstrained minimization problems [15], and then extended in its generalized form for problems with bound and linear constraints [16]. GPS is a non-gradient-based algorithm therefore, it is not as strongly affected by random noise in the cost function. However, it requires more function evaluations than gradient-based algorithms to find the true minimum. A more detailed description of the method can be found in [17]. The MATLAB environment is used to perform the GPS algorithm with its included solver *patternsearch*.

In order to verify if the algorithm and cost function are adequate for solving the optimization problem, a simple case where the closed-form solution is available is investigated.

Considering the case of undamped primary structure $\zeta_e = 0$ and white noise excitation, Yalla and Kareem [10] solved the mean variance problem to obtain analytical expressions for optimal parameters. Figure 2 shows the optimized parameters from Yalla and Kareem and the proposed algorithm as a function of the mass ratio for different length ratio. It can be noticed a very good agreement. A slight difference can be seen between the two methods in the optimum damping ratio for large values of mass ratio which can be attributed to tolerance error in the optimization algorithm and the chosen cost function's interval of integration that, in this case, were chosen between 2 and 5 rad / s. However, the optimized damping ratio sensibility does not influence the characteristics of the system. The wind excitation PSDs considered in this paper are detailed in the next section.

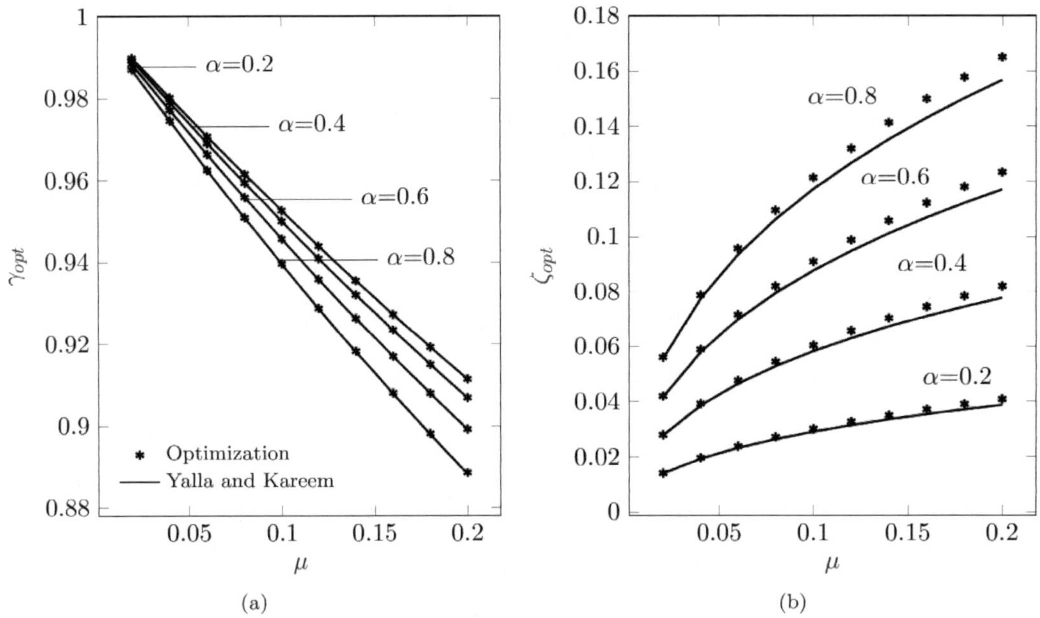

Figure 2: Comparing optimized (a) tunning ratio and (b) damping ratio subject to White Noise spectrum as a function of the mass ratio for different length ratio.

4. Wind profile power spectral density

Wind excitations are highly dynamic, irregular external loads. This section discusses how these could be simulated through the different PSDs such as white noise, Kanai-Tajime, Kaimal and Davenport whose expressions are given

White Noise:

$$S_{WN}(\omega) = S_0 \tag{6}$$

Kanai-Tajimi:

$$\begin{aligned} S_{KT}(\omega) = (1 + 4\xi_g^2(\omega / \omega_g)S_0) \\ /[1 + (\omega / \omega_g)^2 + 4\xi_g^2(\omega / \omega_g)^2] \end{aligned} \tag{7}$$

Kaimal

$$\begin{aligned} S_{Kai}(\omega) = [4S_0^2[L_k / v_{hub})) \\ /[1 + (6\omega(L_k / v_{hub}))]^{5/3} \end{aligned} \tag{8}$$

Davenport

$$\begin{aligned} S_{Dav}(\omega) = 4\kappa L v_{hub}^2 \chi / (1 + \chi^2)^{4/3}, \\ \chi = \omega L / v_{hub} \end{aligned} \tag{9}$$

For Kaimal PSD, $L_k = 340.2\,m$ is a scale parameter that involves the wind turbine high and $v_{hub} = 16\,m/s$ is the mean wind velocity. For Davenport PSD, κ is the drag coefficient referred to the mean velocity and L is the hub height. For Kanai-Tajimi PSD, $\omega_g = 10.5\,rad/s$ and $\xi_g = 0.317$ can be interpreted as characteristic frequency and characteristic damping ratio respectively.

Figure 3 shows the obtained optimized TLCD tunning ratio γ_{opt} and damping ratio ζ_{opt} subject to different wind spectra as a function of the mass ratio μ for a fixed length ratio $\alpha = 0.8$ and 1% primary structural damping. The optimum TLCD damping ratio ζ_{opt} increases for increasing mass ratio μ, and it is only slightly affected by the choice of wind spectrum. On the other hand, the optimum tunning ratio γ_{opt} decreases for increasing mass ratio and it is significantly affected by the choice of wind spectrum. Kaimal and Davenport spectra present almost the same values of tunning ratio, which are overall increasingly smaller for increasing mass ratio, when compared to the white noise spectrum. Furthermore, the choice of wind spectrum can influence the response magnitude and therefore it is relevant for the appropriate choice of optimum TLCD parameters.

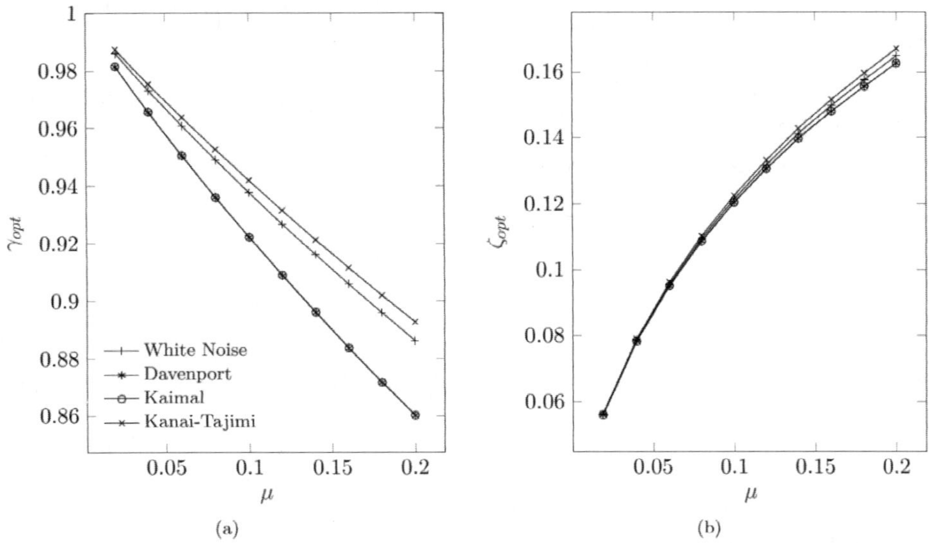

Figure 3: Optimum (a) tunning ratio and (b) damping ratio subject to different wind spectra as a function of the mass ratio for a fixed length ratio $\alpha = 0.8$ and 1% primary structural damping.

5. Concluding remarks

In this paper, it is proposed an optimization approach to find the optimum TLCD parameters for reducing vibration levels in slender structures such as wind turbines, when subject to an arbitrary stationary random wind excitation. The proposed optimization criteria is chosen such that it minimizes the area under the response PSD. Four different wind models, given by a PSD profile, are investigated and it is shown that they can significantly affect the choice of the optimum parameters. A verification of the proposed approach is made considering the case of undamped primary structure under white noise excitation, which has an analytical solution available. The TLCD design parameters can be given by its damping ratio ζ and tunning ratio γ. It is shown that the optimum TLCD damping ratio ζ_{opt} increases for increasing mass ratio and it is only slightly affected by the choice of wind spectrum. Besides, the optimum tunning ratio γ_{opt} decreases for increasing mass ratio and it is significantly affected by the choice of wind spectrum.

Acknowledgements
The authors would like to thank FAPDF, CNPq and MCTI for the financial support (Process number 406895/2013-9).

References

[1] P. Murtagh, A. Ghosh, B. Basu, and B. Broderick, "Passive control of wind turbine vibrations including blade/tower interaction and rotationally sampled turbulence," *Wind Energy*, vol. 11, no. 4, pp. 305–317, 2008.

[2] R. J. McNamara, "Tuned mass dampers for buildings," *J. Struct. Div.*, vol. 103, no. 9, pp. 1785–1798, 1977.

[3] F. Sakai, S. Takaeda, and T. Tamaki, "Tuned liquid column damper-new type device for suppression of building vibrations," in *International Conference on Highrise Buildings, Nanjing, China, Mar*, 1989, pp. 25–27.

[4] S. Colwell and B. Basu, "Tuned liquid column dampers in offshore wind turbines for structural control," *Eng. Struct.*, vol. 31, no. 2, pp. 358–368, 2009.

[5] A. Farshidianfar, "Tuned Liquid Column Dampers with Period Adjustment Equipment for Earthquake Vibrations of High-rise Structures," in *6th National Congress on Civil Engineering, 6NCCE*, 2011.

[6] C. Roderick, "Vibration reduction of offshore wind turbines using tuned liquid column dampers," University of Massachusetts Amherst, 2012.

[7] K.-W. Min, J. Kim, and H.-R. Lee, "A design procedure of two-way liquid dampers for attenuation of wind-induced responses of tall buildings," *J. Wind Eng. Ind. Aerodyn.*, vol. 129, pp. 22–30, 2014.

[8] J. B. Roberts and P. D. Spanos, *Random vibration and statistical linearization*. Courier Corporation, 2003.

[9] G. N. Vanderplaats, *Numerical optimization techniques for engineering design: with applications*, vol. 1. McGraw-Hill New York, 1984.

[10] S. K. Yalla and A. Kareem, "Optimum absorber parameters for tuned liquid column dampers," *J. Struct. Eng.*, vol. 126, no. 8, pp. 906–915, 2000.

[11] O. Altay, C. Butenweg, S. Klinkel, and F. Taddei, "Vibration mitigation of wind turbine towers by tuned liquid column dampers," in *Proceedings of the IX international conference on structural dynamics, Porto, Portugal*, 2014, vol. 12.

[12] H. Gao, K. Kwok, and B. Samali, "Optimization of tuned liquid column dampers," *Eng. Struct.*, vol. 19, no. 6, pp. 476–486, 1997.

[13] K. Shum, "Closed form optimal solution of a tuned liquid column damper for suppressing harmonic vibration of structures," *Eng. Struct.*, vol. 31, no. 1, pp. 84–92, 2009. dampers," *J. Sound Vib.*, vol. 272, no. 3, pp. 893–908, 2004.

[14] D. E. Newland, *An introduction to random vibrations, spectral & wavelet analysis*. Courier Corporation, 2012.

[15] R. Hooke and T. A. Jeeves, "Direct Search Solution of Numerical and Statistical Problems," *J. ACM JACM*, vol. 8, no. 2, pp. 212–229, 1961.

[16] V. Torczon, "On the convergence of pattern search algorithms," *SIAM J. Optim.*, vol. 7, no. 1, pp. 1–25, 1997.

[17] T. G. Kolda, R. M. Lewis, and V. Torczon, "Optimization by direct search: New perspectives on some classical and modern methods," *SIAM Rev.*, vol. 45, no. 3, pp. 385–482, 2003.

IOP Conf. Series: Materials Science and Engineering **280** (2017) 012008 doi:10.1088/1757-899X/280/1/012008

A New Design of the Test Rig to Measure the Transmission Error of Automobile Gearbox

Yixuan Hou[1], Xiaoqin Zhou[1,*], Xiuzhi He[1], Zufei Liu[1] and Qiang Liu[1]

[1] School of Mechanical Science and Engineering, Jilin University, Changchun, 130022,China

* Email adress: xqzhou@jlu.edu.cn

Abstract. Noise and vibration affect the performance of automobile gearbox. And transmission error has been regarded as an important excitation source in gear system. Most of current research is focused on the measurement and analysis of single gear drive, and few investigations on the transmission error measurement in complete gearbox were conducted. In order to measure transmission error in a complete automobile gearbox, a kind of electrically closed test rig is developed. Based on the principle of modular design, the test rig can be used to test different types of gearbox by adding necessary modules. The test rig for front engine, rear-wheel-drive gearbox is constructed. And static and modal analysis methods are taken to verify the performance of a key component.

1. Introduction

Gearbox is a critical component in mechanical transmission, which is widely used in automobile to transform the power generated from engine. It can be regarded as a self-excited vibration source[1], which affects the performance of the entire vehicle. Generally, gearbox is a complicated assembly including gears, shafts, bearings, synchronizers and housings. Above all components, gears plays an significant role in the process of power transmission. Gear transmission error (TE) has been considered as a main source of noise and vibration in gear system, which has been corroborated by several investigators[2-5]. TE is defined as the deviation between the theoretical position and the actual position of the output shaft if the gear drive were perfectly accurate and rigid[6]. However, because of the existence of manufacturing error and teeth deformation, TE is unavoidable in gear system. Current research is mainly concentrated on the measurement of the TE for single gear drive[7-9]. And TE measurement in a complete automobile gearbox under actual operating conditions were rarely conducted[4,10].

To accurately measure the TE of automobile gearbox and investigate the origin of TE, a kind of electrically closed test rig is developed. Due to the adoption of energy-recycling technology, the cost of operation and maintenance is relatively more economic compared with open test rig. The test rig can be used for TE and vibration measurement of gear pairs or complete gearbox. Besides, the system can work flexibly on testing for several types of gearbox by adding necessary modules. The overall structure and operation principle of the test rig are described subsequently.

2. Design of the test rig

Generally, the gearbox test rig can be divided into two types: open system and closed system, according to the energy cycling methods. The limitation of open system is that the energy generated by power source is consumed without recirculation, and loading on gearbox requires additional energy. Although the expenditure of construction for open test rig is relatively lower, subsequent operation and maintenance need further investment. As for the closed system, the energy can be recirculated via feedback device. The energy consumption is much lower than the open system, especially for high-power and long-time testing. Therefore, the closed testing method is chosen, more specifically, electrically closed method.

Based on the principle of modular design, the test rig can satisfy the test requirement for various types of gearbox. And the features of the measurement object determine the structure of the test rig.

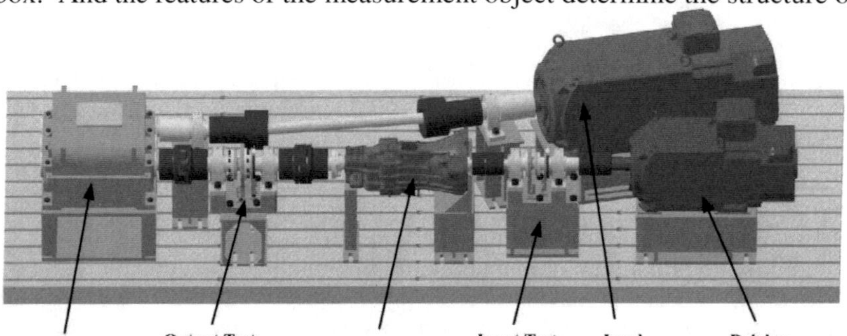

Figure 1. The structure of the test rig for TE measurement of FR gearbox

The front engine, rear-wheel-drive (FR) gearbox includes one input shaft and one output shaft. Accordingly, the overall arrangement of the traditional FR gearbox test rig is linear type, which means occupying a larger floor space. And the output power of the gearbox has high speed and low torque. In order to match the output power of the gearbox and the mechanical characteristic of the load motor, a single speed reducer or back-to-back slave gearbox is always employed to transmit the power or choose a high-power load motor directly.Thanks to the using of self-designed double speed reducer and universal coupling, the TE detection test rig can be installed on a smaller iron floor, which measures 4 by 2 meters. Specifically, the input and output shaft of the reducer are on the same side, so the transmission route of the test rig can be folded. And the gear ratio of the reducer is designed to meet the needs for power matching.

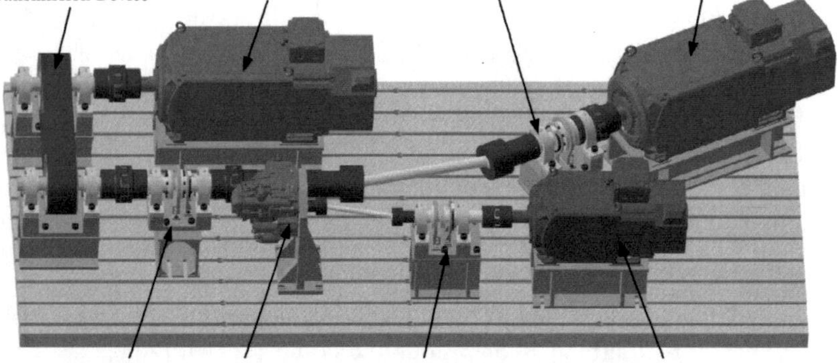

Figure 2. The structure of the test rig for TE measurement of FF gearbox

The front engine, front-wheel-drive (FF) gearbox includes one input shaft and two output shafts. The input shaft and a output shaft locate on one side, and the other output shaft situates on the contrary side. Thus, the structure of FF gearbox test rig is 'h' type. Compared with FR gearbox, the output power of FF gearbox has relatively lower speed and higher torque because of the integration of a main reducer. In order to shorten the length of the test rig, a synchronous belt transmission device is used to fold the transmission route. The synchronous belt transmission device can transmit the output power of the gearbox with an accurate transmission ratio. Therefore, the FF gearbox test rig can be installed on the same iron floor, which increases the universality of the gearbox test rig.

The structures of the test rig for TE measurement of FR and FF gearbox are shown in figure 1 and figure 2 respectively. And the test rig for TE measurement for FR gearbox is built based on the existing conditions of laboratory, which is shown in figure 3. The detailed structure and working principle of the FR test rig will be introduced subsequently.

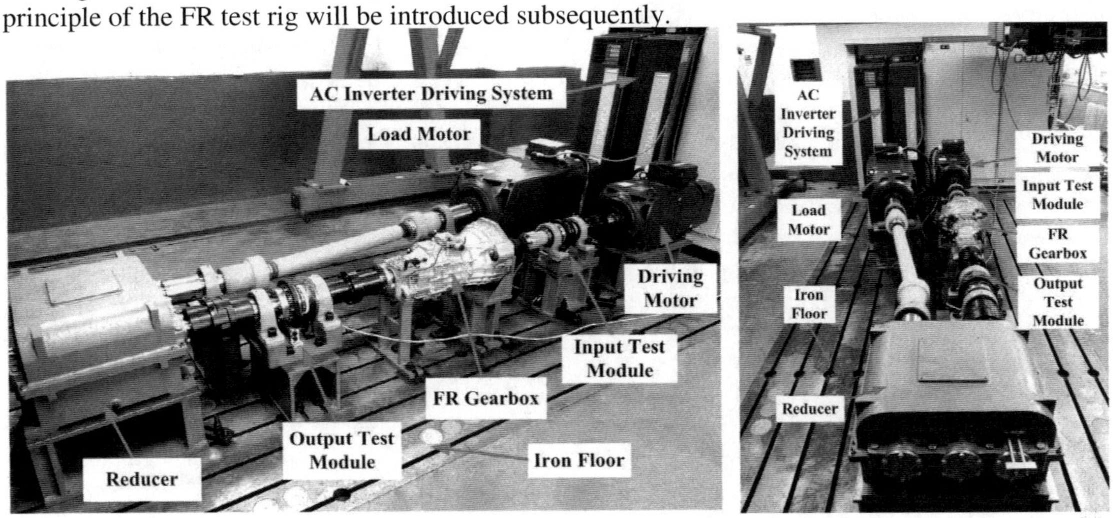

Figure 3. The FR gearbox test rig for TE measurement

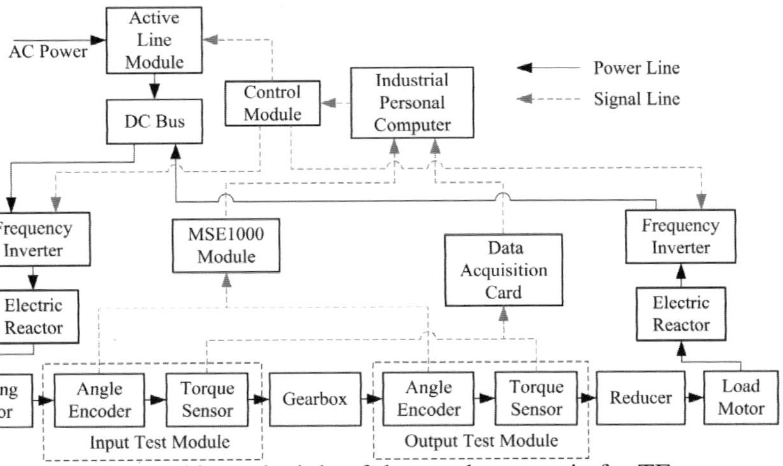

Figure 4. The working principle of the gearbox test rig for TE measurement

The working principle of the test rig are shown in figure 4. The test rig includes two alternating current (AC) motors, serving for driving and loading respectively. The gearbox is powered by the driving motor capable of speeds up to 3000 rpm, and the input torque can arrive at about 150 Nm. The load motor serves as a dynamometer to simulate the load. The driving speed and loading torque can be regulated with in a certain range. Electricity comes from AC power is transformed into direct current (DC) by active line module (ALM) , and transmitted into the DC bus. The power of the driving motor comes from the DC bus via frequency inverter and electric reactor, and the electricity generated by the load motor is fed back to the DC bus through electric reactor and frequency inverter. In this way, the energy can be recycled in the system. The overall power consumption of the test rig depends on the transmission efficiency, and the test rig can be operated much more economically. The ALM and frequency inverters are controlled by the industrial personal computer (IPC) via control module.

The detection of angular motion and torque is achieved by the input and output test module. The structure of the two modules are identical in structure, as is shown in figure 5. Taking the output test module as an example, the angular motion and torque of the output shaft of the gearbox are measured by the Heidenhain angle encoder (ERA4280C) and HBM T40B torque sensor respectively. The Heidenhain angle encoder is consisted of reading head, circular grating and protective cover. The circular grating is mounted on the flange shaft and rotates with the shaft. The reading head and protective cover are fixed on the encoder support. In order to simplify the structure and save space, HBM T40B torque sensor is chosen to measure the torque. The torque sensor is composed by a rotor and a stator, and they are non-contact. The rotor is a torque flange with strain gauges inside, while the stator is used to receive the signal transmitted by the rotor. All the rotating components are supported by two bearing blocks to enhance the rigidity of the module and prevent sensors from damaging. The output signal of angle encoder is received by MSE1000 data acquisition module, and transmitted to MSE software to get angle motion data. Voltage or frequency signal can be chosen as the output signal of torque sensor. Limited to the sampling frequency of the data acquisition card, voltage signal is chosen. The angle motion and torque signal are imported into LabVIEW software for post-processing.

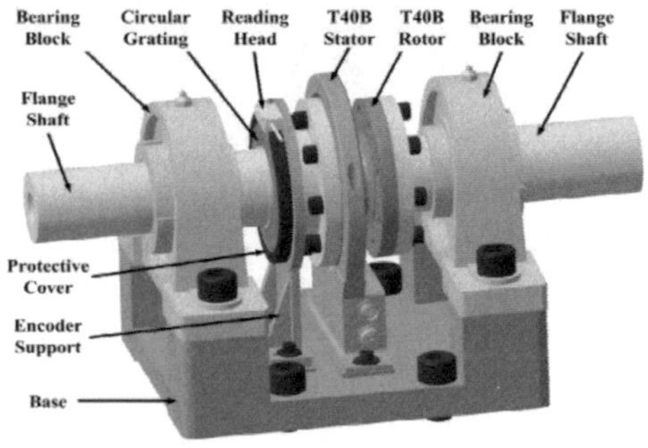

Figure 5. The structure of the TE test module

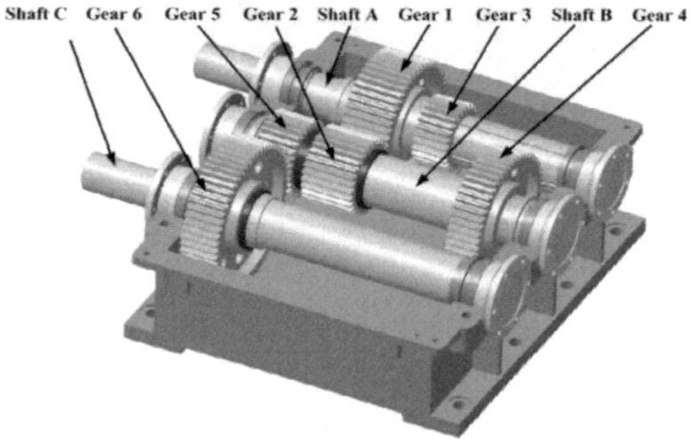

Figure 6. The structure of the reducer

A reducer is designed to match the output speed of the gearbox with the speed of load motor, figure 6 presents the structure of the reducer. The reducer includes two speeds which is realized by using double step sliding gear. And the double step sliding gear is driven by the shifting fork and guide bar, which is shown in figure 7. The transmission ratio of the reducer can can be selected manually according the output speed of the gearbox to match the mechanical characteristic of the load motor. Thanks to the using of the reducer, the load motor can operate at the best working point. Gear 5 and 6 are always in constant mesh. In low-speed mode, gear 1 and 2 are engaged, and the overall reduction ratio is 1.2. In high-speed mode, gear 3 and 4 are engaged, and the overall reduction ratio is 4.

3. Analysis of the reducer

The reducer is the key power conversion device of the test rig, and the performance of the reducer affects the measurement of the TE. As is shown in figure 8, the reducer is modeled in Romax Designer software to realize static analysis and modal analysis under actual loads. And the finite model of reducer housing is shown in figure 9.

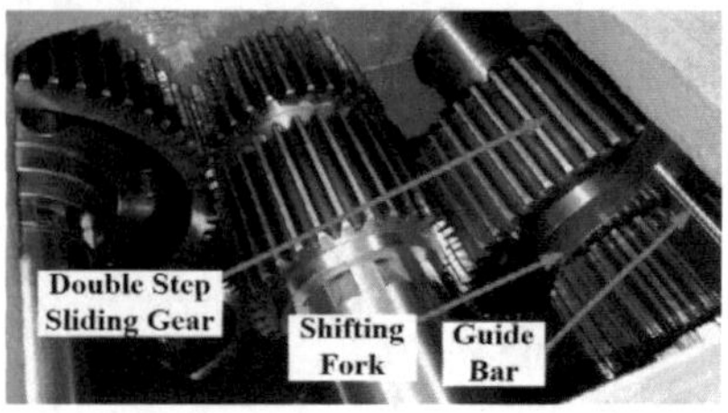

Figure 7. The shifting mechanism of the reducer

Figure 8. Romax model of the reducer

Figure 9. FE model of the reducer housing

3.1. Static analysis

Figure 10 and figure 11 present the deflection of transmission components and the distribution of von Mises stress on the housing respectively. The peak value of shaft deflection magnitude is located on shaft B at 1st speed, and the peak stress is 5.75 Mpa, which is much less than allowable stress. The maximum of von Mises stress is located on the ribs of bearing hole for shaft B, and the peak stress is about 5.81 Mpa, which is also a safe value compared with the yield limit 200Mpa. Thus, the structural strength of the reducer can satisfy the requirement for gearbox loading.

(a) 1st speed (b) 2nd speed

Figure 10. Deflection of transmission components

(a) 1st speed (b) 2nd speed

Figure 11. Distribution of von Mises stress on the housing

3.2. Modal analysis

The inherent characteristic of the housing has an important influence on the vibration and noise performance of the reducer. The modal analysis can be used to analyze the inherent frequency and modal shape of the housing to avoid resonance.The modal analysis of the reducer housing result in 20 inherent frequencies in the frequency range 510-1699Hz. Boundary conditions for housing are fixed. Generally, the low-order modes are more commonly used in modal analysis than high-order modes[11]. Because the high-order modes would attenuate sharply with the effect of damping. And it means that high-order vibrations can hardly generate resonance.

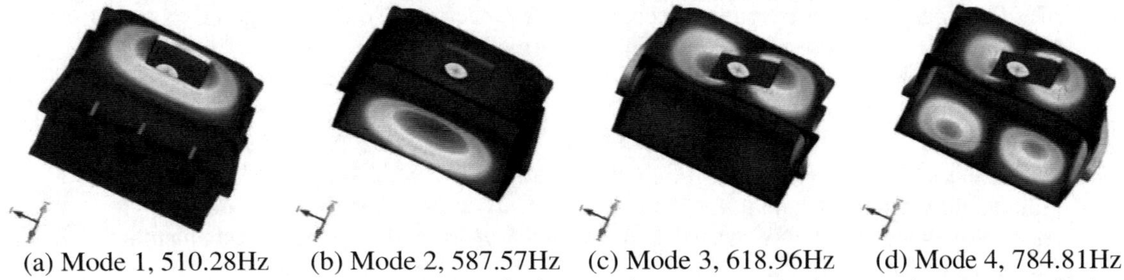

(a) Mode 1, 510.28Hz (b) Mode 2, 587.57Hz (cutaway section) (c) Mode 3, 618.96Hz (cutaway section) (d) Mode 4, 784.81Hz (cutaway section)

Figure 12. First four order modes of the housing

The first four order modal shape and inherent frequency of the housing are shown in figure 12. In the first mode, the upside and the bottom of the housing vibrates along Y axis in positive direction. In the second mode, the bottom of the housing vibrates along Y axis in positive direction, too. In the third mode, the upside of the housing twists around X axis, and the left and right sides of the housing vibrate along Z axis in positive direction. In the fourth mode, the upside and bottom of the housing twists around X axis, and the left and right sides of the housing vibrate along Z axis along negative direction.

The mesh frequencies (MF) of the gears in the reducer are listed in table 1. And the MF of the six gears do not coincide with the inherent frequencies of the housing, thus avoiding housing resonance. Therefore, the performance of the reducer can satisfy the dynamic matching task.

Table 1. Mesh frequencies of the gears in the reducer

	Gear1/Gear2	Gear3/Gear4	Gear5/Gear6
1st speed	1116.9 Hz	1003.8 Hz	
2nd speed		1350.0 Hz	762.5 Hz

4. Conclusion

In this paper, a test rig is designed to measure the TE of automobile gearbox, which is based on electrically closed method. The detailed structure and working principle of the test rig is introduced. The test rig can be used to measure different kinds of gearbox via adding necessary modules. And FR gearbox test rig is constructed based on the existing conditions of the laboratory. The FR gearbox test rig includes a two speed reducer to match the output speed of gearbox and the input speed of load motor. Static and modal analysis methods are taken to validate the performance of the reducer. The analysis result indicates that the reducer can satisfy the dynamic matching task for TE measurement.

Acknowledgement

This work was funded by the Jilin Province Industrial Technology Research and Development Project of China (Grant No.2014Y092 Automobile gearbox transmission error tracing technology research) & Changchun Major Science and Technology Research Project of China (Grant No.14KG043 Automobile gearbox dynamic transmission error measurement under actual operating conditions)

Reference

[1] Gwangmin P, Kyusik P and Bonghyun L 2015 *Correlation of transmission error prediction with vibration measurement for helical gears*, Proc. of the 22nd International Congress on Sound and Vibration, 12-16 July, Florence, Italy

[2] Smith J D 2003 *Gear noise and vibration*, Marcel Dekker Inc, New York

[3] Houser D R, Oswald F B, Valco M J, Drago R J and Lenski J W 1994 *Comparison of transmission error predictions with noise measurements for several spur and helical gears*, NASA Technical Memorandum, n 106647, p 12

[4] Henriksson M 2004 *Analysis of gear noise and dynamic transmission error measurements*, Proc. of 2004 ASME International Mechanical Engineering Congress and Exposition: Applied Mechanics, 13-20 November, Anaheim, California USA, pp 229-237

[5] Velex P and Ajmi M 2006 *On the modelling of excitations in geared systems by transmission errors*, Journal of Sound and Vibration, Vol. 290, Issues 3-5, pp 882-909

[6] Gregory R W, Harris S L and Munro R G 1963 *A method of measuring transmission error in spur gears of 1:1 ratio*, Journal of Scientific Instruments, Vol. 40, pp 5-9

[7] Hotait M A and Kahraman A 2013 *Experiments on the relationship between the dynamic transmission error and the dynamic stress factor of spur gear pairs,* Mechanism and Machine Theory, Vol 70, pp 116-128

[8] Tamminana V K, Kahraman A and Vijayakar S 2007 *A study of the relationship between the dynamic factors and the dynamic transmission error of spur gear pairs,* Journal of Mechanical Design, Vol 129, pp 75-84

[9] Tuma J 2006 *Simple gear set dynamic transmission error measurement*, Proc. of the 13rd International Congress on Sound and Vibration, 2-6 July, Vienna, Austria

[10] Henriksson M 2009 *Transmission Error As Gear Noise Excitation*, Proc. of the ASME International Design Engineering Technical Conferences & Computers and Information in Engineering Conference, 30 August-2 September, San Diego, California USA

[11] Zhu X and Xiong H 2013 *Finite element modal analysis of reducer box based on romax*, Journal of Mechanical Transmission, Vol. 37, Issue 2, pp 94-96

IOP Conf. Series: Materials Science and Engineering **280** (2017) 012009 doi:10.1088/1757-899X/280/1/012009

Study on Gap Flow Field Simulation in Small Hole Machining of Ultrasonic Assisted EDM

Yu Liu, Hao Chang, Wenchao Zhang, Fujian Ma, Zhihua Sha and Shengfang Zhang*

School of Mechanical Engineering, Dalian Jiaotong University, Dalian,China 116028

liuyu_ly12@126.com, C2669112070@163.com, chaochao_0627@163.com, mafujianyx@163.com, shirlysha@hotmail.com, zsf@djtu.edu.cn*

Abstract. When machining a small hole with high aspect ratio in EDM, it is hard for the flushing liquid entering the bottom gap and the debris could hardly be removed, which results in the accumulation of debris and affects the machining efficiency and machining accuracy. The assisted ultrasonic vibration can improve the removal of debris in the gap. Based on dynamics simulation software Fluent, a 3D model of debris movement in the gap flow field of EDM small hole machining assisted with side flushing and ultrasonic vibration is established in this paper. When depth to ratio is 3, the laws of different amplitudes and frequencies on debris distribution and removal are quantitatively analysed. The research results show that periodic ultrasonic vibration can promote the movement of debris, which is beneficial to the removal of debris in the machining gap. Compared to traditional small hole machining in EDM, the debris in the machining gap is greatly reduced, which ensures the stability of machining process and improves the machining efficiency.

1. Introduction

Electrical discharge machining (EDM) as a non-traditional machining method uses the electrothermal effect of pulse discharge between two electrodes to remove materials in dielectric fluid. As there is no mechanical contact between the tool and workpiece during the whole process, it isn't restricted to mechanical properties of the strength of material and the hardness of material, etc. thus EDM has been successfully extended to the field of small hole[1]. However, a lot of debris is generated in the small hole EDM machining process. If the debris in machining gap is not removed in time, it will be accumulated and linked between the electrode and workpiece, which easily leads to the arc discharge and short circuit, affecting machining quality and material removal rate. Flushing is a common method for promoting debris removal in EDM operation, it is a process that makes dielectric fluid get into the machining gap, which will improve fluid mobility and be beneficial to debris removal. But the research indicates that when machining the hole with high aspect ratio, the large machining depth makes flushing difficult to enter the machining gap and to remove the debris[2]. When the depth to diameter ratio of hole is large, the problem of debris removal can be solved with the help of ultrasonic vibration.

Aimed at the problem of debris removal ultrasonic assisted EDM, Y Imai et al. made researches on the effect of the high-frequency response of piezoelectric actuator on the speed of EDM and summarized that the high-frequency vibration can increase the machining speed in fine machining, high smooth surface and $\phi 240\mu m$ micro-hole machining approximately 1.5-2.5 times faster than before[3]. Garn et al. conducted experiments to analyse and concluded that ultrasonic vibration can weaken the effect of systematic delay on processing, reduce arc discharge and improve machining efficiency and surface quality[4]. SA Mastud et al. set up a simulation model of debris movement with ultrasonic vibration. The study found that the amplitude and frequency make the effect of debris movement and accelerate the debris movement[5]. T Zhu et al. researched the effect of ultrasonic vibration on the gap flow field in electrical-discharge milling and showed that the pressure field of cyclical ultrasonic frequency enhances cavitation effect, accelerates dielectric fluid circulation and improves machining stability and machining efficiency[6]. W Chang et al. studied the pressure field, velocity field and the concentration of debris particles in machining gap during a period of ultrasonic vibration with Fluent. Simulation results show that ultrasonic vibration is beneficial to the removal of machining debris from inter-electrode gap[7].

Given many available models, due to the debris movement in high aspect ratio hole EDM machining with ultrasonic vibration and side flushing is not yet fully understood. This paper developed a model to simulate the distribution and removal of debris when the depth to diameter ratio is 3 in ultrasonic assisted EDM with side flushing in fluid dynamical simulation software Fluent and analyse the influence of different parameters during the operation.

2. Establishing for the Model of Inter-Electrode Gap Field

The debris removal process in ultrasonic assisted EDM process for machining small hole is shown in figure 1 The flushing circulates the dielectric liquid in the bottom gap. Meanwhile, amplitude transformer drives the tool to do up and down reciprocating ultrasonic vibration. In this process, due to enhancement of cavitation effect, the electrode causes pressure difference and debris that are generated between two electrodes along with the working liquid is pushed to the region with lower pressure from the region with higher pressure. As the greater mobility of debris and the disturbance of tool vibration in the bottom gap, debris is removed from the bottom gap and side gap, finally removed out of small hole.

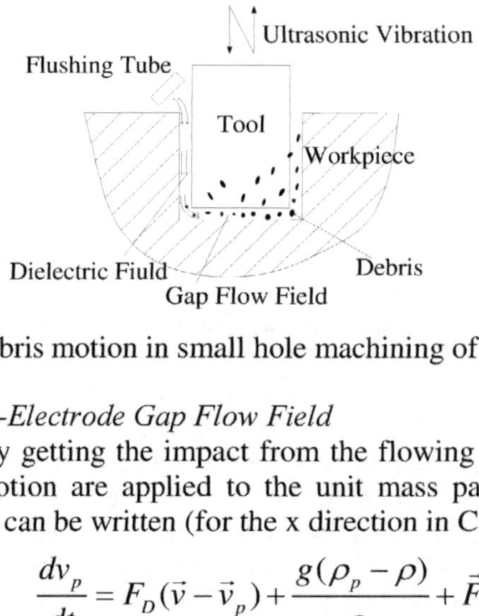

Figure 1. Process of debris motion in small hole machining of ultrasonic assisted EDM.

2.1. Mathematical Model of Inter-Electrode Gap Flow Field

Debris gains speed and moves by getting the impact from the flowing of dielectric liquid in the gap flow field. Newton's laws of motion are applied to the unit mass particles and derive differential kinetic equation[8]. The equation can be written (for the x direction in Cartesian coordinates) as

$$\frac{dv_p}{dt} = F_D(\vec{v} - \vec{v}_p) + \frac{g(\rho_p - \rho)}{\rho_p} + \vec{F} \tag{1}$$

where $F_D(\vec{v} - \vec{v}_p)$ is the drag force per unit particle mass, $\vec{v}_p$ is the particle velocity, $\vec{v}$ is the fluid phase velocity, F_D is expressed as $F_D = \dfrac{18\mu C_D \mathrm{Re}}{24\rho_p d_p^{\,2}}$. μ is the molecular viscosity of the fluid, d_p is the particle diameter, ρ is the fluid density and ρ_p is the particle density. $\vec{F}$ is an additional acceleration term and here consider to be the ultrasonic vibration force.

$$\vec{F} = \frac{dv_u}{dt} = \frac{d\left[2\pi fA\cos\left(2\pi ft+\phi\right)\right]}{dt} \tag{2}$$

where, f is the frequency of ultrasonic vibration, A is amplitude, t is processing time, v_u is tool velocity.

2.2. Modeling for the Inter-Electrode Gap Flow Field

The gap flow field of small hole machining process is considered as a 3D cylinder model which is established and meshed in the Gambit. The electrode and flushing are both cylinders. The machining depth is 3mm. The tool electrode diameter is 1mm. Considering the mobility of working fluid, the region above the electrode is appropriately much larger. Figure 2 is the model of the gap flow field in ultrasonic assisted EDM. In figure 2, 1 is the wall of small hole, 2 is the surface of workpiece, 3 is tool electrode, 4 is flushing and 5 is the flow region of workpiece surface. The flushing tube and working fluid need to exchange data so that an interface boundary is applied, pipe orifice is set to velocity-inlet, upper boundary is set to pressure outlet and rest boundaries are set to walls as default shown as figure 2(a) . The debris is generated near the electrode and workpiece. In order to improve the accuracy and efficiency of calculation, the grid near the bottom of electrode is finely meshed and rest regions are roughly meshed, as shown in figure 2(b). According to the previous research, the width of side gap is set to 0.2mm and height of the bottom gap is set to 0.1mm. Hence, the cylindrical region (A1) is the bottom gap, circular region (A2) is the side gap and upper cylindrical region (A3) is the exterior flowing region, as shown in figure 2(c).

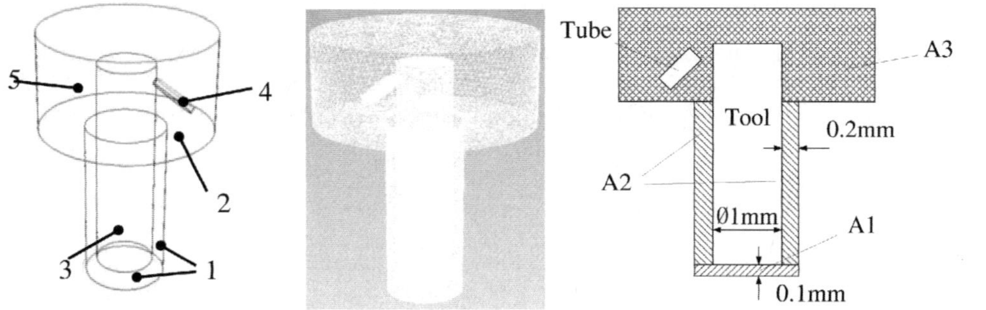

(a) 3D model (b) 3D meshed model (c) Schematics of regions

Figure 2. Model of the gap flow field in EDM.

This paper uses Fluent to simulate the real machining process. The dynamic mesh technique provided by Fluent is used to solve the problem that the shape of flow field is constantly changing in response to the ultrasonic vibration of the tool electrode. The smoothing and remeshing method are applied. The secondary development interface UDF module is also used to simulate the process[9]. In order to simulate the generation of debris in different positions, functions are programmed to generate random numbers. In terms of the fact that the random function on the computer is a fake random, a function *sleep* for delaying is called every time when generating a random number. Then *ranPosition* function is called to generate the positions each discharge. The debris particle materials are copper and Ti alloy respectively. Tool is red copper. Workpiece material is TC4. As the bottom of the electrode is applied with velocity-inlet, laminar-model is used to establish and simulate the flow field.

3. Simulation Results and Analysis

The process of different amplitudes and frequencies within 0.04s is simulated and the debris movement and the velocity of gap flow field are analysed. Pressure solver is set in Fluent. The time is transient. Working fluid is deionized water. The velocity of flushing is 2 m/s. The macro function

DEFINE_CG_MOTION is written and compiled to implement the ultrasonic vibration of electrode. Simulation time is 0.00001 and time step is 4000. The simulation results are shown as follows.

3.1. Impact of side flushing on the removal of debris

Figure 3 and figure 4 are the simulation results without ultrasonic vibration at 0.04s. Figure 3 is debris distribution and figure 4 is velocity field.

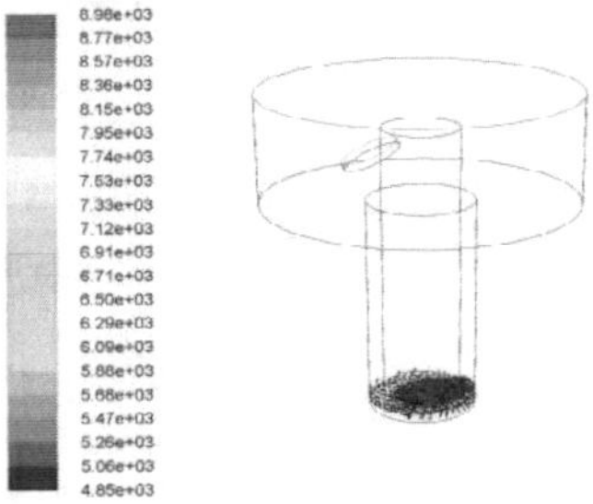

Figure 3. The debris distribution.

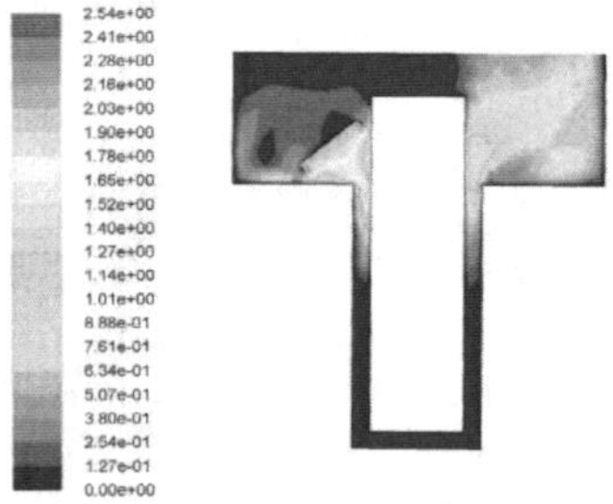

Figure 4. The velocity field.

It is shown in figure 3 that the abundant debris are gathered at the lower right corner and cannot be removed from the machining gap when there is only the effect of side flushing without electrode ultrasonic vibration. Because of the large depth of hole, the machining gap becomes narrow and flowing resistance is increased. It is hard for the flushing liquid overcomes the resistance to enter the side gap and to arrive at the bottom gap and it has a weak effect for the debris movement, which results in the accumulation of debris and could generate the arc discharge and short circuit. It is shown in figure 4 that the velocity of fluid in the bottom gap is approximately 0m/s. The flushing has little effect on the removal of debris.

3.2. Impact of vibration amplitude on the removal of debris

It is shown that the debris distribution with ultrasonic vibration frequency of 20kHz at 0.04s in figure 5 and the (a), (b) and (c) are respectively the debris distribution in the case of the amplitude of as 5μm, 10μm and 20μm.

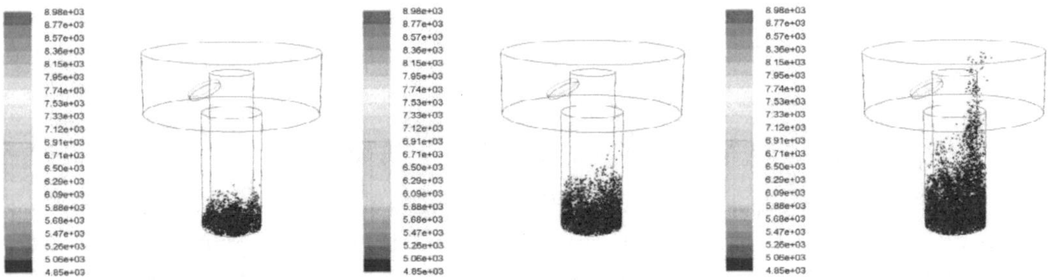

(a) Amplitude of 5μm (b) Amplitude of 10μm (c) Amplitude of 20μm

Figure 5. The debris distribution of the gap flow field.

Compared with the simulation result without ultrasonic vibration as shown in figure 3, the debris movement with ultrasonic vibration is more active, and most of debris in the bottom gap moves into the side gap and even is removed out of the machining gap. The accumulation of debris at bottom is obviously reduced. This is due to the fact that the electrode under the assistance of ultrasonic vibration does the up-and-down piston motion. When the electrode moves up, a negative pressure region is generated in the bottom gap of electrode and the pressure gradient of the entire gap flow field is greater. When the electrode moves down, the working fluid is squeezed by the lower surface of electrode and it is moved from the bottom gap to the side gap and even expelled. This process drives the debris to be removed from the machining gap. figure 6 presents the velocity field with different amplitudes in the gap flow field. It is found that ultrasonic vibration enlarges the effect of working fluid disturbance in the bottom gap and the mobility of fluid is increased. Increasing flow velocity can improve the carrying capacity of working fluid to debris and promote debris at the bottom gap to be removed from the small hole.

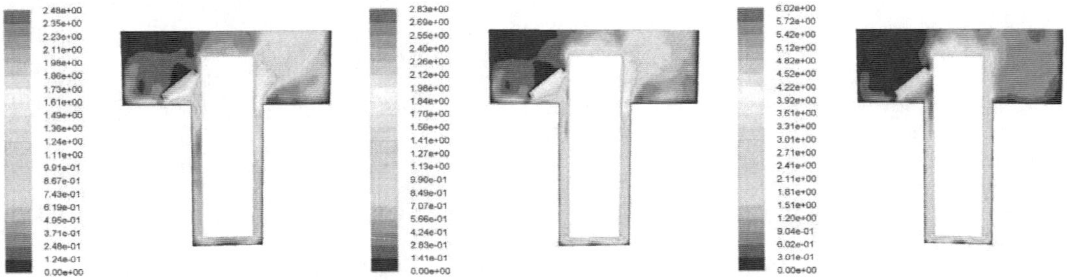

(a) Amplitude of 5μm (b) Amplitude of 10μm (c) Amplitude of 20μm

Figure 6. The velocity field of the gap flow field.

3.3. Impact of vibration frequency on the removal of debris

Figure 7 and figure 8 present the debris distribution with amplitude of 5 and frequencies of 30kHz, 40kHz respectively at the simulation time of 0.04s.

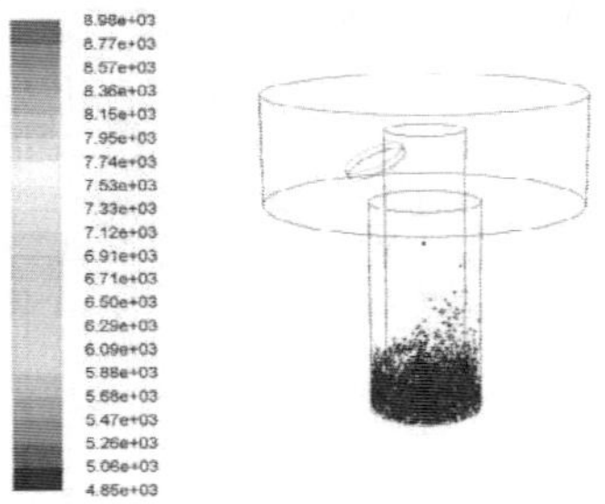

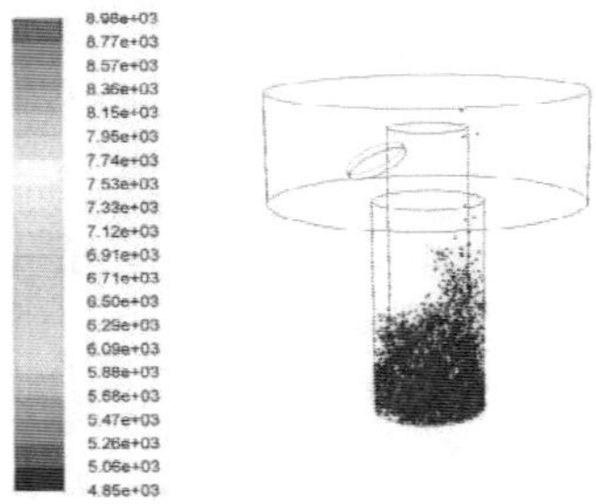

Figure 7. The debris distribution with 30kHz. **Figure 8.** The debris distribution with 40kHz

Compared with debris distribution in figure 5(a), it is found that debris movement become more disorder and violent when increasing frequency and most of debris moves from the bottom gap to the side gap. The velocity of working fluid at the bottom gap is increased and fluid mobility is better. The accumulation of debris is reduced. As the frequency is higher, the number of longitudinal movement of electrode is also increased and the disturbance of working fluid is more violent in the vertical direction. Debris along with the working fluid, are flowed out of the bottom gap. It is beneficial to the excluding of debris.

4. Statistics and Analysis of Debris Distribution

In different amplitudes and frequencies, it is respectively counted that the quantity of debris in A1 region (bottom gap), A2 region (side gap) and A3 region (exterior flowing region) to analyse the results and to find out the relevant laws.

4.1. Impact of vibration amplitude on the debris distribution

When processing 3mm deep hole and frequency with 20kHz, figure 9 presents the quantity variation of debris with the vibration amplitudes of 5μm,10μm and 20μm respectively in A3 region. It is shown that the quantity of debris is gradually increased in A3 region as the process progresses. By observing the velocity of debris removal (slope) with the three kinds of amplitudes, it is found that all of the velocity increases first and then incline to stable value. In addition, as the amplitude increases, the velocity of debris removed is remarkably faster. In unit time, the quantity of debris removed into A3 region is also remarkably increased.

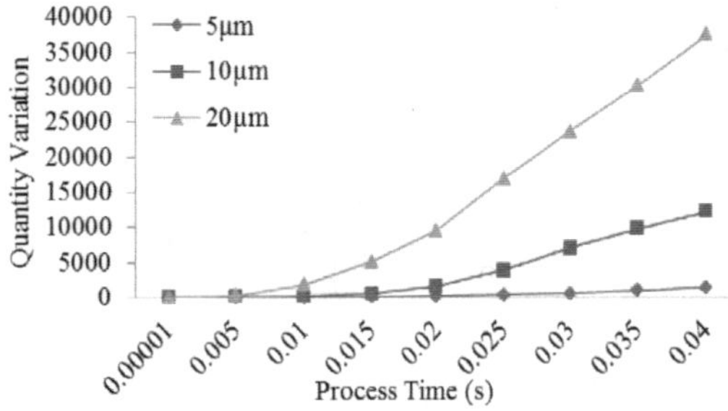

Figure 9. The quantity variation of debris.

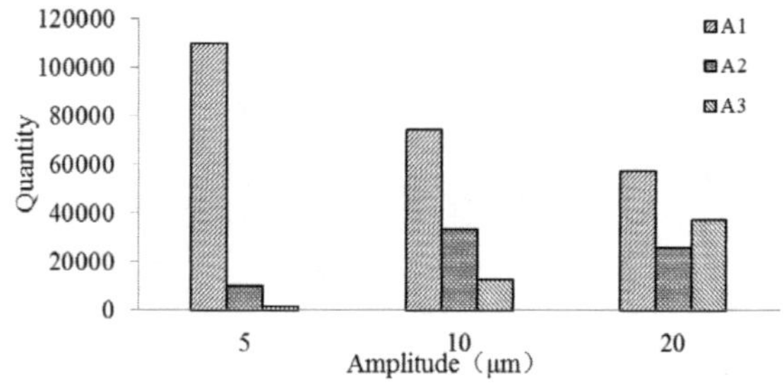

Figure 10. The quantity of debris distribution.

Figure 10 is the quantity of debris distribution with different amplitudes in A1, A2 and A3 regions at 0.04s. It is shown that when the amplitude increases, the quantity of debris in A1 region steadily reduces, the quantity of debris in A2 region increases first and then inclines to stable value and the quantity of debris in A3 region remarkably increases. According to the statistics, 121356 debris is generated in the bottom gap. When the amplitude is 5μm, the quantity of debris in A1 region is 109776, the quantity of debris in A2 region is 109776 that is relatively fewer and it in A3 region is1465 that is the fewest. The removal rate is 1.21% and the residual rate in the bottom gap is 90.46%. As increased to 10um, it is found that the removal rate is 10.1% and the residual rate is 61.24%. When it is 20μm, the quantity of debris in A1 region is less than that of 10μm. The A2 region is declined and the A3 region obviously increases. The removal rate is 30.9% and the residual rate is 47.40%. Therefore, with the growth of amplitude, the debris in exterior flowing region significant increases, the removal rate increases, debris that remains in the bottom gap obviously reduces, and the residual rate reduces. So, when vibration amplitude considerably increases, it is beneficial to the debris removal from the machining gap as well as the improvement of machining efficiency of small hole machining in ultrasonic assisted EDM.

4.2. Impact of vibration frequency on the debris distribution

When ultrasonic vibration amplitude is 5μm, figure.11 respectively presents the quantity variation of debris with the frequencies of 20kHz, 30kHz and 40kHz in A3 region. From figure11, it is observed that the quantity of debris in A3 region continuously grows from 0s to 0.04s. By observing the velocities of debris removal of three frequencies, the velocities increase first and then inclines to stable value. In addition, as the frequency increases, the velocity of debris removal is remarkably faster. When the frequency increases to 40kHz, the velocity and quantity of debris removal is significantly higher than the other two cases.

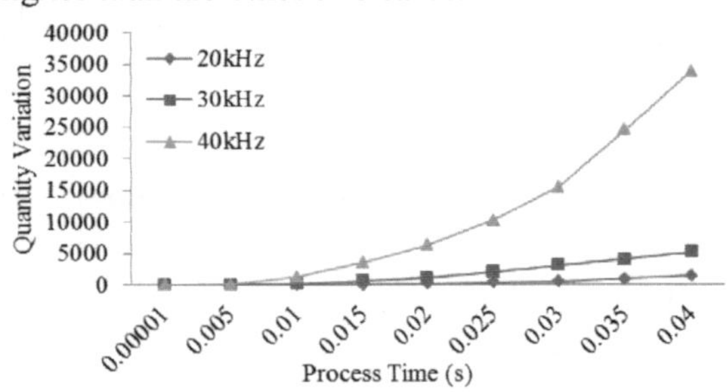

Figure 11. The quantity variation of debris.

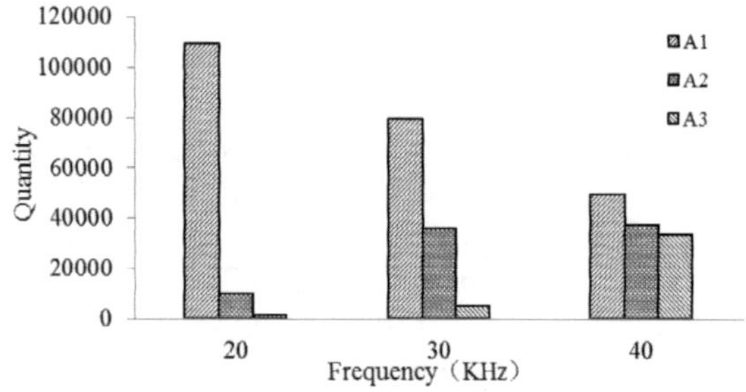

Figure 12. The quantity of debris distribution.

Figure 12 is the quantity of debris distribution with the frequencies of 20kHz, 30kHz and 40kHz in A1, A2 and A3 regions at the simulation time as 0.04s. The laws which the quantity of debris in A1 region, A2 region and A3 region are consistent with figure 10. It is found that most of debris in the bottom gap moves to the side gap and exterior flowing region as the vibration frequency increases. According to the statistical analysis, the removal rate is 1.21% when the frequency is 20kHz and the removal rate increases to 4.37% when it is 30kHz. When it is 40kHz, the removal rate that is 27.9% which is much greater than the former cases. On the contrary, the higher frequency reduces residual rate of debris and they are respectively 90.45%, 65.77%, 41.07%. So it can be seen that higher frequency increases the removal rate increase and reduces the residual rate of debris. Most of debris is removed to the exterior flowing region avoiding the accumulation of debris. The results also indicate that the increasing vibration frequency has a significant effect on the removal of debris.

5. Conclusions

This paper applies the Fluent software to simulate the process of the distribution and movement of debris in small hole machining with ultrasonic assisted EDM. The laws of different amplitudes and frequencies on debris removal, the debris distribution in each region and the flow velocity distribution in the bottom gap are quantitatively researched. Through the analysis, it is found that tool electrode assisted with ultrasonic vibration promotes inter-electrode working fluid to cycle, which is beneficial to the removal of debris in the process. Moreover, as the frequency and amplitude increase, the electrode of disturbance is more violent in the bottom gap, the flow rate of working fluid flowing through the machining gap is faster, and the debris along with the working fluid is more likely to be removed from the machining gap. In addition, the low ultrasonic vibration machining parameters are weak to accelerate the removal of debris. Increasing the vibration parameters obviously improves the removal of debris in machining gap, thus ensures the processing stability and improves the process efficiency.

Acknowledgments

The financial support from the National Natural Science Foundation of China under grant number 51405058, Scientific Research Platform Foundation of Liaoning Province under grant number JDL2016006 and Talent Special Foundation of Dalian City under grant number 2016RQ054 are acknowledged.

References

[1] Pradhan B B, Masanta M, Sarkar B R and Bhattacharyya B 2009 Investigation of electro-discharge micro-machining of titanium super alloy *J. The International Journal of Advanced Manufacturing Technology* **41(11)** pp 1094–1106
[2] Zhang Y, Jing H, Li M, Tang C and Yu Z 2011 Study on Micro Hole Drilling with High Aspect Ratio by Micro EDM Aided with Ultrasonic Vibration *J.Electromachining & Mould* **6** pp 63–66 (in Chinese)
[3] Imai Y, Nakagawa T, Miyake H, Hidai H and Tokura H 2004 Local actuator module for highly accurate micro-EDM *J. Journal of Materials Processing Tech* **149(1)** pp 328–333
[4] Garn R, Schubert A and Zeidler H. 2011 Analysis of the effect of vibrations on the micro-EDM process at the workpiece surface *J. Precision Engineering* **35(2)** pp 364–368
[5] Mastud S A, Kothari N S, Singh R K and Joshi S S 2015 Modeling Debris Motion in Vibration Assisted Reverse Micro Electrical Discharge Machining Process (R-MEDM) *J. Journal of Microelectromechanical Systems* **24(3)** pp 661–676
[6] Zhu T, Zhang J, Wang T, Liu C, Chang W and Song X 2011 Flow Field Simulation During EDM Milling Based on Fluent Software *J. Electromachining & Mould.* (in Chinese)
[7] Chang W, Chen Y, Zhang J, Gu X and Fang M 2015 Ultrasonic Vibration Aided Electrical-discharge Milling Flow Field and Debris Distribution Field Simulation *J. Journal of Basic Science and Engineering* **s1** pp 151–157 (in Chinese)
[8] Zhang W C, Liu Y, Zhang S and Ma F 2015 Research on the Gap Flow Simulation of Debris Removal Process for Small Hole EDM Machining with Ti Alloy *J. Advances in Computer Science Research* **39** pp 2121–2126
[9] Wang J, Wang Y G and Zhao F L. 2009 Simulation of Debris Movement in Micro Electrical Discharge Machining of Deep Holes *C. Materials Science Forum* vol.626–627 pp 267–272

IOP Conf. Series: Materials Science and Engineering **280** (2017) 012010 doi:10.1088/1757-899X/280/1/012010

Study on Abrasive Wear of Brake Pad in the Large-megawatt Wind Turbine Brake Based on Deform Software

Shengfang Zhang, Qiang Hao, Zhihua Sha, Jian Yin, Fujian Ma and Yu Liu*

School of Mechanical Engineering, Dalian Jiaotong University, Dalian, China 116028

zsf@djtu.edu.cn, haoqiang983@163.com, shirlysha@hotmail.com, yinjian429578951@sina.com, mafujianyx@163.com, liuyu_ly12@126.com*

Abstract. For the friction and wear issues of brake pads in the large-megawatt wind turbine brake during braking, this paper established the micro finite element model of abrasive wear by using Deform-2D software. Based on abrasive wear theory and considered the variation of the velocity and load in the micro friction and wear process, the Archard wear calculation model is developed. The influence rules of relative sliding velocity and friction coefficient in the brake pad and disc is analysed. The simulation results showed that as the relative sliding velocity increases, the wear will be more serious, while the larger friction coefficient lowered the contact pressure which released the wear of the brake pad.

1. Introduction

In recent years, the wind turbine industry develops rapidly, which increases the installed capacity of wind turbine. For the high-speed and heavy-load condition of the large-megawatt wind turbine brake, friction of the brake pad/disc causes wear in the brake system which leads to many problems indirectly. It is important to study the friction and wear of the brake pad/disc. Stadler et al. study the tribology behaviour of wear between metal sintered brake and C/C-SiC composite disc[1]. Cho et al. discuss the effect of interface transfer layer of metal-based brake and gray cast iron brake disc on the tribology properties[2]. For rough friction surface, Liu et al. analyse the maximum flash temperature, the contact pressure and real contact area of the rough surface by using the finite element method[3]. Bogdanvoich et al. research the temperature field distribution of the rough surface in the sliding friction process by experiments[4]. Straffelini et al. and Mikael et al. study the internal mechanism of the change of the contact surface of the friction material and the change of the wear rate with the enhanced particle size in the high speed state[5-6]. Uyyuru et al. investigate the effect of strengthening the volume fraction and size distribution on the friction and wear properties of aluminium matrix composites[7]. Zhou et al. take non-smooth surface of circular pits as an example, analyse the wear process of non-smooth surface pits with different diameters and depths[8].

The above researches mainly study the influence of friction material and surface morphology on the friction and wear, and do not analyse the wear rules in abrasion quantitatively. Based on the abrasive wear in the friction mechanism, the Archard wear model is improved in this paper. The abrasive wear model of the large-megawatt wind turbine brake pad is carried out with the finite element analysis by Deform-2D software. The effects of relative sliding velocity and friction coefficient on abrasive wear of brake pad are presented. And the wear rules of the modified Archard wear model are obtained.

2. Calculation model of abrasive wear

For the high-speed and heavy-load condition of the large-megawatt wind turbine brake, based on abrasive wear theory, the generalized linear wear model of Archard is improved, and the variation of the wear depth calculation model can be obtained. It is assumed that the wear depth is related to the relative sliding velocity and the interface pressure, which means that the wear depth can be calculated by the relative sliding velocity and the interface pressure in every unit time. The concrete Archard wear model is as follow[9]:

$$V = K\frac{PL}{H} \tag{1}$$

In which, V is the wear volume, P is the normal pressure of the contact surface in the brake pad/disc, L is the relative sliding distance between abrasive and brake pad, H is the hardness, K is the dimensionless wear coefficient which varies from 10^{-2} to 10^{-7} according to different contact conditions.

However, any interface in the actual project is not smooth absolutely. The actual contacting area is related to the actual interface of the micro-morphology, the point of stress on the brake pad/disc and the wear depth of the interface are closely related to the wear state. The stress changes with the wear depth of the interface, which leads to the variation of the wear degree. Therefore, for a concrete contact point of a simulated wear depth, the original Archard model is no longer applicable. In this case, the Archard model needs to be modified in the specific wear area, which means the stress σ and the relative sliding velocity v are seen as influencing factors. The concrete equation is as follow:

$$h = \frac{K}{H}\int_0^t [v(t)\sigma(t)]dt \tag{2}$$

A single wear is made up of intervals m, and the wear depth is:

$$h_{j,n} = \sum_{i=1}^{m} \Delta h_{j,n} = \sum_{i=1}^{m} \frac{K}{H} v_{j,n}\sigma_{j,n}\Delta t \tag{3}$$

In which, j is the iteration number, n is the contact points number.

So the total wear depth W is:

$$W = \sum_{j=1}^{k} h_{j,n} \tag{4}$$

In which, k is the total iteration number.

According to Equation (4), the quantitative results of abrasive wear can be obtained. The concrete process is divided into two steps: 1) the relative sliding velocity and the stress data which belong to the wear stage are extracted from the contact point. 2) those data are divided and put into the equation to solve.

3. Establishment of abrasive wear finite element model

Based on abrasive wear theory, the micro finite element model of abrasive wear is established by using Deform-2D software. The cone abrasive on the brake pad is 2μm, and the radius of the bottom fillet is 0.5μm. The material is copper-based powder metallurgy. The property is set to be rigid. The brake pad takes 20μm×5μm with rectangle. The material is 45 steel, and the property is set to be plastic. The specific parameters of the material are shown in Table 1.

Table 1. Material parameters of abrasive and brake pad.

	Density (kg/m^3)	Thermal conductivity [W/(m · K)]	Elastic Modulus (GPa)	Poisson's ratio	Linear expansion coefficient	Specific heat capacity [J/(kg · K)]
Abrasive	5250	30	5.2	0.3	1.5×10^{-11}	550
Brake pad	7250	57	138	0.28	1.085×10^{-5}	450

The model can be meshed, and the volume of the deformity body can be compensated by using the adaptive meshing technique. The contact area of the brake pad/disc is grid-encrypted, and the grid is sparse away from the analysed area. The grid model is shown in Figure 1.

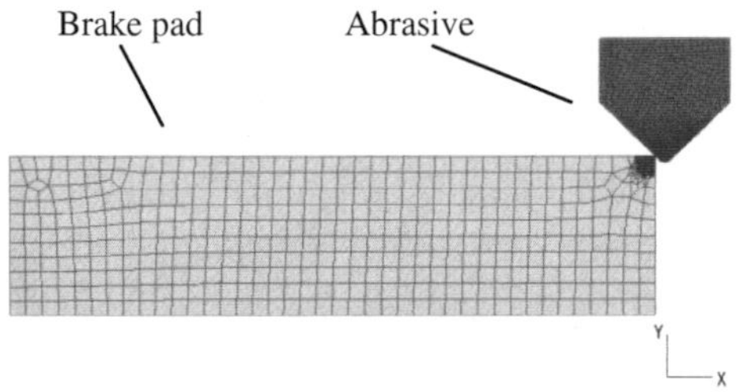

Figure 1. Finite element model of abrasive wear of brake pad.

Regarding to the boundary conditions, the abrasive and the brake pad should be set separately. The full restrain of the brake pad is used to restrain the freedom degree of the bottom unit. The relative motion in the X, Y and Z directions are 0. Abrasive needs to be given an X negative direction of the speed. It is assumed that the heat exchange coefficient between the abrasive and the surface of the brake pad is 45 N/sec/mm/C, and the friction coefficient is assumed unchanged with the temperature. Set the environment temperature to 20℃. In the simulation process, shear failure model and tensile failure model are adopted. The shear model is based on the average plastic strain value at the cell node.

The failure is assumed to occur when the destructive parameter $\omega_1 \geq 1$. The destructive parameter ω_1 is defined as:

$$\omega_1 = \sum \left(\frac{\Delta\varepsilon^{PL}}{\overline{\Delta\varepsilon^{PL}}} \right) \tag{5}$$

In which, $\Delta\varepsilon^{PL}$ is the increment of the average plastic strain, $\overline{\Delta\varepsilon^{PL}}$ is the failure strain.

When the node stress reaches the shear failure criterion, the corresponding material fails at this point. The shear failure criterion is chosen, and the wear equation is used to deduce the modified equation of the Archard wear model. Because the wear is a cumulative process, it takes a long time to work continuously to observe the wear rules. Therefore, to increase the number of simulation steps in the simulation process is needed, the number of simulation steps is controlled in 750 steps, and the data is read and saved with every 10 steps, meanwhile the international unit system is selected. The simulation mode is heat transfer and deformation. For the consideration of the flow stress in the disc material affected by strain, strain rate and temperature, the conjugate gradient method is chosen to be the iterative method.

4. Simulation results and analysis

4.1. Simulation results of wear nephogram analysis

The nephogram of abrasive wear with relative sliding velocity v=10m/s and friction coefficient μ=0.60 is shown in Figure 2.

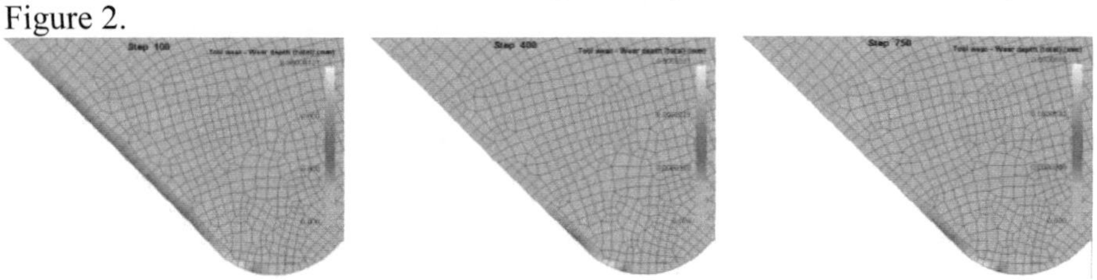

(a) step 100 (b) step 300 (c) step 750

Figure 2. Nephogram of the abrasive wear in different time.

Abrasive wear is a microscopic wear process, and the wear is also a long cumulative process. It can be seen from Figure 2(a) that at the interface of friction, the wear occurs. The abrasive bottom interface wear is more serious, away from the bottom of the surface the wear is less. In the initial stage of wear, the abrasive wear depth is very small, the wear is relatively minor. It is because the abrasive particles and brake pad just come into contact, the contact surface temperature and contact stress are relatively small, and the corresponding wear is relatively small. From Figure 2(b), (c) and (d), it can be found that the wear of the contact surface accumulates and reaches a steady stage gradually with the simulation progresses. Due to the relative sliding velocity and the contact stress, the maximum wear depth reaches h=8.88×10^{-5}mm at the 750th step.

4.2. Effect of relative sliding velocity on abrasive wear

Relative sliding velocity is one of the factors that affects wear. Different relative sliding velocity has a great impact on wear, especially in the high-velocity and heavy-load. The greater relative sliding velocity leads to the higher temperature. Therefore, a single variable method is used to simulate, when μ=0.60 and v=6m/s, v=8m/s and v=10m/s respectively.

The temperature of the interface can be obtained directly from the post-processing results of Deform-2D. At the 750th step, the maximum temperature of the current sub-step is extracted every 10 steps. The interface temperature curve can be obtained at different relative sliding velocity, as shown in Figure 3.

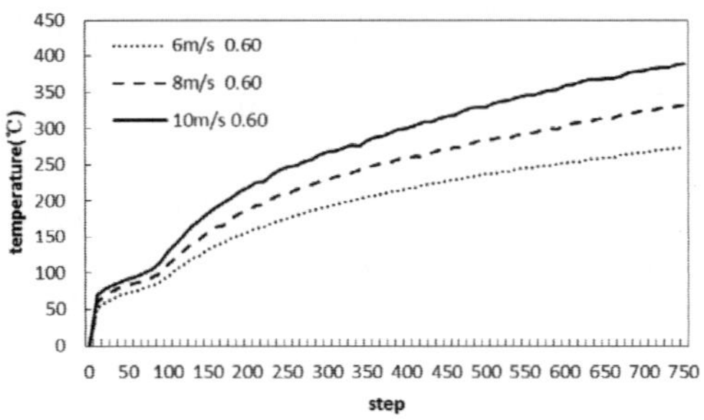

(a) temperature

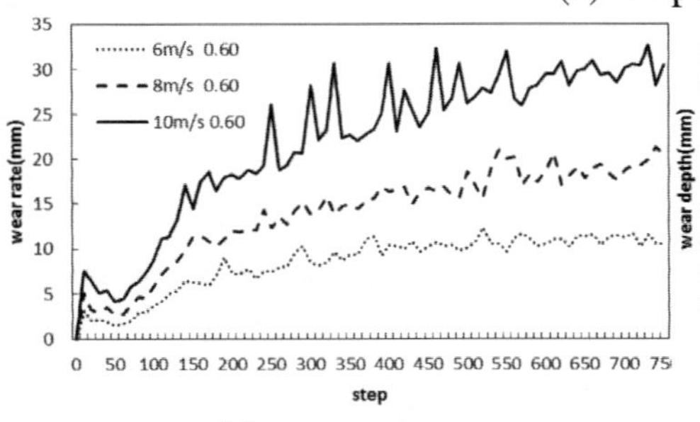

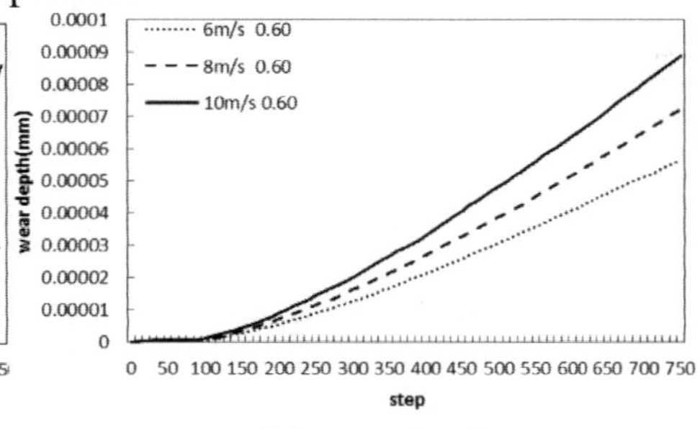

(b) wear rate (c) wear depth

Figure 3. Chart of parameters with different relative sliding velocity.

It can be seen from Figure 3(a) that the interface temperature from the 1st step to the 750th step increases with the wear. At the 750th step, the max interface temperature is at 273.399℃ when v=6m/s. When the relative sliding velocity increases v=10m/s, the max temperature of the interface is at 389.344℃. It can be found that the relative sliding velocity v increases, the interface temperature also increases correspondingly. Mainly because of the relative sliding velocity increasing, the unit time friction and material deformation are generated by the heat increasing, abrasive and brake pad are pressed state, the interface of the heat generated is difficult to spread, the temperature increases correspondently.

Figure 3(b) and (c) show that the relationship between the relative sliding velocity and the wear rate and wear depth. The wear rate of the different relative sliding rate trending and the max wear respectively can be found in Figure 4. It can be seen from Figure 3(b) that the wear rate of abrasive tends to stabilize at v=10mm/s by v=6m/s. When v=10m/s, abrasive wear rate tends to stabilize at v=30mm/s. And the wear rate increases with the increasing velocity, the difference is obvious. The wear rate increases significantly at the beginning of the simulation. Compared with the whole wear process, the wear rate is relatively small in the early stage of wear and shows an upward trend with the increasing of time step. As the relative sliding velocity becomes greater, the wear rate is more obvious, which is due to the impact of vibration on it.

After a period of wear, the corresponding abrasive wear has a relative increase. At the 750th step, we can see a significant difference. When the relative sliding velocity v=6m/s, the max wear depth h=5.63×10^{-5}mm. At v=10m/s, the max wear depth h=8.88×10^{-5}mm. The increase of the relative sliding velocity means that the friction and squeezing of the abrasive increases the wear quantity of the abrasive wear.

4.3. Effect of friction coefficient on abrasive wear

In the friction and wear process, the friction coefficient μ is a non-negligible parameter. It affects the result of wear directly. Take the friction coefficient of μ=0.30, μ=0.45 and μ=0.60 for simulation analysis respectively, when the relative sliding velocity v=10m/s.

The difference of friction coefficient means that the degree of friction is different and has a corresponding effect on the interface temperature. The max temperature of the current sub-step is extracted every 10 steps. The interface temperature curve can be obtained at different friction coefficient, as shown in Figure 4.

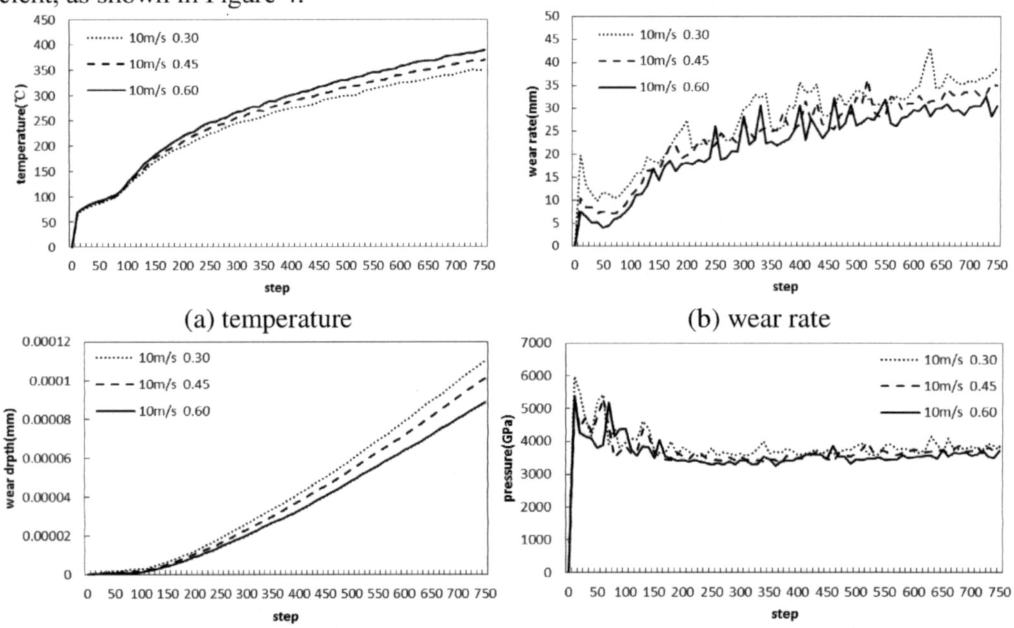

(a) temperature (b) wear rate

(c) wear depth (d) pressure

Figure 4. Chart of parameters with different friction coefficient.

Figure 4(a) shows the relationship between the different friction coefficient and the interface temperature. During the whole process it can be seen the interface temperature increases gradually with simulation steps. At the 750th step, the max interface temperature is at 352.928℃ with the friction coefficient of 0.30. The friction coefficient μ=0.60, the max interface temperature is at 389.344℃. The interface temperature increases as the coefficient of friction increasing, because the source of heat is sheared and squeezed between the materials during the wear process. The increase of the coefficient friction causes the heat generated by friction between the abrasive and the brake pad, so the larger friction coefficient has a certain influence on the temperature rise.

Figure 4(b) and (c) show the relationship between the different friction coefficient and the wear rate as well as wear depth. It indicates the trend and the max of wear in the different friction coefficient respectively. From Figure 4(b) it could be found that when the friction coefficient becomes smaller, the wear rate is greater. It can be seen from Figure 4(c) that the max wear depth h=1.10×10^{-4}mm when μ=0.30. The max wear depth h=8.88×10^{-5}mm at μ=0.60. As the friction coefficient increasing, the wear quantity reduces accordingly. According to the Archard wear model, the wear is a function of the contact pressure and the relative sliding velocity. It can be seen from Figure 4(d) that the effect of the friction coefficient on the contact pressure is not obvious, but the amplification analysis can still find that the contact pressure decreases with coefficient increasing, thus the wear quantity decreases with the increase of friction coefficient.

5. Conclusions

This paper is based on abrasive wear theory of tribology, as well as the variation of the velocity and loads are considered in the micro friction and wear process. The Archard wear calculation model is modified. The micro finite element model of abrasive wear is established, the influence rules of relative sliding velocity and friction coefficient are analysed in the brake pad and disc. The simulation result shows that the wear is not stable at the initial stage, but the accumulation of wear depth tends to be moderate as time went on. The interface temperature rises with the increase of the relative sliding velocity. Although the larger friction coefficient elevates the interface temperature, the wear depth decreases with the influence of interface pressure and the relative sliding velocity.

Acknowledgments

The research work is supported by National Natural Science Foundation of China under Grant No. 51475066 and No. 51675075 and Natural Science Foundation of Liaoning Province under Grant No. 2015020114.

References

[1] Zmago S, Kristoffer K and Tomaz K 2007 Friction behavior of sintered metallic brake pads on a C/C-SiC composite brake disc *J. Journal of the European Ceramic Society* **27** pp 1411–1417

[2] Cho M H, Cho K H and Kim S.J 2005 The role of transfer layers on friction characteristics the sliding interface between friction materials against gray iron brake disks *J. Tribology Letters* **20** pp 101–108

[3] Liu G, Wang Q and Ao Y 2002 Convenient Equations for modeling three dimensional thermo-mechanical asperity contacts *J. Tribology International* **35** pp 411–424

[4] Bogdanovicha P N and Tkachu D V 2006 Temperature distribution over contact area and “hot spots” in rubbing solid contact *J. Tribology International* **39** pp 1355–1360

[5] Straffelini G, Pellizzari M and Molinari A 2004 Influence of load and temperature on the dry sliding behavior of Al-based metal-matrix-composites against friction material *J. Wear* **256** pp 754–763

[6] Eriksson M, Bergman F and Jacobson S 2002 On the nature of tribological contact in automotive brakes *J. Wear* **252** pp 26–36

[7] Uyyuru R K, Surappa M K and Brusethaug S 2005 Effect of reinforcement volume fraction and size distribution on the tribological behavior of Al-composite/brake pad tribo-couple *J. Wear*, **257** pp 39–44

[8] Zhou Z, He Y and Wang S 2014 Effect of Dimple Morphology on Line Contact Wear Resistance of Friction Pairs *J. Journal of Hunan University of Technology* **03** pp 24–29(in Chinses)

[9] Archard J F 1953 Contact and rubbing of flat surfaces *J. Journal of Applied Physics* **24** pp 981–988

IOP Conf. Series: Materials Science and Engineering **280** (2017) 012011 doi:10.1088/1757-899X/280/1/012011

A Numerical and Experimental Study of Ejector Internal Flow Structure and Geometry Modification for Maximized Performance

Mehdi Falsafioon*, Zine Aidoun and Michel Poirier

1615 boul. Lionel-Boulet, Varennes, Quebec, Canada, J3X 1S6

*E-mail : mehdi.falsafioon@canada.ca

Abstract. A wide range of industrial refrigeration systems are good candidates to benefit from the cooling and refrigeration potential of supersonic ejectors. These are thermally activated and can use waste heat recovery from industrial processes where it is abundantly generated and rejected to the environment. In other circumstances low cost heat from biomass or solar energy may also be used in order to produce a cooling effect. Ejector performance is however typically modest and needs to be maximized in order to take full advantage of the simplicity and low cost of the technology. In the present work, the behavior of ejectors with different nozzle exit positions has been investigated using a prototype as well as a CFD model. The prototype was used in order to measure the performance advantages of refrigerant (R-134a) flowing inside the ejector. For the CFD model, it is assumed that the ejectors are axi-symmetric along x-axis, thus the generated model is in 2D. The preliminary CFD results are validated with experimental data over a wide range of conditions and are in good accordance in terms of entrainment and compression ratios. Next, the flow patterns of four different topologies are studied in order to discuss the optimum geometry in term of ejector entrainment improvement. Finally, The numerical simulations were used to find an optimum value corresponding to maximized entrainment ratio for fixed operating conditions.

1. Introduction

Numerous experimental and numerical studies have been conducted to provide a better understanding of mixing and compression processes in ejectors. Among these studies, many have tried to improve the performance of ejectors by modifying the dimensions and/or the configuration of the internal components. One of the items that has a significant effect on the efficiency of an ejector is the distance between the primary nozzle and the entrance of mixing chamber (NXP). Therefore, numerous studies have been carried on this factor of ejector design. Eames et al. [1] observed that nozzle exit position (NXP) and primary nozzle geometry play a strong role on the system performance. Pianthong et al. [2] investigated the effect of operating conditions, NXP and throat area on ejector performance using CFD techniques. Mazzelli and Milazzo [3] applied the CRMC method (Constant Rate of Momentum Change) for the design of an ejector. Their analyses showed that it is necessary to consider the surface roughness of the ejector in order to have a good agreement between the experimental data and CFD results. Zhu et al. [4] developed a 2-D CFD model to investigate the effect of the NXP and the mixing section converging angle. They showed that ejector performance is very sensitive to the converging angle of mixing section. They also reported that the optimum NXP and converging angle cannot be predefined to meet all working conditions. A numerical study of optimum mixing length of gas–gas and gas–liquid ejectors from Li et al. [5] showed that the effect of NXP is not so prominent when the mixing length is much longer than their optimum mixing range. An optimization study of ejector geometry parameters by Lin et al. [6] showed that the nozzle diverging angle and the length of the constant-pressure mixing section were two important geometry parameters that should be carefully considered in the adjustable ejector design. Another investigation by Maghsoodi et al. [7] numerically assessed the effect of four ejector parameters: NXP, mixing tube length, diffuser length and its divergence angle in order to optimize its performance for a fuel cell system. They found that the mixing section length had a significant effect on their ejector performance. The authors also recommended an optimum range for the four parameters studied. Some authors have tried to study the effect of nozzle design on jet performance. Hedges and Hill [8] did an experiment in order to study the influence of nozzle design on jet ejector performance. They tested two conically diverging nozzles with different divergence angle. The exit and throat diameters of the nozzle were fixed in both case and the results show that the overall jet ejector performance was not influenced by the nozzle design. Similarly, a study done by Engdahl and Holton [9] confirms that the nozzle, which was designed by conventional methods for a specific pressure, performed only slightly better than a simple straight-hole nozzle at pressure up to 1170 kPa. Also, a machined nozzle with a convergence section and a 10 degree angle of divergence was only 3 to 6% better than a 690 kPa small pipe-cap nozzle made by drilling a hole in a standard pipe cap. However, altering the nozzle design affects the motive-stream velocity. Kroll [10] also reported that a poorly shaped nozzle results unnecessary shocks and lateral expansion, which consequently decreases jet ejector efficiency significantly. On the other hand, the position of the nozzle (NXP) has a greater effect on jet ejector performance than its design. A number of researchers investigated the optimum position of the nozzle in a jet ejector. Croft and Lilley [11]; and Kim et al. [12] showed that the turbulence effects are decreased when the nozzle is placed right at the entrance of the throat section. However, Croft and Lilley [11] discovered that when the nozzle moves closer to the mixing tube, the entrainment ratio decreases. Most recently, Wang et al. [13] conducted a CFD simulation to improve the ejector performance by varying the ejector primary nozzle's geometries. The simulation results showed that the throat diameter and divergent portion of the primary nozzle play significant role in ejector design since the entrainment ratio of the ejector is rather sensitive to the length and surface roughness of these two portions. Recently, HakkakiFard et al. [14] proposed a computational methodology for designing an optimized ejector with a minimum number of simulation runs.

2. Theory

A gas/gas ejector system can produce cold from heat. Compared to a conventional mechanical compression cycle, in an ejector cycle the compressor is replaced by the trio pump-generatorejector. Integrating the ejector in a conventional refrigeration system in order to boost its performance or in other cases, simply replacing to a compressor with a circulation pump and a generator offers the opportunity to reduce the overall system electricity consumption. Ejector operation consists of using a high primary energy stream mass flow rate (m_1) from the generator, expanding in a supersonic nozzle to entrain and exchange energy with a secondary stream mass flow rate (m_2) from the evaporator, thus producing a cooling effect. Its performance is qualified by the entrainment ratio of the secondary to the primary streams ($\omega = m_2/m_1$) and the compression ratio of the outlet pressure to secondary pressure ($pr = p_{cond}/p_2$). A thermal jump, identified as the" lift", can also be defined as a difference between the condensation temperature (T_{cond}) or the saturated temperature corresponding to the condensation pressure, and the evaporation temperature (T_{evap}) or the saturated temperature corresponding to the evaporation pressure (thermal jump = lift = $T_{cond}-T_{evap}$). The lift characterizes the quality of the cold produced by the ejector.

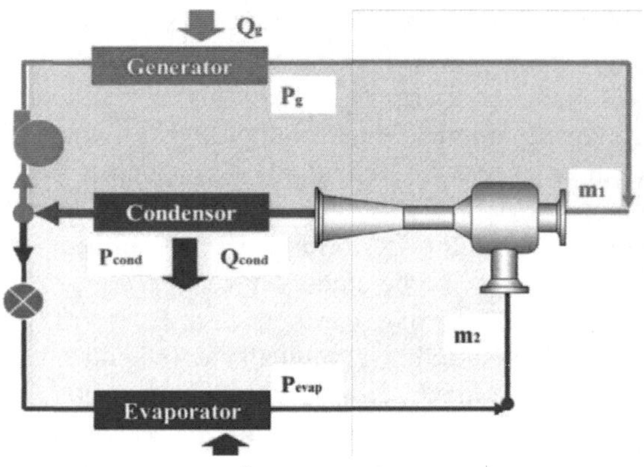

(a) A schematic of the experimental set up (b) View of the constructed prototype for the study

Figure 1. Experimental set up

3. Experimental Setup

A simplified diagram of the designed prototype is shown in Fig. 1(a). It shows the main components: ejector, generator, evaporator, condenser and refrigerant pump. A view of the constructed prototype is given in Fig. 1(b).

The prototype includes several main components:

-A generator, which is a heat exchanger used to vaporize the high pressure liquid refrigerant;

-A re-generator, which recovers part of the sensible energy of the outgoing gaseous refrigerant from the ejector to preheat the liquid refrigerant to the generator;

-A superheater, which allows more flexibility for adjusting the refrigerant superheat temperature before its arrival into the ejector;

-A boiler fueled by natural gas, provides steam at 113°C as an energy source for the generator and the super-heater;

-A fluid cooler and its pump, which use a glycol water loop to reject the condensation energy of refrigerant;

-A tank recovering the liquid refrigerant at the condenser outlet;

-An expansion valve used to adjust the pressure of the secondary flow

The prototype consists of three loops; (I) a loop of ejector cycle in a low-pressure part; II) a loop bringing heat from boiler to the generator and the super-heater; And III) a loop used for heat rejection to the condenser. A ejector initially installed on the prototype was designed to work under different operating conditions in order to study the improvement of entrainment ratio. To determine the prototype performance with different primary nozzle's positions (NXPs) this lenght was modified from 14.0 mm to 36.5 mm with an interval of 7.5 mm. Figure 2 shows a drawing of the ejector with its dimensions. The main body and dimensions are retained for all experiments except NXPs. In this figure NXP's length is equal to 29 mm.

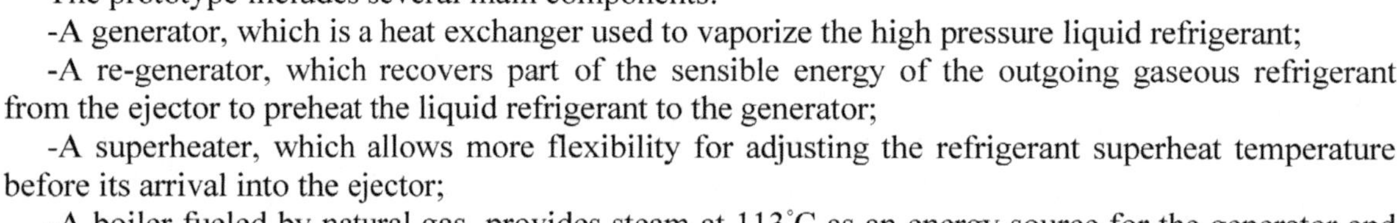

Figure 2. A scheme of the ejector including dimensions (All dimensions are in millimeter. The drawing is not scaled)

Table 1. Working Conditions and Entrainment Ratios from CFD

Condition	P_{gen}	T_{gen}	P_{evap}	T_{evap}	$P_{cond(sat)}$	$T_{cond(sat)}$	$\omega\%$			
							NXP = 14.0 mm	21.5 mm	29.0 mm	36.5 mm
1	2213	87	265	50	650	24	18.2	22.8	25.7	25.1
2	2633	100	425	25	810	32	32.0	36.9	40.3	39.0

4. Numerical Model

CFD simulation is used to determine the effect of the NXP parameter on the performance of a common ejector and consequently to find the optimum value for this parameter. Because of the nature of ejectors, the conservation equations of continuity, momentum and energy are used in their compressible and steady state forms and non-linear sets of discretized equations are solved using the SIMPLEST procedure of Spalding [15] in PHOENICS software, a general commercial CFD package (Flowsolve, London). REFPROP (NIST standard reference data base 23, version 9.1) is coupled with the CFD package to extract R-134a thermophysical properties during the calculations. The $k - \epsilon$ turbulence model is chosen based on the previous studies reported in the literature, e.g. [13, 16, 17].

The ejector is modeled in 2-D and is assumed to be axi-symmetric along the x-axis. A multiblock structured grid is used with a total number of 18,350 quadrilateral elements. A" Pressure inlet" boundary condition is applied for both primary and secondary inlets and a" Pressure outlet" boundary condition is imposed at the outlet. All walls are considered to be adiabatic with no-slip boundary condition. It is also assumed that the primary and secondary flows are fully mixed in the mixing chamber and the vapor expansion in the nozzle is isentropic.

5. Results and Discussion

The CFD code was first validated using experimental results for condition number 2 and NXP = 29 mm (Table 1) in different exit temperatures (T_{cond}). Figure 3(a) shows the entrainment ratios in different ejector exit saturated temperatures for a lift equal to $20^{\circ}C$ and for both numerical and experimental results. In this figure, there is a good accordance between the results that confirms the validity of the CFD code. A difference between experimental and numerical results appears close to the maximum entrainment ratio. In this case, CFD results show a slightly higher entrainment ratio than experimental results. The reason of differences between two curves in this area is currently under investigation. At the next step, the experiments were done for two different working conditions shown in Table 1. Because of a few limitations of the prototype, only four different nozzle positions were chosen to carry out the experiments. For all of these positions, the working conditions were kept fixed in order to evaluate the entrainment ratio as a function of NXP. On the other side, first, CFD simulation was fulfilled with similar NXPs of experiments. Then, six more NXPs were tried using CFD model in order to determine the optimum NXP in terms of the ejector entrainment ratio. The comparison of the experimental and the CFD results for the two different conditions is shown in Fig. 3(b). As is seen in this figure, there is an optimum value for NXP where the highest entrainment ratio can be reached. This figure also shows that this optimum NXP is not fixed for all operating conditions and may slightly from case to case. For instance, for condition 1, the optimum value for the NXP is close to 31 mm, but for the other condition this amount is close to 32.5 mm. Figure 4 shows Mach numbers resulting from numerical simulation for two different NXPs for condition number 2. This diagram reveals the effect of NXP on the shocks occurring inside the ejector and consequently the role that this element plays on the efficiency of an ejector. In this figure, for the case with NXP = 29.0 mm, the flow is always supersonic in the area between the nozzle and the mixing chamber. This allows for better mixing of primary and secondary flows and results an improvement of entrainment ratio.

IOP Conf. Series: Materials Science and Engineering **280** (2017) 012011 doi:10.1088/1757-899X/280/1/012011

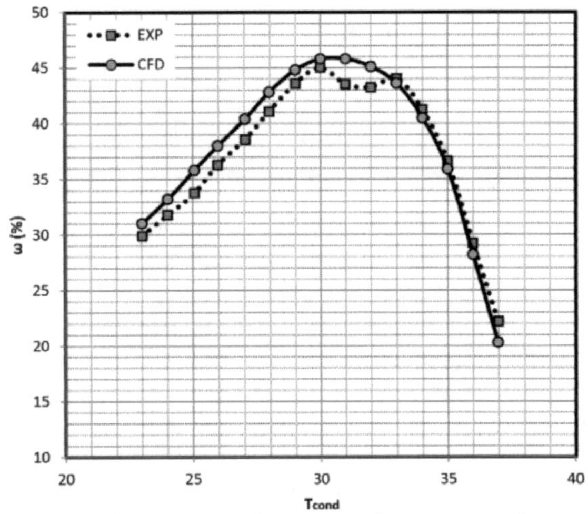

(a) Results for entrainment ratios versus ejector exittemperature in saturated state and for a lift of 20°C

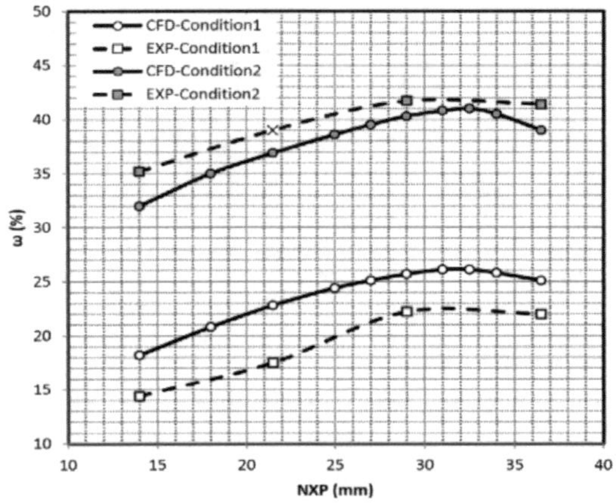

(b) Results for the entrainment ratio versus nozzle'sposition toward mixing chamber (NXP). Conditions are defined in Table 1[1]

Figure 3. Numerical (CFD) and experimental (EXP) results

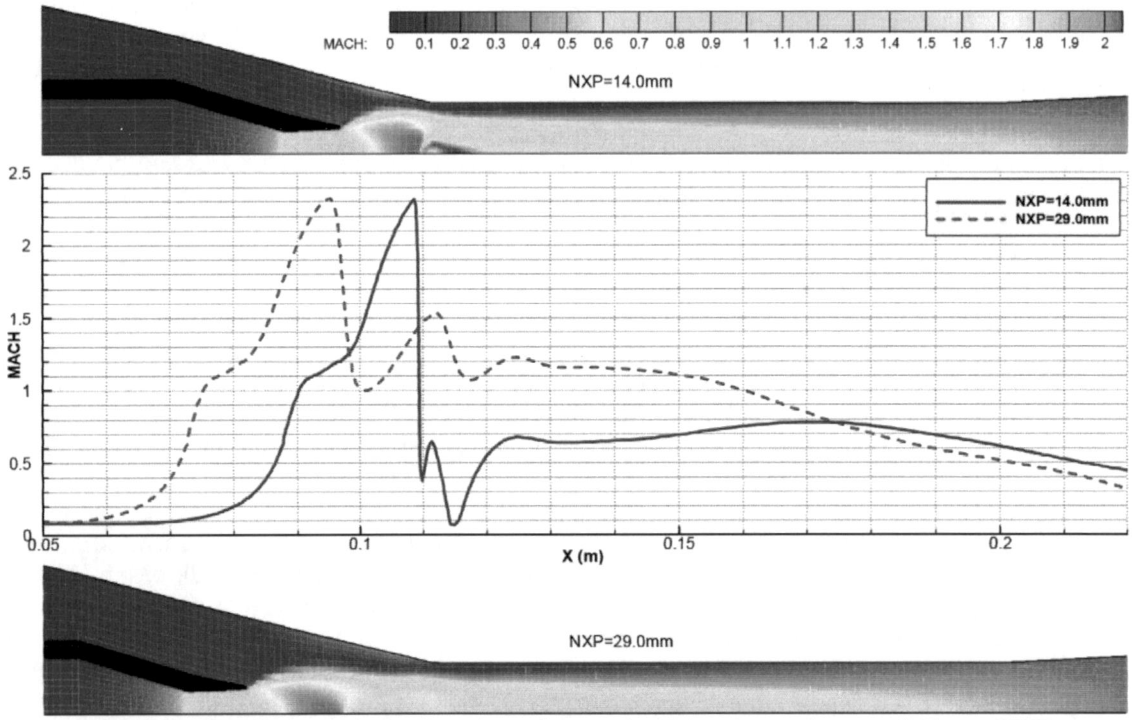

Figure 4. Variation of the Mach number along the x-axis of the ejector for two different NXPs

[1] The cross point shows a missing data for this condition and is not a real output from the designed prototype

IOP Conf. Series: Materials Science and Engineering **280** (2017) 012011 doi:10.1088/1757-899X/280/1/012011

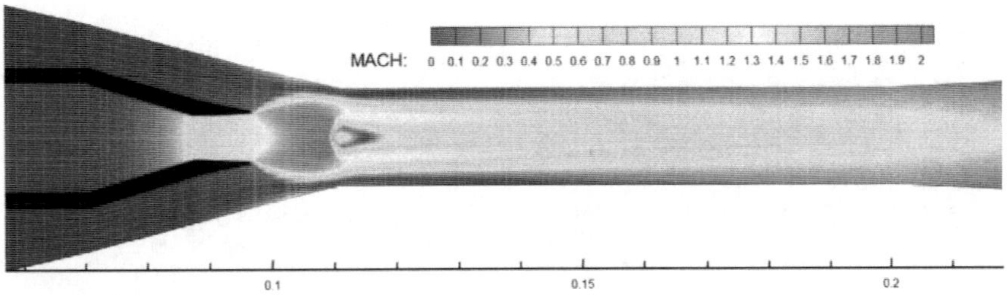

NXP = 14.0 mm

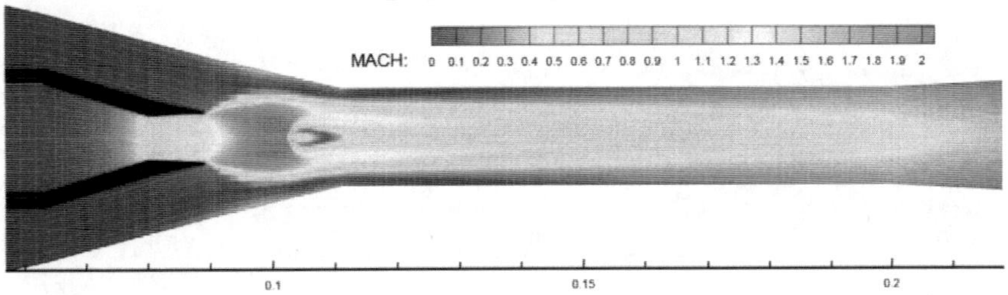

NXP=21.5mm

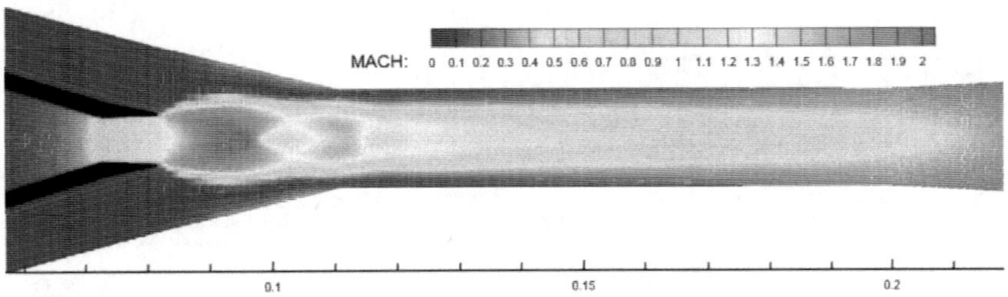

NXP=29.0mm

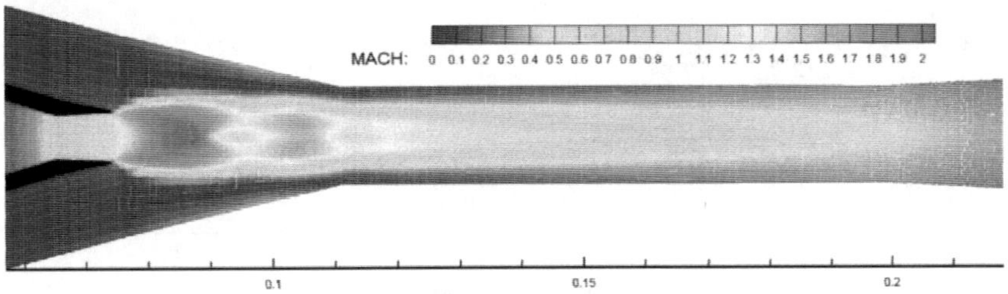

NXP=36.5mm

Figure 5. Mach contours for four different NXPs at the working condition 2

IOP Conf. Series: Materials Science and Engineering **280** (2017) 012011 doi:10.1088/1757-899X/280/1/012011

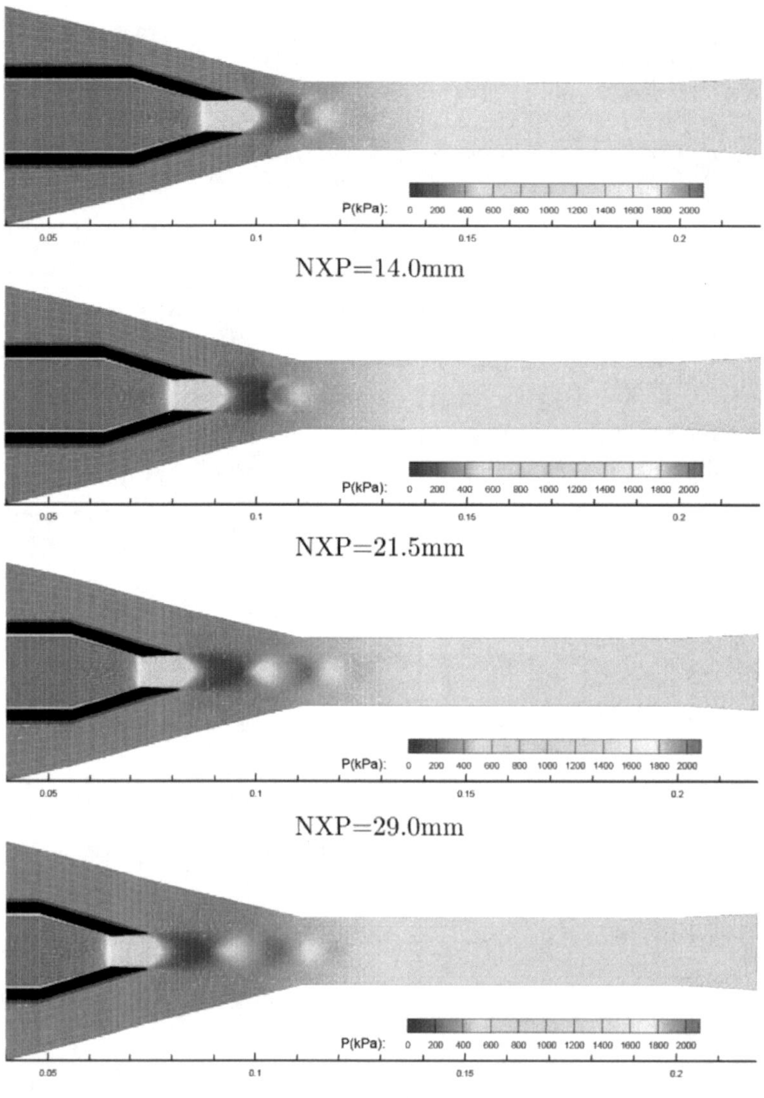

Figure 6. Pressure and velocity contours for four different NXPs at the working condition 2

Nevertheless, for cases where the nozzle is too far from the mixing chamber (NXP is too large) the efficiency starts to reduce because the primary flow is choked right away after the nozzle and has lost its energy to perform the suction of secondary flow. On the contrary, for a case with a short NXP (a.e. *NXP* = 14 *mm*) secondary flow does not have enough area to exit and to interact with primary flow. This case may even result in reverse flow and negative entrainment. Briefly, there is an optimized value for NXP which is not a constant value, differing from case to case with different configurations and different working conditions. With the help of CFD simulation, the phenomena which is happening inside the ejector can be visualized. This also can give a better understanding of the fluid flow pattern and the pressure field. To this end, the Mach and Pressure contours are drawn respectively in Fig. 5 and Fig. 6 for four different NXPs at working condition 2 (see Table 1). The CFD results presented in Fig. 5 accompanied with those results in Fig. 4 show that for those length with a better entrainment ratio the supersonic area is longer along the ejector. This figure is shown in more detail in Fig. 4 for two NXP = 29.0 mm and NXP = 14.0 mm. It can be seen that for NXP = 29.0

mm more shocks happen in the ejector, which helps to have better mixing as well as a better compression ratio at the end of ejector.

Figure 6 shows that for the configurations with longer NXP, more swirls appear between nozzle and mixing chamber area. This could be interpreted as a reason for having better mixing of primary and secondary fluids and consequently having better entrainment ratio.

6. Conclusion

In this study the effect of nozzle position with respect to the entrance of the mixing chamber has been studied numerically and experimentally. The study was first conducted with similar experimental and numerical conditions that showed a good accordance between two studies and also validated the CFD model. At the next step, further numerical studies were carried out in order to determine an optimum value for NXP for each different case. The results revealed that this optimum value changes slightly from case to case. While comparison between theory and experiment shows a relatively good agreement, the numerical predictions clearly show an optimum value corresponding to maximized entrainment ratio for fixed operating conditions while experiment show no maximized value, likely because of a limited number of data points obtained experimentally as opposed to a larger number of simulated values over the same range of conditions.

References

[1] Eames I W, Ablwaifa A E and Petrenko V 2007 *Applied Thermal Engineering* **27** 2833–2840
[2] Pianthong K, Seehanam W, Behnia M, Sriveerakul T and Aphornratana S 2007 *Energy Conversion and Management* **48** 2556–2564
[3] Mazzelli F and Milazzo A 2015 *International journal of refrigeration* **49** 79–92
[4] Zhu Y, Cai W, Wen C and Li Y 2009 *Applied Thermal Engineering* **29** 898–905
[5] Li C, Li Y and Wang L 2012 *Applied Thermal Engineering* **48** 237–248
[6] Lin C, Cai W, Li Y, Yan J, Hu Y and Giridharan K 2013 *Applied Thermal Engineering* **61** 649–656
[7] Maghsoodi A, Afshari E and Ahmadikia H 2014 *Applied Thermal Engineering* **71** 410–418
[8] Hedges K and Hill P 1974 *American Society of Mechanical Engineers and Canadian Society for Mechanical Engineering, Fluids Engineering Conference, Montreal, Canada* p 1974
[9] Rao N S and Kremer H 1993 *Chemical engineering & technology* **16** 94–101
[10] Kroll A E 1947 *Chemical Engineering Progress* **43** B21–B24
[11] Croft D and Lilley D 1976 *14th Aerospace Sciences Meeting* p 183
[12] Kim H D, Setoguchi T, Yu S and Raghunathan S 1999 *Journal of Thermal Science* **8** 79
[13] Wang L, Yan J, Wang C and Li X 2017 *International Journal of Refrigeration* **76** 219–229
[14] Hakkaki-Fard A, Aidoun Z and Ouzzane M 2015 *Energy Conversion and Management* **105** 1291–1302
[15] Patankar S 1980 *Numerical Heat Transfer and Fluid Flow* (McGraw Hill)
[16] Watanawanavet S 2008 *CFD optimization study of high-efficiency jet ejectors* Ph.D. thesis Texas A&M University
[17] Bartosiewicz Y, Aidoun Z, Desevaux P and Mercadier Y 2005 *International Journal of Heat and Fluid Flow* **26** 56–70

IOP Conf. Series: Materials Science and Engineering **280** (2017) 012012 doi:10.1088/1757-899X/280/1/012012

Efficient Algorithm for Sidewall Machining of Aircraft Structural Part

Zujie Zheng[a], Guolei Zheng[a], Zheng Jiang[b] and Xiaofeng Wu[b]

[a]Department of Mechanical Engineering & Automation, Beihang University, Beijing 100191, China;
[b]Shanghai Aerospace control technology institute,shanghai, 201109 , China;

E-mail: zhengzujie@buaa.edu.cn

Abstract. Traditionally, the boundaries of the pocket bottom face are often used as the guide elements of the machining process of sidewall machining. However, for complex pocket, overcutting often occurs due to the incorrectness of the guide elements. To solve this problem, the sidewall machining model is presented at first. Then, terms such as rim convex edge and visible property are introduced in detailed. Finally, the rim convex edges are projected onto the bottom face and the visible guide lines are constructed by sub-lines selection based on the visible property. At last a case study is given to verify the feasibility and effectiveness of the proposed algorithm. The result shows that the algorithm can avoid overcutting for complex pocket.

1. Introduction

Sidewall machining is one of the most common and important machining stage in the aircraft fabrication process of structural parts and turbines which have high accuracy requirements and are easy to deform [1]. Since most sidewall faces of aircraft structural part are ruled faces, 3-axis flank machining[2] is still widely adopted in the finishing machining process. Moreover, for the complex pocket composed of inclined faces, 3-axis flank machining is often used to remove the residual material left by rough machining before 5-axis flank machining[3-6]. Generally, 3-axis sidewall machining uses the offsetting curves of the boundaries of the related bottom face as the tool-path elements[7]. The 2D-curve offsetting problem has been regarded as the key issues for tool-path generation of sidewall machining[8,9]. However, under the shelter of the inclined faces, some boundaries are invisible along the opposite direction of the machining direction. Therefore, interference occurs when the cutting tool locates in the invisible area.

The research reported in this paper is motivated by the above observation. The objective of this work is to provide an efficient algorithm that can receive precise machining region for sidewall machining process, which would avoid overcutting in the machining process. In other word, the main task of the current work is to define machining region from the perspective of geometry, therefore machining parameter is not considered here. Section 2 describes the proposed algorithm in detailed and the section 3 presents the case study. Conclusions are given in Section 4.

2. The proposed algorithm

2.1 Machining model

Sidewall machining is used to remove the radial allowance left by rough machining process, and the machining model of 3-axis sidewall machining can be expressed as follows:

$$MV = (d_r, C, H_b, H_t)$$
$$TP = f(MV) \tag{1}$$

Where MV is a single machining volume that can be described by the four parameters. d_r represents the radial allowance, C means the guide line that limits the movement range of the machining tool, H_b and H_t are the bottom and top height of the machining volume. In this paper, the end-milling cutter is used to generate the tool path TP, which can be generated by the tool path generation function f, and the main task of the function is to compute the offsetting curve of the guide curve.

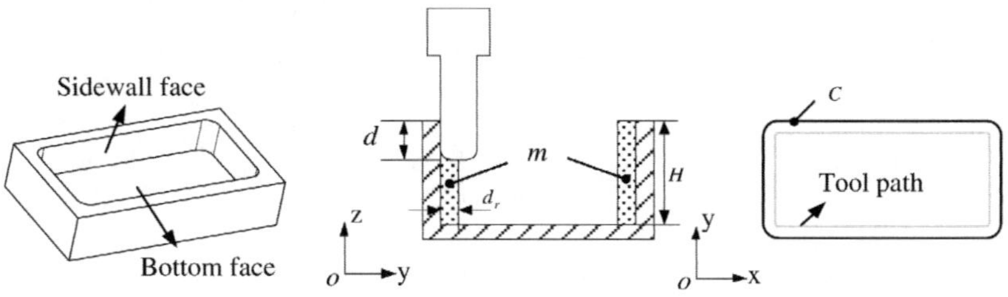

(a) Pocket (b) Machining volume (c) Tool path

Figure 1. Sidewall machining model

Fig.1 shows the machining model of a simple pocket, where m is the machining volume and C is the guide line. Generally, the boundaries of the bottom face are regarded as the guide lines for the sidewall machining volume. However, for complex pocket composed inclined faces, part of the bottom face is inaccessible for the cutting tool. Since the cutting tool direction is often fixed during machining time, overcutting often occurs during the machining process. Therefore, in order to avoid overcutting phenomenon, the guide line should be visible.

2.2 The proposed method

Base on the machining model, the most important task is to compute the precise guide line. Fig. 2 presents the flowchart of the proposed algorithm.

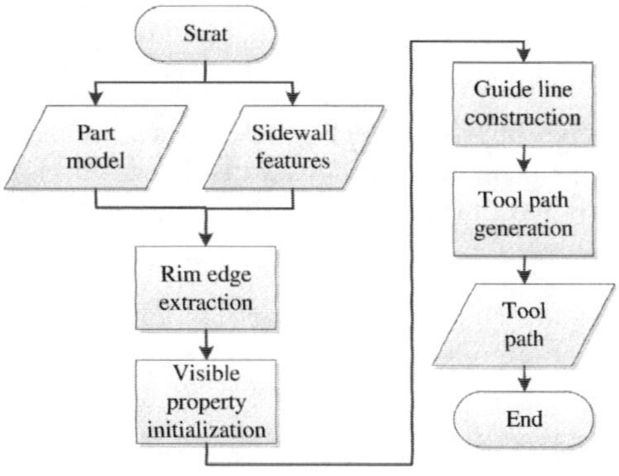

Figure 2. Flowchart of the proposed algorithm

2.2.1 Rim convex edge extraction. **Rim convex edge.** Rim convex edge (RCE) is the convex edge that is visible along the opposite direction of the machining direction (z-axis), and it can be extracted from CAD model by the following rule.

Rule 1. Suppose $\boldsymbol{e}$ is a common convex edge of face f_1 and f_2, $\vec{n}_1$ and $\vec{n}_2$ are normal vector of face f_1 and f_2 respectively, which point out of the volume. $\theta_1(\theta_2)$ is the angle between $\vec{n}_1$ ($\vec{n}_2$) and z-axis. If $\theta_1 \geq 90^\circ$ & $\theta_2 \leq 90^\circ$ or $\theta_1 < 90^\circ$ & $\theta_2 \geq 90^\circ$, then $\boldsymbol{e}$ is a RCE.

2.2.2 Visible property initialization. Traditionally, the edges of out loop on the inclined faces are projected onto bottom face to construct precise guide line[7]. However, for complex pocket, the number of projection lines may be large and the relationships between each other are complicated, as shown in Fig. 3.

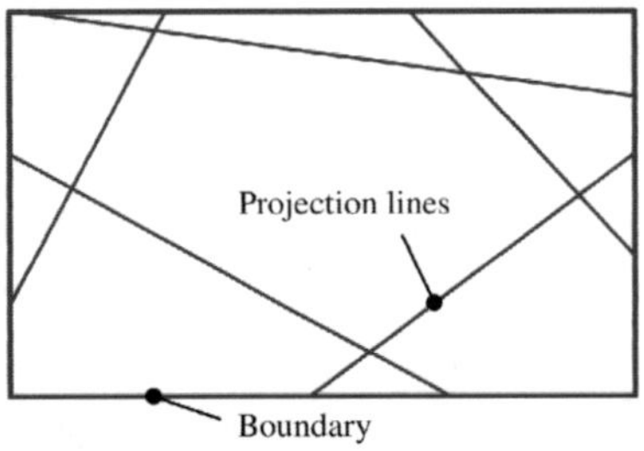

Figure 3. Projection lines onto the bottom face

In order to simplify the guide line construction, a visible property is introduced to RCE. Before the property initialization, it is necessary to sort the edges on the inclined faces. The edges in outer contour loop of inclined faces are sequenced by the right-hand rule, while it is unnecessary to initialize the visible property for the edges in inner contour loop since the edges cannot limit the travel range of the cutting tool. Table 1 shows the detailed visible property values for different edges, where positive edge means the edge whose ending vertex is the vertex of next edge in the contour loop, else the edge is called negative edge.

Table 1. Visible property initialization

Belonging face	Edge type	Visible property	
		Outer Loop	Inner Loop
Inclined face	Positive Edge	-1	*
	Negative Edge	1	*

2.2.3 Visible guide line construction. Visible guide line. Set C is a set of continuous lines on bottom face of the pocket, $C = (L_1, L_2, \cdots, L_n)$, F is a set of inclined faces, $F = (f_1, f_2, \cdots, f_m)$. $\forall L_p$, L_p is the z_ray on a random point of L_i, $L_i \in C$, if $\exists p, p = L_p \cap F$, and the point p lies on an outer loop of the inclined faces, then C is visible guide line.

Fig. 4 illustrates three types of two projection lines of RCEs. For the part intersecting and full intersecting case, sub-lines can be easily generated by the intersecting point and some of the sub-lines should be removed. The visible property of the RCE is used to remove the invalid sub-lines.

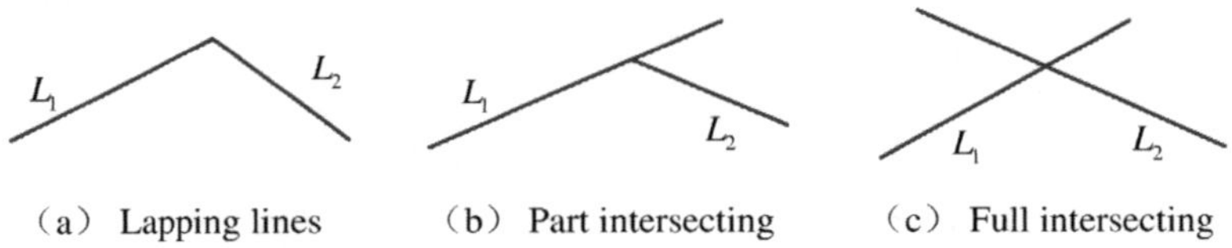

Figure 4. The relationship between projection lines

Take the full intersecting lines L_1 and L_2 for example, the steps to remove invalid sub-line are described as follows.

Step 1: Suppose $\overline{l_1}$ and $\overline{l_2}$ are two sub-lines of L_1, and $\vec{T}_2$ represents the vector from the starting point to the end point of L_2.

Step 2: Compute the vector $\overrightarrow{p_{st}, p_{mid1}}$ and $\overrightarrow{p_{st}, p_{mid2}}$, where p_{st} is the starting point of L_2 while p_{mid1} and p_{mid2} are the middle points of $\overline{l_1}$ and $\overline{l_2}$ respectively. Then, compute the cross product $\vec{T}_{21} = \vec{T}_2 \times \overrightarrow{p_{st}, p_{mid1}}$, $\vec{T}_{22} = \vec{T}_2 \times \overrightarrow{p_{st}, p_{mid2}}$.

Step 3: Set θ_i means the angle between $\vec{T}_{2i}$ and z-axis, $1 \le i \le 2$. When the visible property of value of L_2 is equal to 1, then $\overline{l_i}$ should be removed when $\theta_i = 0^\circ$. Else, $\overline{l_i}$ should be removed when $\theta_i = 180^\circ$.

As shown in Fig. 5, there are three projection lines on the bottom face, and α_1, α_2 and α_3 are the visible property of L_1, L_2 and L_3 respectively. At first, we consider the intersecting lines L_1 and L_2, the sub-lines $\overline{l_1}$ and $\overline{l_2}$ can be removed by the method mentioned above. Then, $\overline{l_3}$ and $\overline{l_4}$ can be extracted and removed by the same method. For the open pocket, the visible guide line is an unenclosed loop and it is unnecessary to group the original boundaries of bottom face and the visible guide line. But, for the closed pocket, the visible guide lines are composed of valid projection sub-lines and the original bottom boundaries, as shown in Fig. 5(c).

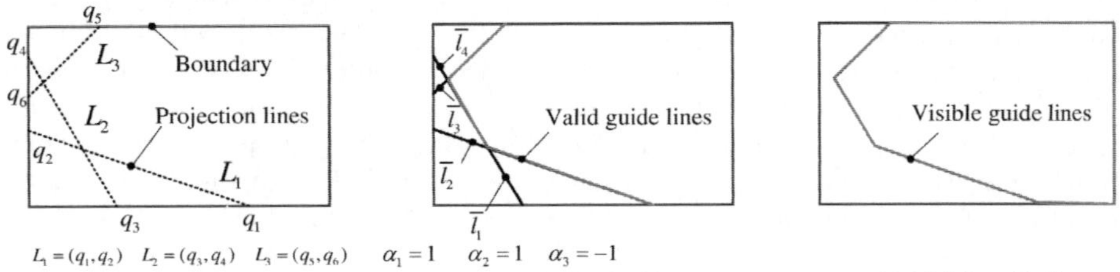

(a) Projection lines (b) Projection lines (c) Visible guide line

Figure 5. Visible guide lines construction example

2.2.4 Tool path generation. Based on the sidewall machining model, the tool path can be easily generated after getting the precise guide lines. At first, the basic tool path elements are generated by the offsetting function[7]. Then, the axial passes number of the tool path is computed by the height of sidewall faces and the maximum axial cutting depth.

3. Implementation

To validate the feasibility and effectiveness of the proposed algorithm, a case study on a simple aircraft structural part is conducted on the platforms of Microsoft Visual Studio 2005 and CATIA. Fig. 6 shows the tool path comparison between the traditional method and the proposed method. By the original method, the bottom boundaries are used as the guide line for sidewall machining (see Fig. 6(b)), which induces the overcutting in the machining process, as shown in Fig. 6 (c). By applying the proposed algorithm, three rim edges on the inclined faces are extracted, as shown in Fig.6 (a). Then, based on the precise guide line (see Fig.6 (d)), the traditional tool path generation function[7] is used to generate the optimized tool path, as shown in Fig.6 (e). The result shows that the proposed algorithm can avoid overcutting for complex pocket with inclined faces.

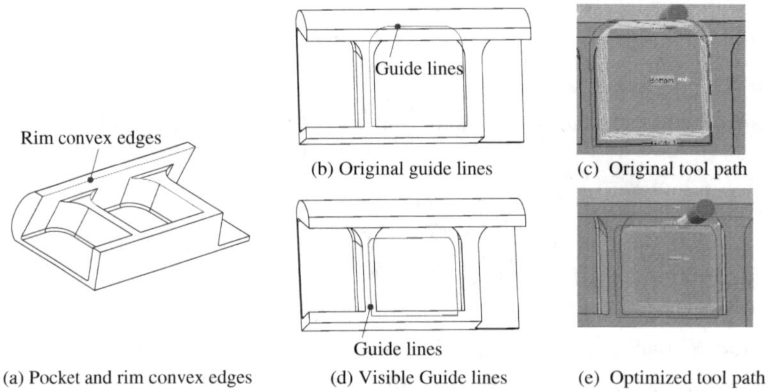

Figure 6. Tool path by the proposed algorithm

4. Conclusion

In this paper, to avoid the overcutting in the sidewall machining process of complex pocket composed of inclined faces, a novel efficient algorithm is proposed. The key of the algorithm is to generate precise guide lines that are visible along the opposite direction of the machining direction. Compared to the traditional tool path generation methods, the proposed algorithm focus on the boundary correctness for the sidewall faces. Rim convex edge and the visible property are developed and used in the invalid projection line removal process, which deeply improve the efficiency and accuracy of guide line generation. Computer implementations are provided, and the results show that the proposed algorithm is feasible and efficient.

References

[1] Chiou John C.J. Floor, wall and ceiling approach for pocket machining. *Computer Aided Design*, 2005.37(4): 373-385.

[2] Held M, Spielberger C. A smooth spiral tool path for high speed machining of 2D pockets. *Computer Aided Design*, (2009) 539-550.

[3] Bo Pengbo, Bartoň Michael, Plakhotnik Denys, et al. Towards efficient 5-axis flank CNC machining of free-form surfaces via fitting envelopes of surfaces of revolution. *Computer Aided Design*, 2016.79:1-11.

[4] Bo Zhou, JinBin Zhao, Lun Li, et al. Double spiral tool-path generation and linking method for complex pocket machining. *Machining Science and Technology*, 2016. 20(2): 262-289.

[5] Zhiwei Lin, Jianzhong Fu, Hongyao Shen, et al. Smooth contour-parallel tool path generation for high-speed machining through a dual offset procedure. *The International Journal of Advanced Manufacturing Technology*, 2015.81(5-8):1233-1245.

[6] Min Zhou, Guolei Zheng, Zhibo Luo, et al. Algorithm for recognizing multi-axis surface feature in aircraft structural parts based on main face extention. *Journal of Computer-Aided Design & Computer Graphics*, 2015.27:2376-83, in chinese.

[7] Kim H C. Tool path generation for contour parallel milling with incomplete mesh model. *The International Journal of Advanced Manufacturing Technology*, 2010.48(5-8):443-454.

[8] Minho Chang, Sang C Park, Y.C. Chuang. Three-dimensional profile curve machining on three-axis machines. *International Journal of Production Research*, 2009.47(10), 2587-2599.

[9] Minho Chang, Chong Man Kim, Sang C Park. Tool-path generation for sidewall machining. *Computers & Industrial Engineering*, 2009.56(4): 1649-1656.

IOP Conf. Series: Materials Science and Engineering **280** (2017) 012013 doi:10.1088/1757-899X/280/1/012013

Unsupervised Learning —A Novel Clustering Method for Rolling Bearing Faults Identification

Li Kai , Luo Bo, Ma Tao, Yang Xuefeng and Wang Guangming

1037 Luoyu Road, Hongshan District, Wuhan(430074) Hubei Province, P. R. China

15527965836@163.com

Abstract. To promptly process the massive fault data and automatically provide accurate diagnosis results, numerous studies have been conducted on intelligent fault diagnosis of rolling bearing. Among these studies, such as artificial neural networks, support vector machines, decision trees and other supervised learning methods are used commonly. These methods can detect the failure of rolling bearing effectively, but to achieve better detection results, it often requires a lot of training samples. Based on above, a novel clustering method is proposed in this paper. This novel method is able to find the correct number of clusters automatically the effectiveness of the proposed method is validated using datasets from rolling element bearings. The diagnosis results show that the proposed method can accurately detect the fault types of small samples. Meanwhile, the diagnosis results are also relative high accuracy even for massive samples.

1. Introduction

The rolling bearing is a key part which is widely used in rotating machinery, ranging from simple electric fans to complex machine tools [1]. The traditional technology of fault diagnosis usually fails in analyzing complex faults with no further fault information, and intelligent diagnosis theory based on machine learning has been central in study and application recently[2,3].Correspondingly, supervised learning methods such as artificial neural networks(ANN), decision trees, support vector machines(SVM), etc have been used widely. Samanta et al. [4] employed ANNs and SVM to diagnose faults of bearings and utilized time domain features to characterize the health conditions of the bearing. Tamilselvan [5] used the deep belief network (DBN) in an aircraft engine fault diagnose is based on health state classification in a small sample with enough fault information. Widodo et al [6]proposed two methods of multi-class classification techniques for fault diagnosis through relevance vector machine(RVM)and support vector machines(SVM),aiming to detect the fault of low speed bearing. B.A. Paya and I.I. Esat[7] preprocessed the real time domain vibration signals of bearing and gear faults by wavelet transforms, then employed the neural network to perform fault detection .Hai dong shao et al[8] presented a novel optimization deep belief network(DBN) for rolling bearing fault diagnosis, this method has a good generalization performance in complex classifications. Feng Jia, Yaguo Lei et al[9] presented a novel intelligent method which overcome the deficiencies of the aforementioned intelligent diagnosis methods for detecting of rotating machinery with massive data, the DNNs are trained by using raw signals in the time domain.Oveall,supervised learning methods have been more mature in the application of fault diagnosis, but it generally requires a massive fault or health samples for training learning model. However, during the actual processing, the lifecycle of rolling bearing is relatively long. So it is difficult for acquiring a large number of fault samples.

IOP Conf. Series: Materials Science and Engineering **280** (2017) 012013 doi:10.1088/1757-899X/280/1/012013

Moreover, the unknown fault types may occur during production, and the learning model can not accurately detect the fault types at that time. In 2014,Alex Rodriguez and Alessandro Laio[10] proposed a novel approach, this algorithm assume that cluster centers are surrounded by neighbors with lower local density and that they are at a relatively large distance from any points with a higher local density. This novel clustering method can make up for the shortcomings of the above two methods. Meanwhile, it is suitable for nonspherical clusters and determine clustering centers automatically.

2. A brief introduction to the novel clustering method

The novel clustering algorithm seems to be able to meet the above requirements. The following is a brief introduction to the novel clustering method in two steps.

2.1. The local density ρ_i and distance δ_i

Cluster centers can be defined as two main features :(1) Cluster centers are surrounded by neighbors with lower local density. (2) They are at a relatively large distance from any points with a higher local density [9].That is a cluster center has higher ρ_i and large value of δ_i as compared with non-center data points.Suppose the dataset to be clustered is $S=\{X_i\}_{i=1}^{N}$ .The cut-off local density ρ_i of data point is defined as[8,9]:

$$\rho_i = \sum_j \chi(d_{ij} - d_c) \qquad (1)$$

Where $\chi(x)=1$ if $x<0$ and $\chi(x)=0$ otherwise. ρ_i is equal to the number of points that are closer than d_c to point i . d_c is a cut-off distance that to be selected, As a rule of thumb, one can choose d_c so that the average number of neighbors is around 1 to 2% of the total number of points in the data set [9]. From equation (1), we find that cut off kernel are discrete numeric values. It is easy to cause the different data points tend to be the same local density, which will have an impact on our searching clustering centers.Thus, we choose Gaussian kernel to calculate the local density ρ_i .Gaussian kernel can obtain continuous value, which reduces the probability of the same local density of different data points. The Gaussian kernel local density ρ_i of data point is defined as:

$$\rho_i = \sum_{j \in I_S / \{i\}} e^{-\left(\frac{d_{ij}}{d_c}\right)^2} \qquad (2)$$

Above equation (1) and equation (2) where d_{ij} is the distance between data points i and j .In machine learning, the similarity measures include: Euclidean distance, Manhattan distance, Minkowski distance, cosine of included angle, etc. For clustering in rolling bearing fault, we find that the Euclidean distance is better than others. δ_i measured by computing the minimum distance between the point i and the nearest point with higher density[9]:

$$\delta_i = \min_{j:\rho_j > \rho_i}(d_{ij}) \qquad (3)$$

3. Fault diagnosis of rolling bearing using the novel clustering method

3.1.Data description

The bearing fault data used here are provided by the Case Western Reserve University (CWRU) collected from the electrical engineering laboratory [10]. The experimental platform consists of two horsepower motors, two acceleration sensors, a torque sensor and an electronic control device, as shown in Fig.1.The single point failure of the bearing is arranged on the bearing using the electric spark machining technology. The fault diameter is 0.007, 0.014, 0.021, 0.028 inch respectively (1 inch=25.4 mm). Each bearing was tested at four different loads (0, 1, 2, and 3), 0,1,2,3 load corresponding to 1797rpm, 1772rpm, 1750rpm, 1730rpm running speed of motor respectively. Experimental data of rolling bearings are acquired from the experimental system under four different operating conditions [11]: (1) the normal conditions; (2) with the inner race fault; and (3) with the ball fault. (4) With the outer race fault. In addition, the rolling bearing with outer race fault has three measurement directions which is 3 o'clock, 6 o'clock and 12 o'clock.

3.2.Data preprocessing

The number of fault sample in 12K drive end is 59,as shown in Table 1, Sampling time of each sample data is 10 seconds. Due to the different length of time domain vibration signal between different bearing fault types.However,we need to measure the similarity between different bearing fault types. However, we need to measure the similarity between different bearing fault types.

Figure 1. Test bed of rolling bearings.

Therefore, the time domain signal data of each fault type need to be preprocessed, and the preprocessing steps are as follows: (1) the time domain signal data interception of each bearing fault. Each sample is taken from the first 12000 data points, as shown in Fig.2(1); (2) Calculating auto-power spectrum for the interception of the fault time domain vibration signal, as shown in Fig.2(2); (3) Smoothing the auto-power spectrum of each bearing fault sample, as shown in Fig.2(3).It is found that the smoothing process is very important to the clustering result. As the clustering process is very sensitive to the noise signal, so it is needed to eliminate the noise by smoothing. In addition, as there are many kinds of fault types in 12K drive end, in order to simplify the expression, we choose one kind of bearing fault types from three different fault positions (the ball fault, the inner race fault and the outer race fault) to represent the pretreatment process, as shown in Fig. 2

3.3. Diagnosis results

After the raw bearing fault vibration signals are preprocessed in 3.2 section. The novel clustering method is used to cluster the 12K drive end bearing fault samples to find out the same type of faults. The similarity between each bearing fault sample is computed by Euclidean distance. The density is estimated by a Gaussian kernel with variance $d_c = 0.03$. The clustering decision graph is shown in Fig.3 (a) when the auto-power spectrum is not smoothed.

Table1. Description of 12K drive end bearing fault datasets

Fault position	Fault diameter(inch)	Fault direction	Load condition
The ball fault	0.007 0.014 0.021 0.028	Non	0,1,2,3
The inner race fault	0.007 0.014 0.021 0.028	Non	0,1,2,3
The outer race fault	0.007	3 o'clock 6 o'clock 12 o'clock	0,1,2,3
	0.014	6 o'clock	0,1,2,3
	0.021	3 o'clock 6 o'clock 12 o'clock	0,1,2,3

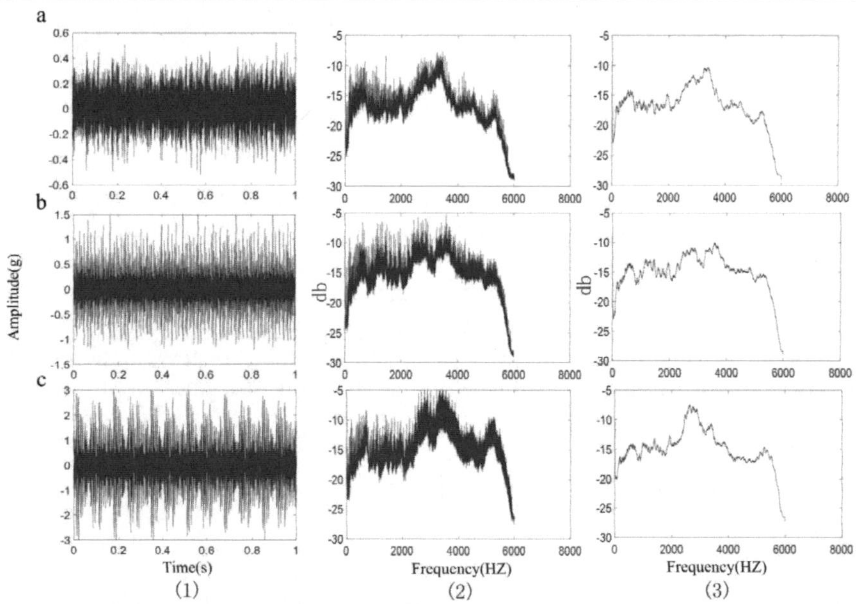

Figure 2. The time domain vibration signal and spectra for three different positions of bearing fault:(a) The ball fault (The fault diameter is 0.007 inch under 0 load condition), (b) The inner race fault (The fault diameter is 0.007 inch under 0 load condition), (c) The outer race fault (The fault diameter is 0.007 inch and Fault direction is 3 o'clock under 0 load condition)

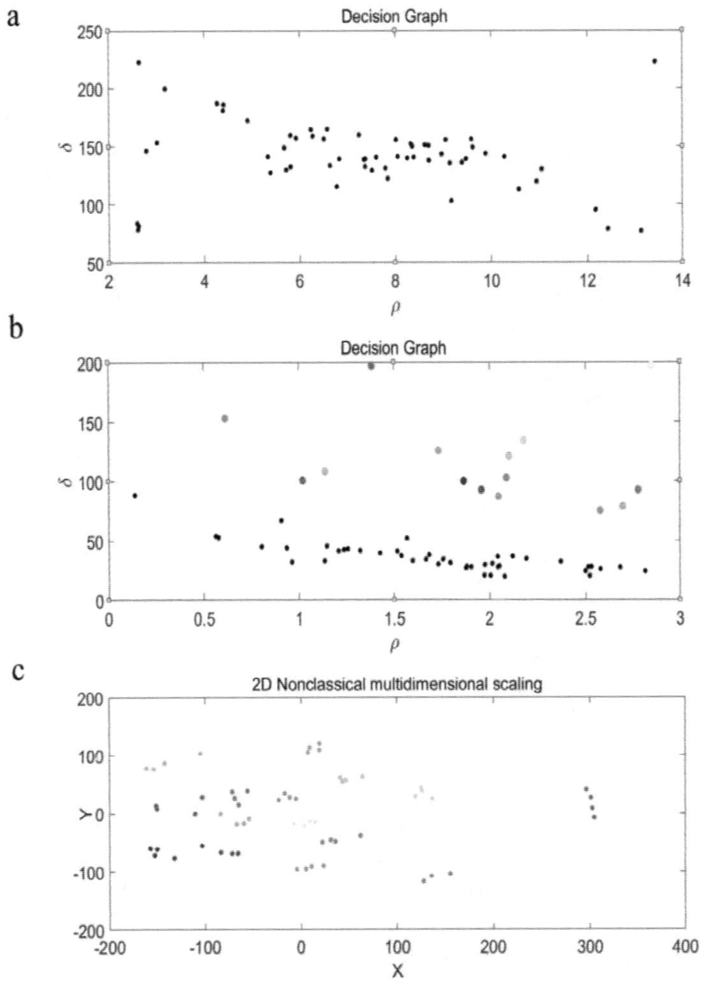

Figure 3. Cluster analysis of 12K drive end bearing fault.(a) The corresponding decision graphs, with the auto- power spectrum is not smoothed,(b)The corresponding decision graphs , with the auto-power spectrum has been smoothed and the centers colored by cluster,(c) The distribution graph of the sample after classifyin

Fig. 3(b) shows the clustering decision graph after smoothing and the clustering results of each sample is depicted in Fig.3(c). Compared with Fig.3 (b), the exact number of cluster centers is not clear and a consequence of the sparsity of the data points in Fig.3 (a).Therefore, it is required to smooth the auto-power spectrum before clustering bearing fault samples. In addition, From Fig. 3 (b) and Fig.3 (c) shows that the 12K drive end bearing fault is divided into 15 clusters. In addition, according to the number of clustering centers we found using the novel clustering method.K-means was used to classify the rolling bearing fault samples again. Compared with the two clustering methods, both of the methods can identify all kinds of fault types, but the novel clustering method does not need to determine the number of clusters in advance and the accuracy rate of the classifying results is 100%.

4. Conclusions

Currently, supervised learning methods are commonly used for the fault diagnosis of bearing and other related parts and the accuracy is often very significant. However, supervised learning methods often require massive the known fault types as training samples, which is not so easy for the actual processing conditions. And as the actual processing conditions are often more complex, so there will be some unknown fault types occasionally. It is a challenge for fault diagnosis using the supervised learning methods. On this basis, this paper proposes a novel method for classifying the fault types of rolling bearings, this clustering method that can find the cluster centers from the samples automatically is different from the traditional K-means and K-medoids, it does not need to determine the number of clusters in advance. The effectiveness of the proposed method is verified using 59 samples .59 rolling bearing fault samples contain a lot of fault types, clustering results show that each of fault types can be separated; the accuracy can even reach 100%. But by looking for suspected clustering centers can distinguish between high similarities of fault types. On the other hand, the novel clustering method is sensitive to noise, thus the auto-power spectrum of the time domain signal of the bearing faults should be smoothed before classifying, so that the classifying effect will be more significant.

Acknowledgments
The research is supported by the National Natural Science Foundation of China under Grant nos. 51375193, the Postdoctoral Science Foundation of China under Grant nos. 2016M602282 and the State key laboratory smart manufacturing for special vehicles and transmission system nos.GZ2016KF008.the National Major Special Projects Foundation of China under Grant nos. 2014ZX04014101. the National Natural Science Foundation of China under Grant nos. 51775210

References
[1] Wang W and Lee H, An energy kurtosis demodulation technique for signal denoising and bearing fault detection Meas. Sci. Technol. 24 (2013)025-601
ransform technique for bearing condition monitoring Meas. Sci. Technol. 24 (2013)085004
[2] Liu J, Wang W, Golnaraghi F and Liu K, Shannon wavelet spectrum analysis for bearing fault diagnostics Meas. Sci. Technol. 19(2008) 015-105
[3] Lei Y G, He Z J and Zi Y Y, Application of an intelligent classification method to mechanical fault diagnosis Expert. Syst. Appl. 36(2009) 9941–8
[4]B. Samanta, Artificial neural networks and genetic algorithms for gear fault detection, Mech. Syst. Signal Process. 18 (2004) 1273–1282.
[5] Tamilselvan P and Wang P F, Failure diagnosis using deep belief learning based health state classification Reliable. Eng. Syst. Saf. 115 (2013)124–35
[6] A. Widodo, E.Y. Kim, J.-D. Son, B.-S. Yang, A.C. Tan, D.-S. Gu, B.-K. Choi, J. Mathew, Fault diagnosis of low speed bearing based on relevance vector machine and support vector machine, Expert Syst. Appl. 36 (2009) 7252–7261.
[7]B.A. Paya, I.I. Esat, Artificial neural network based fault diagnostics of rotating machinery using wavelet transforms as a preprocessor, Mech. Syst. Signal Process. 11 (1997) 751–765
[8] H. Shao, H. Jiang, X. Zhang, and M. Niu,Rolling bearing fault diagnosis using an optimization deep belief network, Meas.Sci.Technol.26(2015)115-002
[9] Feng Jia, Yaguo Lei n, Jing Lin, Xin Zhou, Na Lu, Deep neural networks: A promising tool for fault characteristic mining and intelligent diagnosis of rotating machinery with massive data, Mech. Syst. Signal Process. 72-73 (2016) 303–315
[10] Rodriguez, A. and Laio,A.Clustering by fast search and find of density peaks. Science (New York, N.Y.), 344(2104)1492–1496
[11] X.S. Lou, K.A. Loparo, Bearing fault diagnosis based on wavelet transform and fuzzy inference, Mech. Syst. Signal Process. 18 (2004) 1077–109

IOP Conf. Series: Materials Science and Engineering **280** (2017) 012014 doi:10.1088/1757-899X/280/1/012014

Effect of CT Specimen Thickness on the Mechanical Characteristics at the Crack Tip of Stress Corrosion Cracking in Ni-based Alloys

Cui Yinghao[1], Xue He[1] and Zhao Lingyan[2]

[1] School of Mechanical Engineering, Xi'an University of Science and Technology, Xi'an 710054, China
[2] School of Science, Xi'an University of Science and Technology, Xi'an 710054, China

xue_he@hotmail.com

Abstract. It's important to obtain accurate stress corrosion crack(SCC) growth rate for quantitative life prediction of components in nuclear power plants. However, the engineering practice shows that the crack tip constraint effect has a great influence on the mechanical properties and crack growth rate of SCC at crack tip. To study the influence of the specimen thickness on the crack tip mechanical properties of SCC, the stress, strain and C integral at creep crack tip are analyzed under different specimens thickness. Results show that the cracked specimen is less likely to crack due to effect of crack tip constraint. When the thickness ratio *B/W* is larger than 0.1, the crack tip constraint is almost ineffective. Value of C integral is the largest when *B/W* is 0.25. Then specimen thickness has little effect on the value of C integral. The effect of specimen thickness on the value of C integral is less significant at higher thickness ratio.

1. Introduction

Ni-based alloy 600 is widely used in nuclear primary pressure vessel and dissimilar metal welded joints due to its good high temperature corrosion resistance and mechanical properties. The stress corrosion cracking (SCC) caused by crack tip creep in high temperature water environment is one of the important factors lead to the failure of structures [1]. Therefore, it is very important to obtain accurate creep crack propagation rate for quantitative prediction nuclear structure life [2-3]. Influence of the specimen thickness constraint on creep crack tip mechanical field is considered to correlate the creep rupture mechanics parameters with the structural crack propagation rate, and evaluate the structural integrity and the life prediction. Practice shows that the crack tip constraint has a great influence on the mechanical properties at the creep crack tip.

The crack tip constraint is usually controlled by the specimen geometry, the crack size, and the loading mode [4]. The influence of materials and structures fracture behavior [5] and crack tip constraint, quantitative characterization of constraint parameters [6-8], and the law of influence and mechanism of compound restraint on CGR [9] were studied by many scholars.

Different thickness finite element model is established by compact tension specimen using Ni-based Alloy 600. The micro-mechanical properties of crack tip region and the main parameters of SCC crack propagation rate are studied. The influence of CT specimen thickness on the fracture parameters

IOP Conf. Series: Materials Science and Engineering **280** (2017) 012014 doi:10.1088/1757-899X/280/1/012014

at creep crack tip of Ni-based Alloy is further analyzed, which is of great significance for the study of SCC crack initiation and propagation behavior.

2. Finite element modeling

2.1. Specimen model

Compact tensile specimen with prefabricated crack is used as the research object. 1T-CT specimen geometry is shown in Figure 1, which is coincidence with the American Society for Experimental and Materials Standards ASTM E399-90 standard [10]. The specimen width W=50 mm, the crack length a = 0.5W, the sample thickness is B, and the value of B/W is 0.05, 0.1, 0.2, 0.3 and 0.4, respectively.

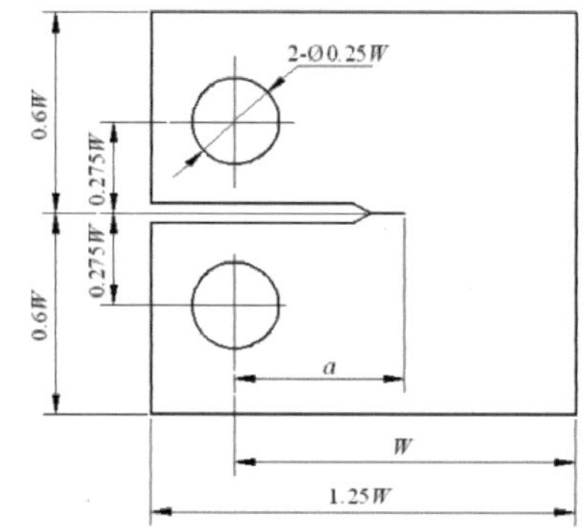

Figure 1. Geometry size of 1T-CT specimen

The finite element mesh of the 1T-CT global model and the local region of the crack tip are shown in Figure 2 when the B/W = 0.05, 8-node linear brick (C3D8) is used in simulation and total number is 138 820. To obtain a more accurate crack tip stress-strain, the mesh is refined at the crack tip and the total number at cracked local area is 42 680.

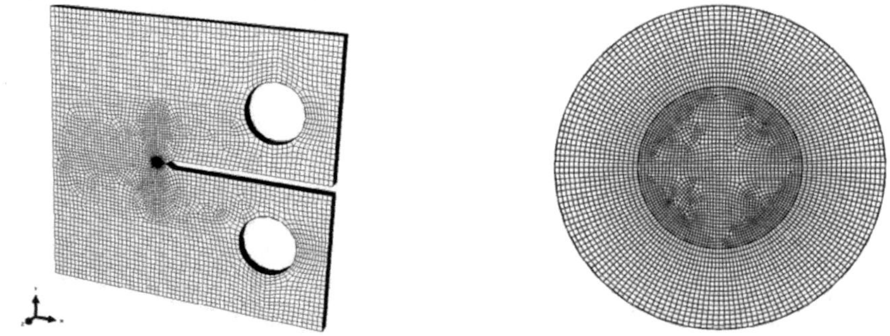

(a) Global mesh model (b) Mesh at crack tip zone

Figure 2. Finite element mesh model of 1T-CT

2.2. Material model

The Ni-based Alloy 600 is a power-hardened material whose constitutive relationship is usually characterized by the Ramberg-Osgood relationship [11], and its form is as follows:

$$\frac{\varepsilon}{\varepsilon_0} = \frac{\sigma}{\sigma_0} + \alpha \left(\frac{\sigma}{\sigma_0} \right)^n \tag{1}$$

Where ε is the strain, including elastic and plastic strain. σ is the total stress; ε_0 is the yield strain of the material, σ_0 is the yield stress of the material, and n is the strain hardening exponent of the material, α is the offset coefficient of the material.

The creep rate is calculated using the power law model [12-13] in the FEM calculation:

$$\dot{\varepsilon}_{cr} = A\sigma^n t^m \tag{2}$$

Where $\dot{\varepsilon}_{cr}$ is the creep strain rate. σ is the applied stress; A is the power law multiplier, the value is 1.24×10^{-23}, n is the creep exponent, and the value is 7.5.

The mechanical properties of the material is given in Table 1.

Table 1. Material mechanical property of Nickel-based alloy 600 at 343℃

Material	E /MPa	ν	σ_0 /MPa	n	α
Alloy600	193 000	0.288	385	4.799	1.0

2.3. Loading and boundary conditions

Due to symmetry of CT specimen, only the half specimen was modelled. Boundary condition is applied to the symmetry plane of the model in Y direction, and to the load application in the X direction, as shown in Figure 2(a). Loading is applied to the load hole by using the multiple point constraints facility in ABAQUS, which joins the hole centre to the nodes of the hole surface. The concentrated force is applied on the reference point to ensure the crack tip stress intensity factor K_I=15 MPa·m$^{1/2}$.

3. Results and Discussion

3.1. Effect of specimen thickness on SCC crack tip stress

The Mises stress distribution in front of the crack tip of different thickness samples is shown in Figure 3. The radius is 0.5 mm, and it can be seen that the Mises stress decreases with increasing of the specimen thickness. The Mises stress decreases slowly as the thickness increases when thickness ratio≥0.1, that is to say, The effect of thickness on the Mises stress at this time is not so great.

3.2. Effect of specimen thickness on SCC crack tip PEEQ

The PEEQ distribution in front of the crack tip of different thickness samples is shown in Figure 4. The radius is also set as 0.5 mm. The PEEQ at crack tip decreases with the increasing of the specimen thickness, but the influence is not so obvious, which indicates that PEEQ at crack tip is not a appropriate parameter to describe crack propagation.

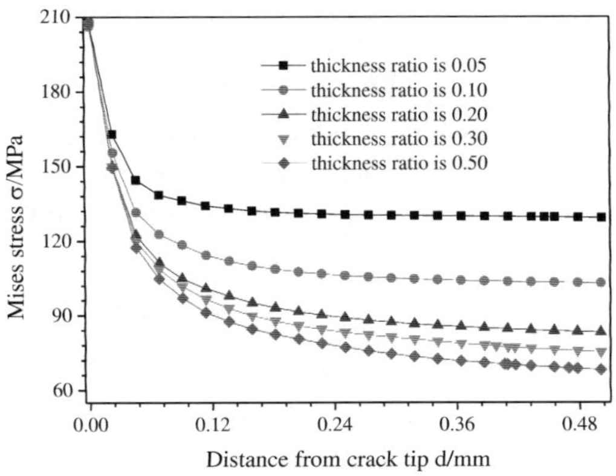

Figure 3. Mises stress ahead of crack tip (MPa)

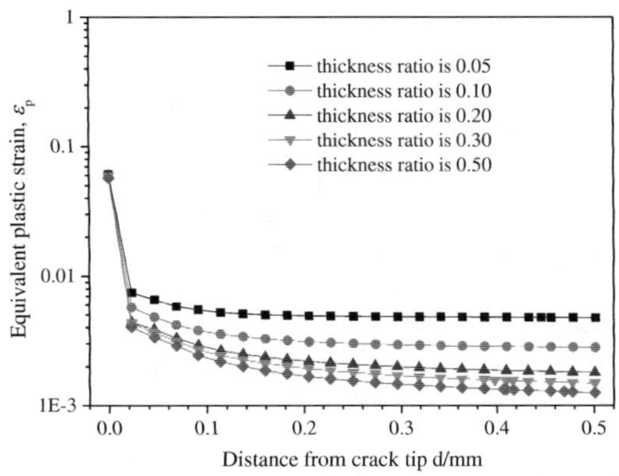

Figure 4. Tensile plastic strain ahead of crack tip

3.3. Effect of specimen thickness on SCC crack tip creep strain

The creep strain distribution in front of the crack tip of different thickness samples is shown in Figure 5. The creep strain at crack tip decreases with the increasing of the specimen thickness, but the influence is not so obvious, which indicates that creep strain at crack tip is not a appropriate parameter to describe crack propagation.

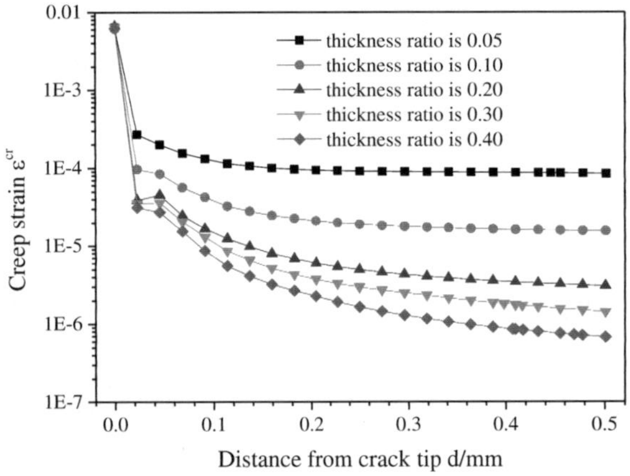

Figure 5. Creep strain ahead of crack tip

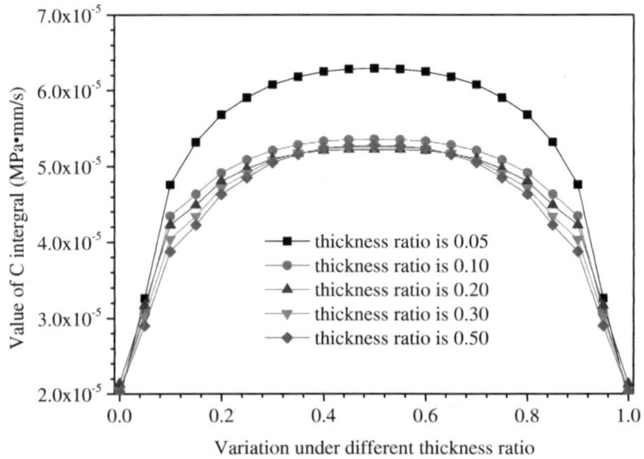

Figure 6. Value of C integral

3.4. Effect of specimen thickness on C integral

The effect of specimen thickness on the crack tip C integration is shown in Figure 6. The value of the C-integral decreases gradually as the thickness of the sample increase, so the constraint is larger when specimen is greater, and the specimen is less likely to crack at this time. Compared with the thinner sample, the crack growth rate is also smaller due to the thicker specimen being closer to the plane stress state. Compared with the stress and strain parameters, C integral is a reasonable crack tip fracture parameter, suitable for describing the crack initiation and growth of stress corrosion cracking.

4. Conclusions

By analyzing the influence of the thickness of the compact tensile specimen on the creep stress-straom field at the crack tip of the Ni-based alloy SCC, the conclusions are as follows:

(1) When the sample thickness ratio $B/W \geqslant 0.1$, the crack tip effect is almost ineffective; when $B/W=0.25$, The crack tip stress, strain and C integral is the maximum, then the specimen thickness has little effect on the C integral value.

(2) Compared with the stress and strain parameters, C integral is a reasonable crack tip fracture parameter, suitable for describing the crack initiation and expansion of stress corrosion cracking.

(3) To accurately predict the crack propagation rate of the low-constraint nuclear welding structure, it is necessary to consider the influence of the constraint level of the specimen on the crack propagation rate.

Acknowledgments

This work is financially supported by the National Natural Science Foundation of China (Grant nos. are 51475362 and11502195).

References

[1] Ma Cheng, Peng Qunjia, Han Enhou. Review of stress corrosion cracking of structural materials in nuclear power plants [J]. Journal of Chinese Society for Corrosion and Protection, 2014, 34(1): 37-45.

[2] Xue He, Gong Xiaoyan, Zhao Lingyan. Effect of mechanical parameter selection in Quantitative estimation of the growth of environmentally assisted cracks at flaws in light water

reactor components with complex mechanical condition[C]. ASME 2011 Pressure Vessels and Piping Conference, 2011: 893-899.

[3] PENG Qunjia, XUE He, HOU Juan. Role of water chemistry and microstructure in stress corrosion cracking in the fusion boundary region of an Alloy 182-A533B low alloy steel dissimilar weld joint in high temperature water [J]. Corrosion Science, 2011, 53(12): 4309-4317.

[4] Tan Jianping, WANG Guozhen, XUAN Fuzhen, et al. The effect of CT specimen thickness on high temperature creep crack growth rate of 25Cr2NiMo1V steel [J]. Materials for Mechanical Engineering, 2012, 36(10): 42-46.

[5] Kim Y J, Kim J S, CHO S M, et al. 3-D constraint effects on J testing and crack tip constraint in M(T), SE(B), SE(T) and C(T) specimens: numerical study [J]. Engineering Fracture Mechanics, 2004, 71(9): 1203-1218.

[6] Dodds Jr R H, Shih C F, Anderson T L. Continuum and micro mechanics treatment of constraint in fracture [J]. International Journal of Fracture, 1993, 64(2): 101-133.

[7] Mostafavi M, Smith D J, Pavier M J. A micromechanical fracture criterion accounting for in-plane and out-of-plane constraint [J]. Computational Materials Science, 2011, 50: 2759-2770.

[8] Yand Jie, Wand Guozhen, Xuan Fuzhen. Unified characterisation of in-plane and out-of-plane constraint based on crack-tip equivalent plastic strain [J]. Fatigue & Fracture of Engineering Materials & Structures, 2013, 36(6): 504-514.

[9] He Bin, Liu Jun, Chen Jianen, et al. Finite element simulation of the growth rate of high temperature creep crack of P92 steel with different constraints [J]. Materials Review, 2016, 30(16): 145-149.

[10] ASTM Standard E399-90. Annual book of ASTM standards[M]. USA: ASTM International, 2002.

[11] Ueda Y, Shi Y, Sun S. Effect of crack depth and strength mis-matching on the relation between J-integral and CTOD for welded tensile specimens(mechanics, strength & structure design) [J]. Transaction of JWRI, 1997, 26(1): 133-140.

[12] Zhao Lingyan, Cui Yinghao, Xue He, et al. Effect of crack propagation initiation angle on the stress corrosion cracking behavior in welding transition zone of Ni-based alloys [J]. China Sciencepaper, 2016, 11(16): 1864-1866.

[13] Quintero H, Mehmanparast A. Prediction of creep crack initiation behaviour in 316H stainless steel using stress dependent creep ductility[J]. International Journal of Solids & Structures, 2016, s 97–98:101-115.

IOP Conf. Series: Materials Science and Engineering **280** (2017) 012015 doi:10.1088/1757-899X/280/1/012015

Parameterized LMI Based Diagonal Dominance Compensator Study for Polynomial Linear Parameter Varying System

Xiaobao Han, Huacong Li and Qiusheng Jia

School of Power and Energy Northwestern Polytechnical University Xi'an , China

ranger_qing@nwpu.edu.cn

Abstract. For dynamic decoupling of polynomial linear parameter varying(PLPV) system, a robust dominance pre-compensator design method is given. The parameterized pre-compensator design problem is converted into an optimal problem constrained with parameterized linear matrix inequalities(PLMI) by using the conception of parameterized Lyapunov function(PLF). To solve the PLMI constrained optimal problem, the pre-compensator design problem is reduced into a normal convex optimization problem with normal linear matrix inequalities (LMI) constraints on a new constructed convex polyhedron. Moreover, a parameter scheduling pre-compensator is achieved, which satisfies robust performance and decoupling performances. Finally, the feasibility and validity of the robust diagonal dominance pre-compensator design method are verified by the numerical simulation on a turbofan engine PLPV model.

1. Introduction

For MIMO control problem, the decoupling performances are important because coupling interaction among different control loops can significantly worst the control characteristics. An promising methodology for multi-input multi-output(MIMO)control systems are convert the control system into single-input single-output(SISO) control system by using decoupling precompensation, then design controllers for every decoupled control loop[1,2]. In diagonal dominance precompensation study for decoupling purpose, Hawkins has proposed a pseudo-diagonalization method which minimize the off-diagonal effects at certain frequencies[3].Neil Munro obtains optimal reduction in the off-diagonal elements based on Perron-Frobenius theory and the conception of generalised diagonal dominance[4].Nobakhti uses genetic algorithms to achieve diagonal dominance without choosing a design frequency in previous methods[5]. In[6] Chughtai etc. give a new method for determining a constant precompensator for reducing the interactions in MIMO system by solving LMI constrained optimization problem. B. LABIBI presents a methodology for diagonal dominance of large-scale systems via Eigen structure assignment[7]. Based on the theory of generalized diagonal dominance, S. Jamebozorg gives a design method of static precompensator for the reduction of interaction in linear multivariable systems[8]. For LPV systems, the decoupling compensation is important and difficult in control system design with the change of scheduling parameters[9]. In [10], A LPV system output regulation is achieved by nonlinear compensation approach based on PLF. J. Mohammadpour proposes a method using parameter-dependent static inversion or SVD decomposition of the system to minimize the effects of the off-diagonal terms in the MIMO system[11,12]. Based on minimization of the H_2 norm, Amin Nobakhti proposes a LMI method for the design of these precompensators which improve the efficiency in speed of computation and design conservatism[13]. In this study a LMI based decoupling compensator is designed to obtain the diagonal dominance of LPV system using parameterized Lyapunov function, then the global solutions are achieved by solving LMI restrained optimization problems on constructed convex polyhedron vertex set. Finally, the given decoupling compensator design method is verified by numerical simulations.

2. Problen formulation

For LPV decoupling compensator design problem, consider the following PLPV system(1), which depends on the parameter σ with bounded parameter variation rate.

$$\Sigma p \begin{cases} \dot{x}_p = A_p(\sigma)x_p + B_p(\sigma)u \\ y_p = C_p(\sigma)x_p + D_p(\sigma)u \end{cases} \tag{1}$$

Where $x_p \in \mathrm{R}^n$ is the state vector, $u \in \mathrm{R}^n$ is the input vector, $y_p \in \mathrm{R}^n$ is the output vector. The coefficient matrices of system (1) have the same form, which is defined as follows:

$$\ell(\sigma) = \sum_{i=0}^{g} \sigma^i \ell_i \,, \sigma \in [\sigma_{\min}, \sigma_{\max}] \,, |\dot{\sigma}| < D_{\max} \tag{2}$$

Where ℓ_i are known real constant matrices with appropriate dimensions. Moreover, we consider the following form of PLF needed in system analysis and synthesis.

$$\mathrm{V}(x,\sigma) = x^T P(\sigma) x \,, P(\sigma) = \sum_{i=0}^{\mathrm{m}} \sigma^i P_i \tag{3}$$

Where, $P(\sigma) \in R^{n\times n}$ is a symmetric positive definite polynomial Lyapunov matrix of degree m, P_i are symmetric real constant matrices with appropriate dimensions. The LPV decoupling compensator design purposes are design compensator $\mathrm{K}_p(\sigma)$ which can decoupling the interaction between different control loops under the varying of scheduling parameter σ, then the MIMO control problem can be converted into several SISO control problems easily.

3. LMI based decoupling compensator design

The decoupling compensator design purpose is to find a static compensator that optimize diagonal dominance condition of compensated system under certain constrains[6]. Firstly, the following parameterized regular splitting definition is given based on the conceptions of parameterized regular splitting and parameterized fundamental dominance which extended from the basic conceptions in[6]:

Parameterized regular splitting: Let $\mathrm{A}(s,\sigma) \in C^{\mathrm{n\times n}}$ be a parameterized complex matrix, then, $\mathrm{A}(s,\sigma)$ is said to have a regular splitting if

$$\mathrm{A}(s,\sigma) = \mathrm{B}(s,\sigma) + \mathrm{D}(s,\sigma) \tag{4}$$

such that $\mathrm{B}(s,\sigma)$ is non singular. Then the fundamental dominance is defined as:

Parameterized Fundamental dominance: Let $\mathrm{A}(s,\sigma) \in C^{\mathrm{n\times n}}$ be a complex matrix, and let $\mathrm{A}(s,\sigma)=\mathrm{B}(s,\sigma)+\mathrm{D}(s,\sigma)$ be a regular splitting, where $\mathrm{B}(s,\sigma) \in C^{\mathrm{n\times n}}$ and $\mathrm{D}(s,\sigma) \in C^{\mathrm{n\times n}}$. Then, the term $\mathrm{B}(s,\sigma)$ is said to be the dominant term of $\mathrm{A}(s,\sigma)$ if and only if

$$\rho(\mathrm{B}(s,\sigma)^{-1}\mathrm{D}(s,\sigma)) < 1 \tag{5}$$

where $\rho(.)$ is the spectral radius and is used to scale the magnitude of the maximum eigenvalue.

If any square parameterized transfer function matrix(PTFM) $G(s,\sigma)$ has a regular splitting of the form

$$G(s,\sigma) = \mathrm{X}(\sigma) + Y(s,\sigma) \tag{6}$$

and the parameterized precompensator $\mathrm{K}_p(\sigma) = \mathrm{X}(\sigma)^{-1}$; then the compensated system would be

$$G(s,\sigma)\mathrm{K}_p(\sigma) = \mathrm{I} + Y(s,\sigma)\mathrm{K}_p(\sigma) \tag{7}$$

Thus, if $X(\sigma)$ is the dominant term of $G(s,\sigma)$, then the identity will be the dominant term of the compensated system. Since all eigenvalues of a matrix are bounded above by its maximum singular value or $\rho(,)<\bar{\sigma}(\cdot)$ and $\|\cdot\|_\infty = \sup_\omega \bar{\sigma}(.)$; the inequality (5) could be presented as

$$\|B(s,\sigma)^{-1}D(s,\sigma)\|_\infty \le 1 \tag{8}$$

Then, for the system having the regular splitting as in (6) the problem of achieving diagonal dominance can be presented as an optimization problem: Find a compensator $K_p(\sigma)$ can minimize.

$$\|Y(s,\sigma)K_p(\sigma)\|_\infty = \|G(s,\sigma)K_p(\sigma)\text{-}I\|_\infty \le \gamma \tag{9}$$

The minimization of γ can be presented in the form of LMIs by using the Bounded Real Lemma (BRL)[6]. According to BRL, the state space system (1) corresponding PTFM form is $G(s,\sigma)=C_p(\sigma)(sI\text{-}A_p(\sigma))^{-1}B_p(\sigma)+D_p(\sigma)$, then A_p is stable and $\|G(s,\sigma)\|_\infty<\gamma$, if and only if there exists a positive defined polynomial parameterized symmetric matrix $P(\sigma)$ satisfied LMI constrains(10).

$$\begin{bmatrix} A_p(\sigma)P(\sigma)+P(\sigma)A_p(\sigma)^T+\dot{P}(\sigma) & B_p(\sigma) & P(\sigma)C_p(\sigma)^T \\ * & -\gamma I & D_p(\sigma)^T \\ * & * & -\gamma I \end{bmatrix} < 0 \tag{10}$$

Let $G(s,\sigma)$ is the plant without compensation and define $\tilde{G}(s,\sigma)=G(s,\sigma)K_p(\sigma)-I$ as the compensated plant. Based on (10) the diagonal dominant constant precompensator $K_p(\sigma)$ will be the solution of optimization problem(11)restrained by LMIs.

$$\min_\omega \gamma \text{ s.t.} \begin{bmatrix} A_p(\sigma)P(\sigma)+P(\sigma)A_p(\sigma)^T+\dot{P}(\sigma) & B_p(\sigma)K_p(\sigma) & P(\sigma)C_p(\sigma)^T \\ * & -\gamma I & K_p(\sigma)^T D_p(\sigma)^T - I \\ * & * & -\gamma I \end{bmatrix} < 0 \tag{11}$$

4. Convex polyhedron construction based compensator design

For PLPV model, the PLF reduces the control system analysis and synthesis problem into an optimization problem with the PLMI constraints. To guarantee the global characters of compensator design, we convert it into a normal convex optimization problem with LMI constraints based on a methodology of convex polyhedron construction[14].

Suppose $\Phi_P=(P_0,\cdots,P_m)^T$, considering (2) and (3), then the constraints (11) can be written as:

$$F(\sigma)=\sum_{i=0}^{m}\sigma^i F_i(\Phi_P,\gamma)<0 \tag{12}$$

Where $F_i(\Phi_P,\gamma)$ is a symmetric matrix function that affinely depends on Φ_P and γ. For (12), we have the following theorem:

Theorem 1[15]: In m dimensional linear space, assume that there exists a convex polyhedron H_0 including curve Γ which is defined as:

$$\Gamma=\{(\sigma,\cdots,\sigma^m)^T \mid \sigma\in[\sigma_{min},\sigma_{max}]\} \tag{13}$$

The $p^j\in R^m$ and the vertex set of H_0 is $V_0=\{p_i \mid i=1,\cdots,q\}$.If there exist Φ_P and γ that satisfy (14):

$$\bar{F}(p_i)=\sum_{j=0}^{m}p_{i,j}F_j(\Phi_P,\gamma)<0,\forall p_i\in V_0 \tag{14}$$

Where $p_{i,j}$ is the j th dimensional coordinate of p_i, then (12) will be satisfied on Γ.

Based on theorem 1, the global compensator design can be achieved by satisfying (14) on the constructed convex polyhedron vertex set. Then the convex polyhedron construction method in [14]is used and the construction method is not given in detail because the scope of this article. For LPV system(1), if the constructed convex polyhedrons vertex set is $V=\{v_i \mid i=1,\cdots,q\}$,then the optimization problem(11) can be formulated as(15).

$$\min_{\omega}\gamma \text{ s.t. } \sum_{q=0}^{k} v_{iq}P_q>0,\ \begin{bmatrix} \sum_{q=0}^{s=2k} v_{iq}\sum_{n+j=q}^{n\le k,j\le k} \mathrm{A}_{pj}^{\mathrm{T}} P_n + \sum_{q=0}^{s=2k} v_{iq}\sum_{n+j=q}^{n\le k,j\le k} P_n \mathrm{A}_{pj} + \sum_{n=1}^{k} n v_{i(n-1)} D_{\max} P_n & \sum_{q=0}^{s=2k} v_{iq}\sum_{n+j=q}^{n\le k,j\le k} B_{pj}\mathrm{K}_{pn} & \sum_{q=0}^{s=2k} v_{iq}\sum_{n+j=q}^{n\le k,j\le k} P_j C_{pn}^{\mathrm{T}} \\ * & -\gamma \mathrm{I} & \sum_{q=0}^{s=2k} v_{iq}\sum_{n+j=q}^{n\le k,j\le k} \mathrm{K}_{pj}^{\mathrm{T}} D_{pn}^{\mathrm{T}} - \mathrm{I} \\ * & * & -\gamma \mathrm{I} \end{bmatrix} <0,\ \forall v_i \in \mathrm{V} \quad (15)$$

Where, $v_{i0}=1$, v_{iq}, $q\in(1,\cdots,s)$ is the q th dimensional coordinate of vertex v_i. Then the compensator design problem for PLPV system with constrains(11)can be reduced into a normal LMI constrained convex optimization problem.

5. Numerical simulation

To verify the LMI based decoupling compensation method, the following PLPV model (16) is considered and the coefficient matrices satisfied the definition (1) and (2) with degree 2[14].

$$\Sigma_0:\begin{cases} \dot{x}(t)=A(\sigma)x(t)+B(\sigma)w(t) \\ y(t)=Cx(t) \end{cases} \quad (16)$$

Where the coefficient matrices $[A_0|A_1|A_2]$, $[B_0|B_1|B_2]$ and C are:

$$[A_0|A_1|A_2]=\left[\begin{array}{ccc|ccc|ccc} -4.365 & -0.672 & -0.336 & -0.561 & 0.855 & 0.589 & 0.670 & -1.375 & -0.991 \\ 7.088 & -6.557 & -4.601 & 2.533 & -1.040 & -7.737 & -2.896 & -1.529 & 10.516 \\ -2.410 & 7.584 & -14.310 & 3.192 & 1.797 & -2.589 & -3.578 & 2.839 & 1.909 \end{array}\right] [B_0|B_1|B_2]=\left[\begin{array}{cc|cc|cc} 2.374 & 0.749 & -0.160 & -0.352 & 0.156 & 0.131 \\ 1.366 & 3.444 & 0.116 & -2.484 & -0.496 & 4.038 \\ 0.946 & -9.619 & -0.111 & -4.606 & -0.031 & 0.895 \end{array}\right],\ C=\begin{bmatrix} 0 & 1 & 0 \\ 0 & 0 & 1 \end{bmatrix} \quad (17)$$

The scheduling parameter $\sigma\in[0,1]$ and $D_{\max}=10$.

In simulation, the pseudo-diagonalization is used as a comparative method with the given LMI based compensator. The coefficient matrices of designed compensator $\mathrm{K}_p(\sigma)$ is given in(18).

$$\mathrm{K}_{p0}=\begin{bmatrix} 16.67 & 4.60 \\ 1.56 & -1.63 \end{bmatrix},\ \mathrm{K}_{p1}=\begin{bmatrix} -12.25 & -5.72 \\ -2.07 & 1.30 \end{bmatrix},\ \mathrm{K}_{p2}=\begin{bmatrix} -3.30 & 9.69 \\ 0.03 & 0.12 \end{bmatrix} \quad (18)$$

In figure 1 and figure 2, the dominance ratio and Inverse Nyquist Array (INA) with row Gershgorin discs of Σ_0 are given respectively where the $\sigma=0.1$. Figure 3 and figure 4 show the dominance ratio and INA with row Gershgorin discs of pseudo-diagonalization compensated system Σ_1 respectively where the key compensation frequency is 0Hz and $\sigma=0.1$. Figure 5 and figure 6 show the dominance ratio and inverse Nyquist array (INA) with row Gershgorin discs of LMI based compensated system Σ_2 respectively where the parameter $\sigma=0.1$.

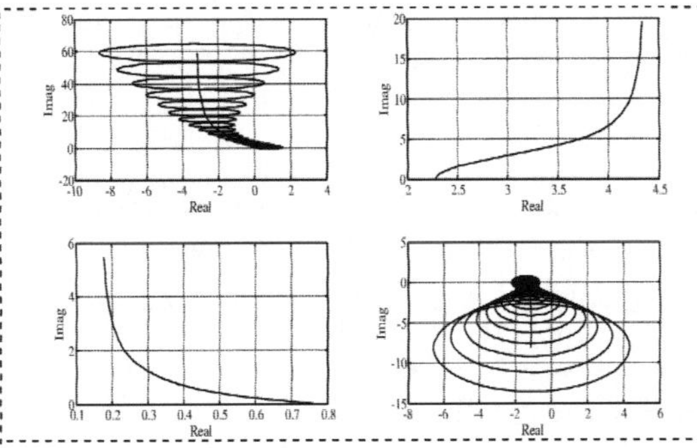

Figure 1. INA with Gershgorin discs of Σ_0.

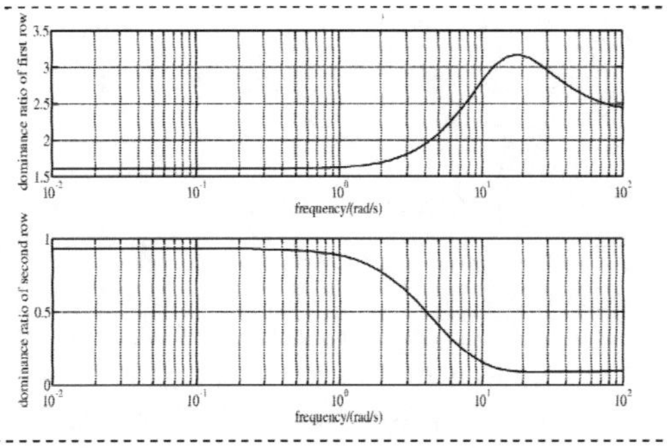

Figure 2. Dominance ratio of Σ_0.

IOP Conf. Series: Materials Science and Engineering **280** (2017) 012015 doi:10.1088/1757-899X/280/1/012015

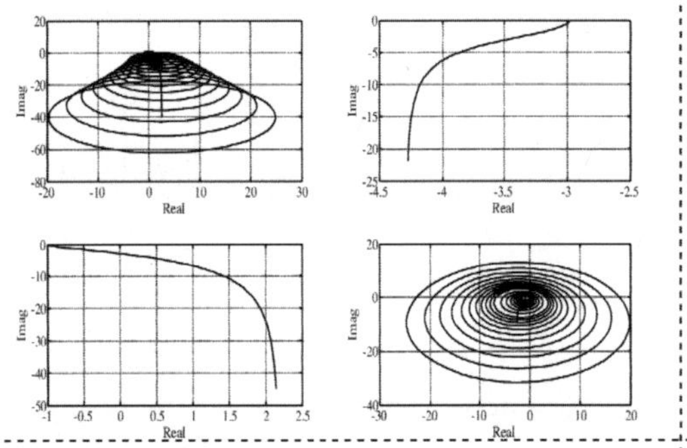

Figure 3. INA with Gershgorin discs of pseudo-diagonalization compensated Σ_1.

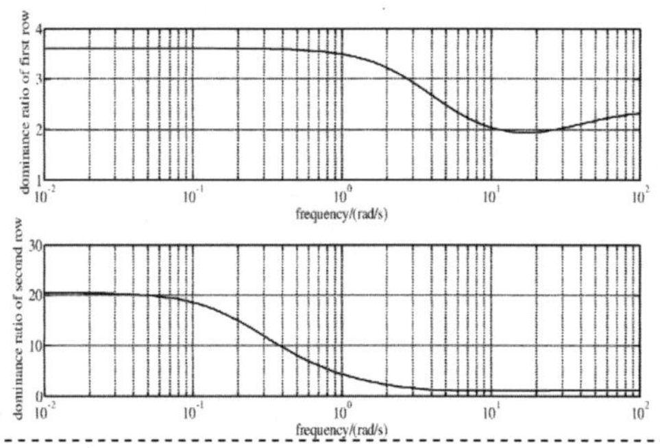

Figure 4. Dominance ratio of pseudo-diagonalization compensated Σ_1.

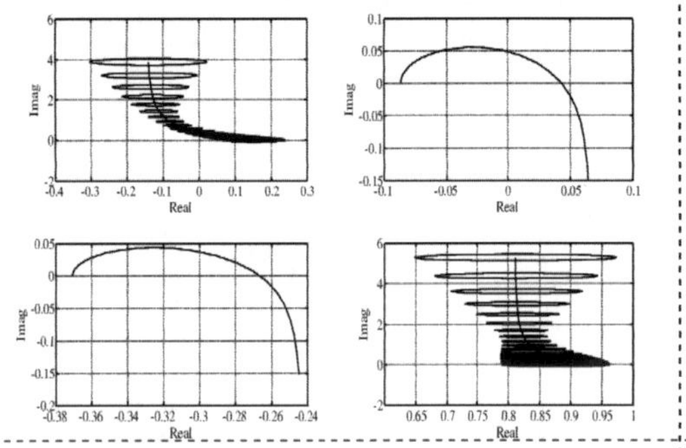

Figure 5. INA with Gershgorin discs of LMI based compensated Σ_2.

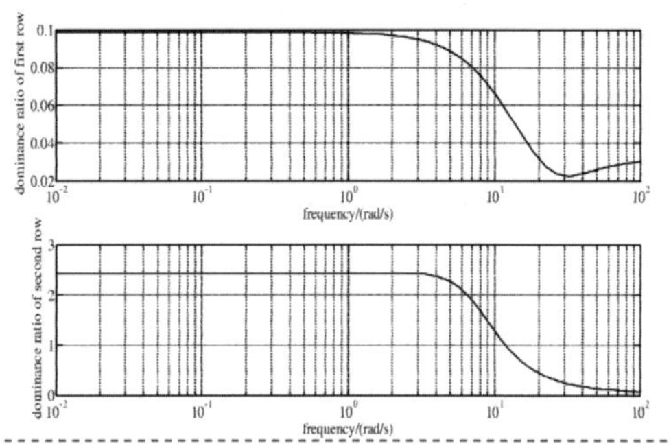

Figure 6. Dominance ratio of LMI based compensated Σ_2.

In figure 1 and figure 2, the first row of Σ_0 is almost non diagonal dominance with large scale of Gershgorin discs size and the second row shows better diagonal dominance condition. Figure 3 and figure 4 shows that the first row of Σ_1 is almost non-diagonal dominance, the second row is non-diagonal dominance and both rows with large scale of Gershgorin discs size. As shown in figure 5 and figure6, the LMI based compensation model show good diagonal dominance characteristics in working frequency domain and get satisfying decoupling performance with small scale of Gershgorin discs size. For LPV model, the LMI based compensation method also have global characteristics of diagonal dominance compensation without specifying certain frequency domain which the classical method can not satisfy. The more simulations on $\sigma \in [0,1]$ also show satisfying diagonal dominance performance results of compensated models.

6. Conclusion

In MIMO LPV control system analysis and synthesis, the interaction between different control loops can dramatically influence the control performance which changed with the varying schedule parameters. For a PLPV system a global robust dominance pre-compensator design method is proposed. Based on parameterized Lyapunov function, this parameterized compensator design is converted into an optimal problem constrained with parameterized linear matrix inequalities; To get the global solution, the design problem was reduced into a normal convex optimization problem with normal LMI constraints on constructed convex polyhedron. Finally, the feasibility and validity of the parameterized robust diagonal dominance pre-compensator design method are verified by the numerical simulation on a PLPV model.

Acknowledgments
This research was supported by the National Natural Science Foundation of China (51506177).

References

[1] Gao Yahui, Zang Jun, YaoHua, et al. Decoupling PI controller design for aeroengine, Journal of Bei jing university of Aeronautics and Astronautics, 40(4),pp:449-454, 2014.

[2] Zhu Yubin, Fan Siqi, Ren Xinyu, et al. A multivariable decoupling control based on DRNN for aeroengine, Journal of Propulsion Technology, 27(2), pp:150-153, 2006.

[3] D.J. Hawkins, Pseudo-diagonalisation and the inverse Nyquist array method, IEEE Proceedings Part D, Control Theory and Applications, 119(3), pp. 337–342, 1972.

[4] Neil Munro , Recent extensions to the inverse Nyquist array design method, Proceedings of 24th Conference on Decision and control, pp.1852-1857, 1985.

[5] A. Nobakhti, N. Munro and B. Porter, ''Evolutionary achievement of diagonal dominance in linear multivariable plants'', Electronics Letters, 39, pp. 165–166, 2003.

[6] S.S. Chughtai and N. Munro, Diagonal dominance using LMIs, IEEE Proceedings Part D, Control Theory and Applications, 151(2), pp. 225–233, 2004.

[7] B. LABIBI, H. J. MARQUEZ and T. CHEN, Diagonal dominance via eigenstructure assignment ,International Journal of Control,Vol. 79, No. 7, pp:707–718, 2006.

[8] S. Jamebozorg, A. Shahmansoorian, Multivariable Decoupling Control System Based on Generalized Diagonal Dominance, International Journal of Electronics Communication and Computer Engineering, Volume 6, Issue 2, pp:199-204,2015.

[9] Mingxing Li, Yinmin Jia, and Junping Du, LPV Control With Decoupling Performance of 4WS Vehicles Under Velocity-Varying Motion, IEEE Transactions on Control Systems Technology, Vol. 22, NO.5, 2014.

[10] Fen Wu, Xun Song and Zhang Ren , A Nonlinear Gain-Scheduling Compensation Approach Using Parameter-Dependent Lyapunov Functions, Journal of Dynamic Systems Measurement and Control, Vol.138, 2015.

[11] J. Mohammadpour, K. Grigoriadis, M. Franchek, Yue-Yun Wang and I. Haskara, LPV Decoupling and Input Shaping for Control of Diesel Engines, 2010 American Control Conference, Baltimore, MD, USA, PP:1477-1482, 2010.

[12] J. Mohammadpour, K. Grigoriadis, M. Franchek, and Y.Y. Wang, LPV decoupling for control of multivariable systems, International Journal of Control Vol. 84, No. 8, August, PP:1350–1361, 2011.

[13] Amin Nobakhti, Hong Wang, An LMI formulation for H_2 based design of multivariable precompensators, Proceedings of the 2007 American Control Conference, New York City, USA, PP:6163-6168, 2007.

[14] Xiaobao Han, Zhenbao Liu, Huacong Li and Xianwei Liu, Output feedback controller design for polynomial linear parameter varying system via parameter-dependent Lyapunov functions, Advances in Mechanical Engineering, Vol. 9(2), PP:1–9, 2017.

[15] Azuma T, Watanabe R, Uchida K, et al, A new LMI approach to analysis of linear systems depending on scheduling parameter in polynomial forms, Automatisierungstechnik, Vol. 48,: PP:199–205, 2000.

IOP Conf. Series: Materials Science and Engineering **280** (2017) 012016 doi:10.1088/1757-899X/280/1/012016

Numerical Analysis of Thermal Stresses around Fasteners in Composite Metal Foils

S K Nammi, J Butt, J –L Mauricette and H Shirvani

Engineering and Built Environment, Anglia Ruskin University, Bishop Hall Lane, Chelmsford CM1 1SQ, UK.

sathish.nammi@anglia.ac.uk, javaid.butt@anglia.ac.uk, jean-luc.mauricette@pgr.anglia.ac.uk, hassan.shirvani@anglia.ac.uk

Abstract. The process of composite metal foil manufacturing (CMFM) has reduced a number of limitations associated with commercial additive manufacturing (AM) methods. The existing metal AM machines are restricted by their build envelope and there is a growing market for the manufacture of large parts using AM. These parts are subsequently manufactured in fragments and are fastened together. This paper analyses the thermal stresses around cylindrical fasteners for three layered metal composite parts consisting of aluminium foil, brazing paste and copper foil layers. The investigation aims to examine the mechanical integrity of the metallurgically bonded aluminium/copper foils of 100 micron thickness manufactured in a disc shape. A cylindrical fastener set at an elevated temperature of 100 °C is fitted in the middle of the disc which results in a steady-state thermal distribution. Radial and shear stresses are computed using finite element method which shows that non-zero shear stresses developed by the copper layer inhibit the axial slippage of the fastener and thereby establishing the suitability of rivet joints for CMFM parts.

Symbols

a radius of fastener	t_t thickness of top (aluminium) layer
b outer radius of composite	t_m thickness of middle (brazing) layer
ϑ Poisson's ratio	t_b thickness of bottom (copper) layer
α Coefficient of Thermal Expansion	σ_r radial stress
E Elastic Modulus	σ_θ tangential stress
T Increase in Fastener Temperature	E_{al} Elastic Modulus of aluminium
α_{al} Coefficient of Expansion of aluminium	E_{br} Elastic Modulus of brazing
α_{br} Coefficient of Expansion of brazing	E_{co} Elastic Modulus of copper
α_{co} Coefficient of Expansion of copper	r distance from centre of composite

Notations

CalculiX CrunchiX: CCX	CalculiX GraphiX: CGX
Composite Metal: CM	Coefficient of Thermal Expansion: CTE
Composite Metal Foil: CMF	Composite Metal Foil Manufacturing: CMFM
Finite Element: FE	Finite Volume: FV

1. Introduction

Additive Manufacturing (AM) has been receiving unprecedented attention from the academia, mainstream media, investment and national governments around the world. A number of commercial systems exist based on AM technologies for the manufacture of products using a wide variety of materials such as plastic, photopolymers, metal, ceramics etc. These processes are capable of manufacturing products with good mechanical properties and are used widely in industries such as automotive, aerospace and medical. The most important product from these technologies is made out of metal powder, as metal parts can be used directly in industrial applications. The list of reasons for using AM for the manufacture of metal parts is long but the most important one is perhaps its ability to provide design freedom. Complex and intricate geometries that are very difficult and time consuming or sometimes impossible to manufacture using conventional methods can be easily produced by AM technologies. Although methods such as Direct Metal Laser Sintering (DMLS) and Electron beam Melting (EBM) are capable of manufacturing high quality metal parts, there are a number of limitations associated with them. The biggest issue is the cost of manufacture and the pre/post processing, which takes up a lot of time and resources. Composite Metal Foil Manufacturing (CMFM) has been introduced as a cost-effective alternative to these processes and has shown consistent results for certain mechanical properties [1]. It works by combining Laminated Object Manufacturing and brazing technologies. CMFM can join similar as well as dissimilar metal foils for the production of high quality metal and composite products at a fraction of the cost, and does not require extensive pre/post processing to improve properties or aesthetics [2]. The salient feature of CMFM process is that various metals of different grades can be combined to produce composites tailored to an application. The process has only been used to produce parts for testing of properties such as tensile strength, peel strength and shear strength. It has also produced some products of everyday use such as spanners manufactured with multiple layered aluminium alloy foil and a composite from alternate layers of copper and aluminium as shown in Figure 1.

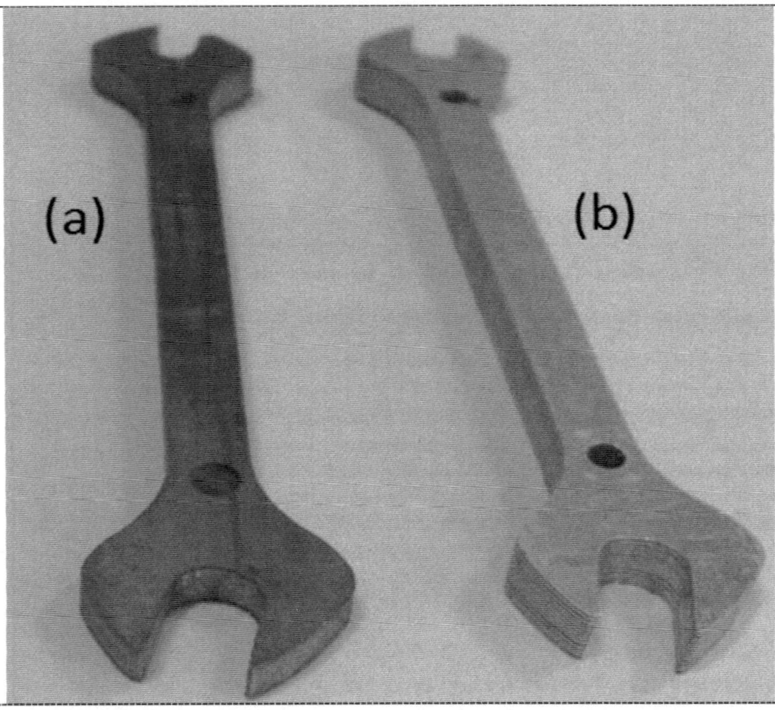

Figure 1. CMFM based composites made of layered (a) copper-aluminium (b) aluminium.

There are a number of applications where large metal parts are produced in fragments (due to constraints on build envelope) and are fastened together with rivets for aerospace, automotive and marine structures. A smart way is the use of composite structures, as they can provide better performance as compared to using single material metal parts. This approach has its issues as well because often mechanical failures occur due to variation in thermal expansion of the two joined materials if poorly designed. CMFM can provide design freedom which will help in making complex shapes and it has already demonstrated higher fracture values for its composite products (copper/aluminium) compared to the parent alloys used [3]. This process, however, has not been tested rigorously for thermal expansion of its joining materials. Therefore, this paper will analyse the stresses induced in the composites made by CMFM due to thermal expansion mismatch between steel fasteners. This knowledge is needed to extend the applications of CMFM process, particularly for allowable stress limits around cylindrical fasteners. There are few studies on stresses in composites but most of them focus on homogeneous materials [4]. However, there are none on CMFM parts and this paper will aim to investigate such effects.

2. Thermal stresses around cylindrical fasteners

The general equations for stress gradients developed in reference [5] for a plate due to a hot-spot is directly relevant to this work. Figure 2 shows the cross-section of a circular plate (radius, b) with a hot-spot (radius, a) at temperature T.

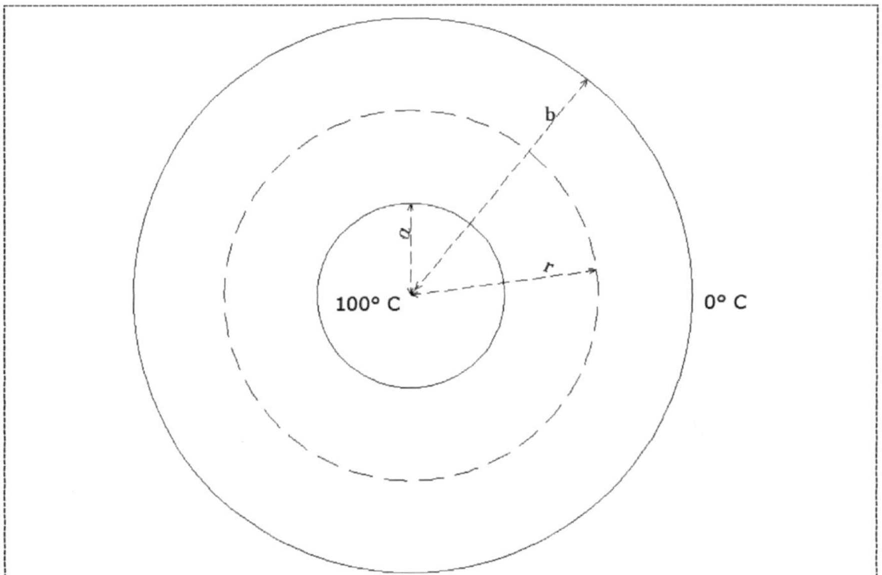

Figure 2. Schematic of circular plate with a hot-spot.

The temperature ($\emptyset$) of plate varies in radial direction according to equation (1):

$$\emptyset = T \log \frac{\left(\frac{b}{r}\right)}{\log\left(\frac{b}{a}\right)} \qquad (1)$$

The non-dimensional radial stress can be expressed according to equation (2):

$$\frac{\sigma_r}{E\propto T}=\frac{1}{2\log\left(\frac{b}{a}\right)}\left[\frac{\left\{1-(1-v)(1-t)\log\left(\frac{b}{a}\right)\right\}\left\{(b^2/r^2)-1\right\}}{\{(1+v)+(1-v)t\}\left\{\frac{b^2}{a^2}\right\}+(1-v)(1-t)}-\log\left(b/r\right)\right] \quad (2)$$

Similarly, the normalised shear stress which is the difference between radial and tangential components can be expressed according to equation (3):

$$\frac{\sigma_r-\sigma_\theta}{E\propto T}=\frac{1}{2\log\left(\frac{b}{a}\right)}\left[\frac{\left\{1-(1-v)(1-t)\log\left(\frac{b}{a}\right)\right\}(2b^2/r^2)}{\{(1+v)+(1-v)t\}\left\{\frac{b^2}{a^2}\right\}+(1-v)(1-t)}-1\right] \quad (3)$$

Both radial and tangential stresses in the neighbourhood of the hot-spot converge to a value of 0.5EαT for a large plate i.e., b/a→∞. Composites are made of at least two or more layers with distinct material properties for each. Analytical equations for two layered systems are relatively simple. However, as the number of layers increase such equations become more complex. No closed form solutions exist for composites of various different layers and hence in this work, FE methods were used. CMFM part investigated here is a composite made of three layers. The top and bottom layers are aluminium and copper foils respectively that are sandwiching a layer of brazing paste. All three layers have different parameters for E, α& ϑ.

3. Experimental Methodology

FE models of a fastener were built with three layered composite consisting of aluminium foil, brazing paste and copper foil. A benchmark study was initially conducted with analytical models available for isotropic structure. Afterwards, full FE analyses were conducted by assigning material data described in Table I for the composite layers and fastener. The variations in radial and tangential stresses for various configurations of models were computed. Parametric investigations were conducted by varying thicknesses of each layer of composite.

Table I. Properties of Fastener, Metal Foils and Brazing Paste Layer.

Properties	Fastener	Aluminium 1050 H14 ½ hard temper	99% Copper	80% Zinc 20% Aluminium Brazing Paste
E (GPa)	205	70.0	115.0	52
ϑ	0.3	0.3	0.3	0.35
α (µm/m .°K)	12.0	23.0	17.0	21.8

3.1. Assumptions

1. There is no clearance between fastener and surrounding composite material.
2. Fasteners are of same thickness as that of composite with no end bolt or head.
3. There is no interference friction between fastener and composite.
4. The increase in fastener temperature was 100℃.
5. Temperature of inner and outer faces of composite was at 100 & 0℃ respectively, radially varying according to equation (1).
6. Faces in the thickness direction were considered to be insulated. The heat entering into the composite via fastener at the inner face leaves at the outer face.

IOP Conf. Series: Materials Science and Engineering **280** (2017) 012016 doi:10.1088/1757-899X/280/1/012016

3.2. Material properties

Material tests were conducted for the grade of copper, aluminium and brazing in CMFM parts. Properties summarised in Table I were assigned for metal alloys and brazing paste in composite. The properties of steel in reference [6] were assigned to the fastener.

4. Finite element analysis

The CalculiX [7] is an open source non-linear FE code that is compatible with operating systems such as Linux, Windows, Mac and Android. It is divided into two independent codes (viz: Calculix GraphiX and Calculix CrunchiX). The Calculix GraphiX, also referred to as CGX, handles pre-processing to define a geometry, mesh, loads, and boundary conditions; it has both command and interactive modes. The CGX has an interactive 3D-environment that uses libraries in the openGL API for post-processing results. The Calculix CrunchiX also referred as CCX is a solver code that can do both linear and non-linear calculations. It can solve using, FE and FV methods. The CCX code is capable of performing many types of analyses including static, dynamic, fluid and thermal. FE models of the fastener and composite were built from scratch. An API script was coded with definitions of the geometry and mesh. This was done to facilitate parametric studies, it was easier to define/update model details with script. Commands were described in a script that is read by CGX. An auxiliary python code was used to update script with parameters to define geometry and meshing. In this way, FE model generation and meshing was automated. A section cut of disc equal to 1/16 of full model was built to reduce computational cost i.e., a wedge shaped sector equal to 22.5° arc cut out of a full cylinder. An illustration of the FE model is shown in Figure 3. Solid continuum elements were used for discretization; 8-noded elements (type C3D8) were used to represent both fastener and composite. Analysis was repeated with both linear and quadratic elements. The nodes of the fastener and composite were tied for their radial displacements. Constraints to specific nodes were assigned with the "*EQUATION" keyword. A finer mesh was used for the composite in the neighbourhood fastener. The radial dimension of elements were increased away from the zone of interest. An illustration of fine and coarse meshes used are shown in Figure 4. Particular attention was given for stress distribution around the fastener.

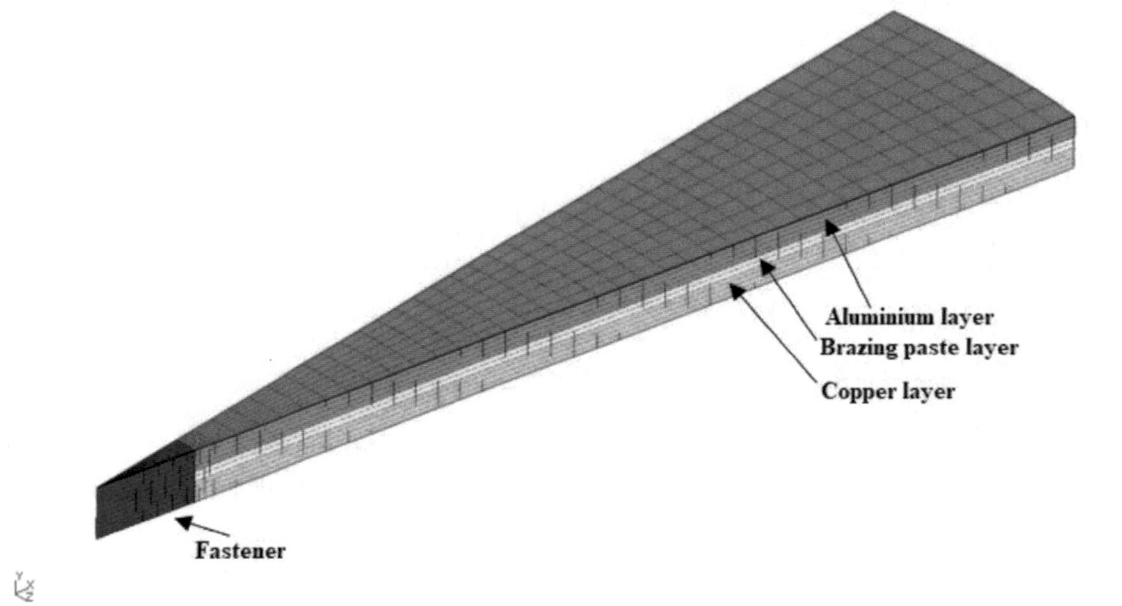

Figure 3. FE model of fastener surrounded by 3-layered composite.

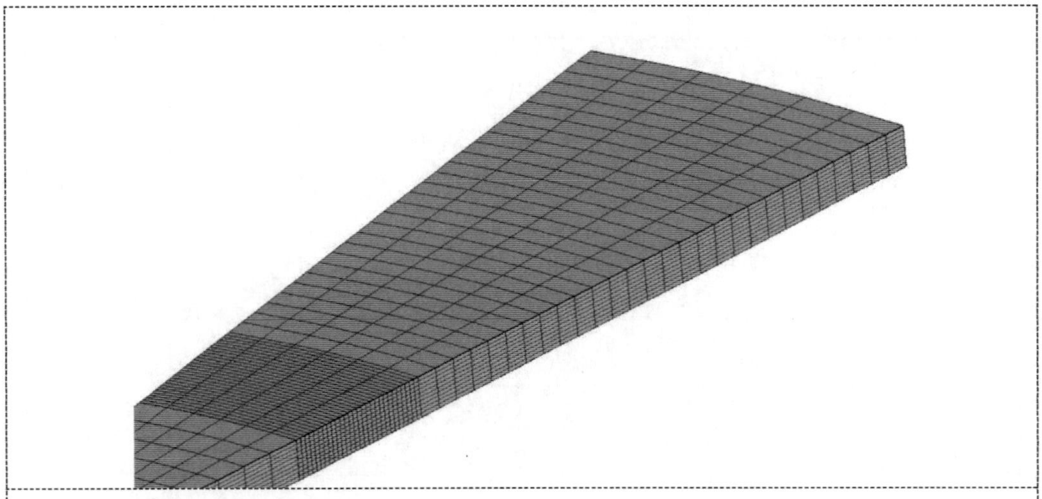

Figure 4. Coarse and fine Mesh.

5. Model validation

The fastener and all three layers of the composite were assigned with identical elastic modulus, Poisson's ratio and CTE. For simplicity, material data [8] for the FE model was assigned with CTE=10^{-05}, E=10^{05} and ϑ=0.3. The model is analogous to a homogeneous plate with a local hot-spot. Thermal stresses predicted by the FE model were compared with closed-form solutions (equations 2 and 3). The fastener was subjected to a 100 ℃ temperature increase and stresses developed at the interface between the fastener and surrounding materials were examined. The radial and tangential stresses becomes 0.5EαT if the extent of surrounding material is large compared to the fastener. The outer radius of the plate was increased successively up to 15 times that of fastener. The FE model results were within 2% of analytical results. Comparisons of FE and analytical results are shown in Figure 5 and Figure 6 for b/a=2.

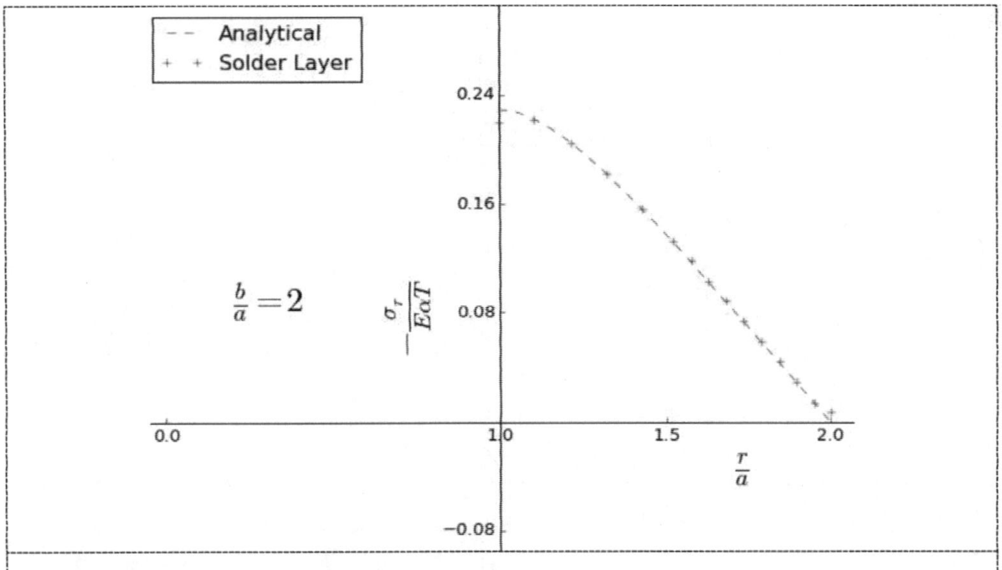

Figure 5. Variation in radial stresses for b/a=2.

IOP Conf. Series: Materials Science and Engineering **280** (2017) 012016 doi:10.1088/1757-899X/280/1/012016

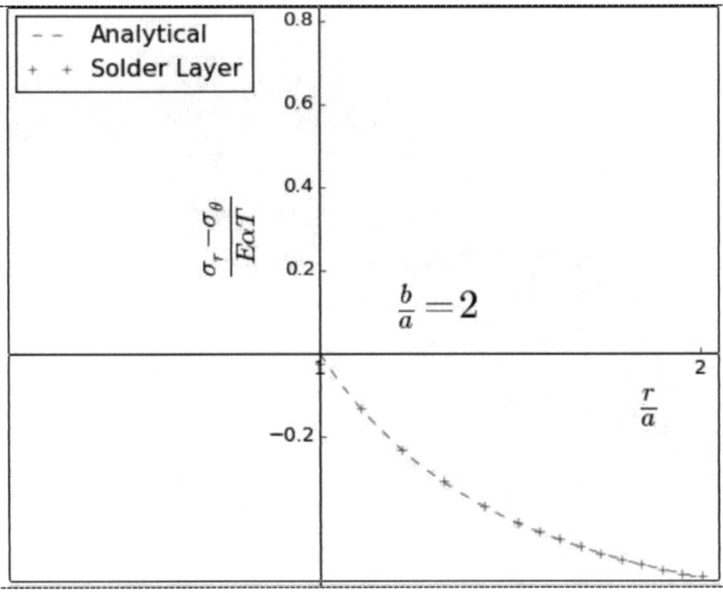

Figure 6. Variation in shear stresses for b/a=2.

6. Results and Discussion

All post-processing images were generated in CGX. Plots for the variations of radial (σ_r) and shear stresses ($\sigma_r - \sigma_\theta$) as a function of position from centre were generated for each layer of composite. Plots for b/a ratios 2 and 3 are presented in Figure 7 and Figure 8. It can be seen that stresses in aluminium and brazing layers is analogous to analytical works. However, shear stresses are non-zero in the neighbourhood of the fastener for the copper layer. The FE stress contours for tangential stresses for b/a ratio 3 are shown in Figure 9.

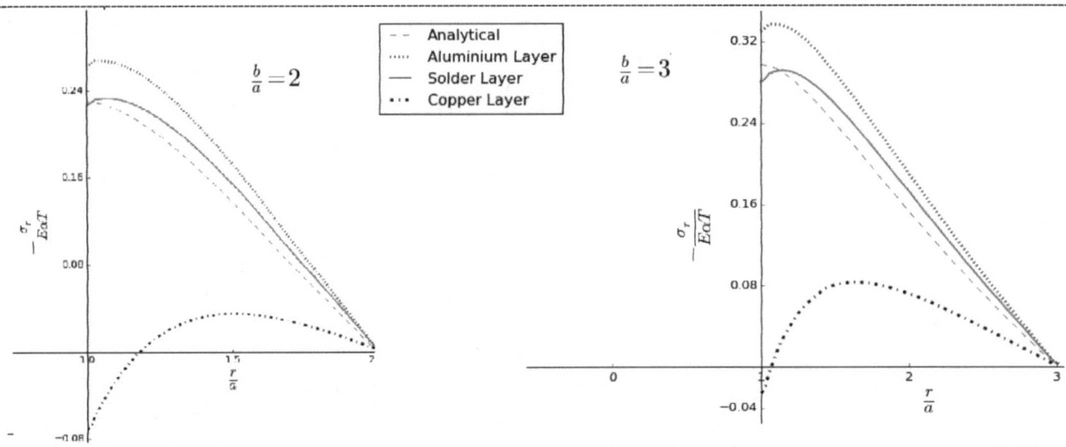

Figure 7. Radial stresses in composites.

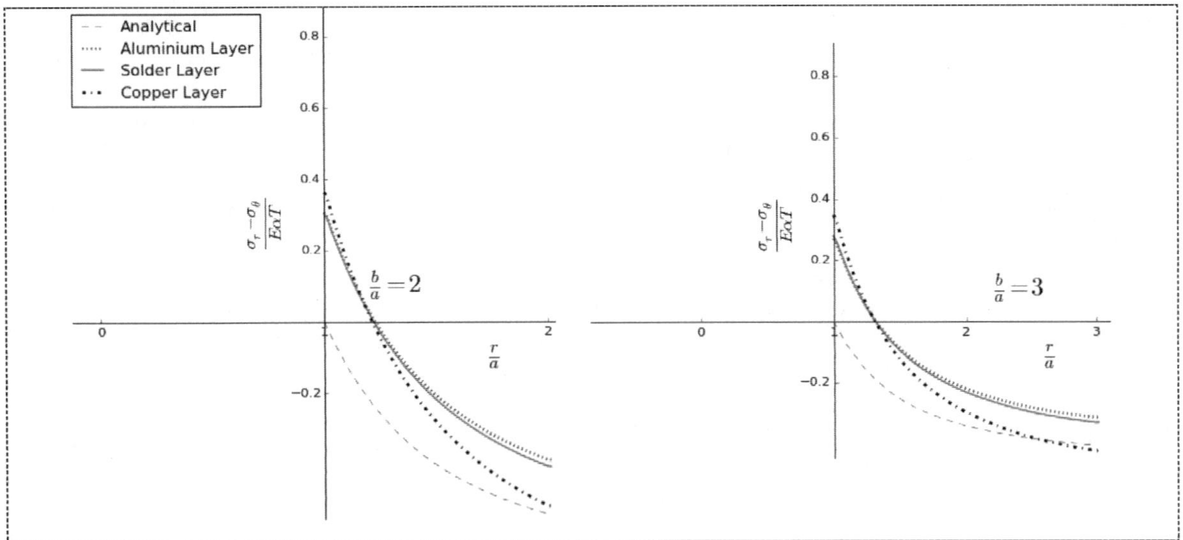

Figure 8. Shear stresses in composites.

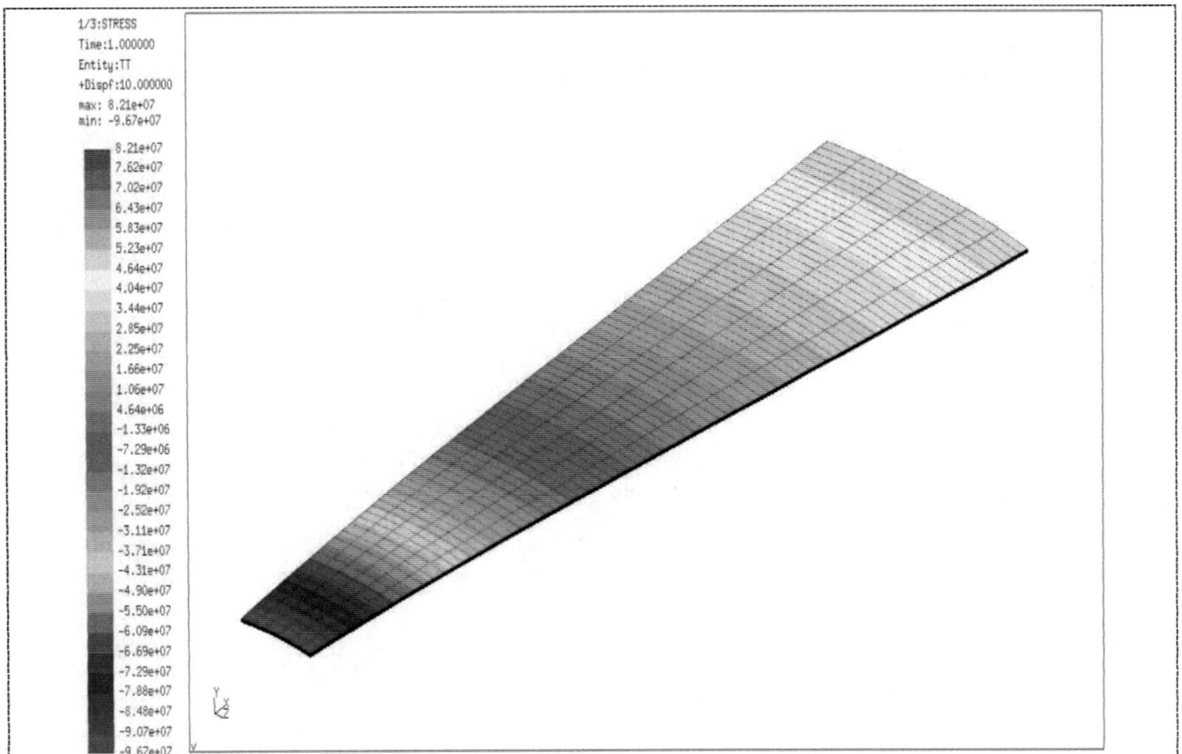

Figure 9. Tangential stresses in composites for b/a ratio of 3.

IOP Conf. Series: Materials Science and Engineering **280** (2017) 012016 doi:10.1088/1757-899X/280/1/012016

7. Conclusions

Numerical investigations were carried out for thermal stresses around cylindrical fasteners in a CMFM composite made from thin foils (100 microns) of aluminium and copper. This study will support the fastening of large parts manufactured by CMFM. The obtained results enable the design of slip-free composite structures that can have bespoke characteristics and advantages. The shear stresses ($\sigma_r - \sigma_\theta$) are zero for homogeneous materials whereas they are always non-zero for Composite Metal Foil (CMF) under elevated temperatures. This is favourable as it will inhibit the fastener loosening axially which will provide a better support to the joint made between large parts. The numerical analysis of the aluminium/copper foil structure has shown the same trait where the copper layer is under compression and thus resulting in an even tighter fitting of the fastener. This added feature combined with the other advantages will make CMFM a very good option for joining large structures using fasteners.

References

[1] Butt, J., Mebrahtu, H. and Shirvani, H., 2016. Microstructure and mechanical properties of dissimilar pure copper foil/1050 aluminium composites made with composite metal foil manufacturing. Journal of Materials Processing Technology, 238, pp.96-107.

[2] Butt, J., Mebrahtu, H. and Shirvani, H., 2015. Peel and tensile test investigation of aluminium 1050 foil parts made with a new additive manufacturing process. International Journal of Rapid Manufacturing, 5(1), pp.95-115.

[3] Butt, J., Mebrahtu, H. and Shirvani, H., 2016. Rapid prototyping by heat diffusion of metal foil and related mechanical testing. The International Journal of Advanced Manufacturing Technology, 84(9-12), pp.2357-2366.

[4] M. L. Blosser. Thermal stress in high temperature cylindrical fasteners. NASA Technical Reports. NASA-TM-100611, NAS 1.15:100611.

[5] E.W. Parkes. The Stresses in a Plate due to a Local Hot Spot: An Analysis of a Practical Thermal Stressing Case. Aircraft Engineering and Aerospace Technology, Volume 29, Issue 3, pp. 67-69.

[6] EN 1993-1-1 (2005): Eurocode 3: Design of steel structures – Part 1 -1: General rules and rules for buildings.

[7] CALCULIX. A free software 3D FEM program. http://www.calculix.de.

[8] D. J. Burrows. Benchmark Tests in Thermal Stress Analysis using the Finite Element Method. NAFEMS-P02.

IOP Conf. Series: Materials Science and Engineering **280** (2017) 012017 doi:10.1088/1757-899X/280/1/012017

Study on Measuring the Viscosity of Lubricating Oil by Viscometer Based on Hele - Shaw Principle

Longfei Li

College of Power Engineering, Naval University of Engineering, Wuhan Hubei 430033, China

Email address:13260641243@163.com

Abstract: In order to explore the method of accurately measuring the viscosity value of oil samples using the viscometer based on Hele-Shaw principle, three different measurement methods are designed in the laboratory, and the statistical characteristics of the measured values are compared, in order to get the best measurement method. The results show that the oil sample to be measured is placed in the magnetic field formed by the magnet, and the oil sample can be sucked from the same distance from the magnet. The viscosity value of the sample can be measured accurately.

1. Introduction

Viscosity is an important indicator of lubricating oil and has an important impact on the performance of the lubricating oil [1]. If the viscosity is too small, it is difficult to form a solid oil film, and then can not achieve the role of reducing friction. However If the viscosity is too large, it will cause the equipment to start difficult, and also makes the oil can not be cycle normally among the equipment, resulting in equipment lubrication failure finally[2]. Therefore, the choice of the correct viscosity of lubricating oil is of great importance to the normal use of the equipment. At the same time, in the oil monitoring process, viscosity is often the test items. In the laboratory, when measured the viscosity of lubricants, the viscometer is often used which is based on the principle of Hele-Shaw, but in the actual use of the process found that the measurement method has a great influence of the result when measuring the viscosity of the lubricating oil .

In this paper, by comparing the results of different ways of using the viscosity measurement, to find the best way to use the viscometer based on Hele-Shaw principle when measuring the viscosity of lubricant viscosity, and ensure the accuracy of measurement of lubricating oil viscosity, provide guidance and method for laboratory use of this type of viscometer to measure viscosity.

2. Experimental equipment

A viscometer based on the Hele-Shaw principle, ordinary droppers, processed droppers (Fig.1), a magnet, an experimental oil sample, a non-woven fabric, etc.

3. Experimental method

Method 1. The sample viscosity is measured directly by viscometer. First, wipe the measuring board of the viscometer with a non-woven fabric. Then drain the sample to be measured with a dropper and drop two drops of oil into the instrument vertically.

Attention please: In order to keep the accuracy of measurement results, attention should be paid to measurements:

(1) When the sample is measured, the first data is discarded as there haven't formed an effective and uniform oil film in the sample detection area before the first time's measurement.

(2) In order to prevent the effect of the initial velocity of the sample on the experimental results, the height of the dropper from the instrument should be the same as when the sample is dropped, and the sample is naturally dropped.

Method 2. Shake the sample before the measurement, and then use the method 1 to measure the sample viscosity value;

Method 3. When measuring the magnetic force of the sample, it is found that the lubricant sample has weak magnetic properties. Therefore, use a magnet to assist in viscosity measurements. The specific method is as follow. First put the oil sample bottle on the magnet do without shaking the sample, and then use the treated straw to suck the sample from the same position at the bottom, which is shown as fig. 2, Finally use the method 1 to measure the viscosity value.

Figure1. Treated dropper

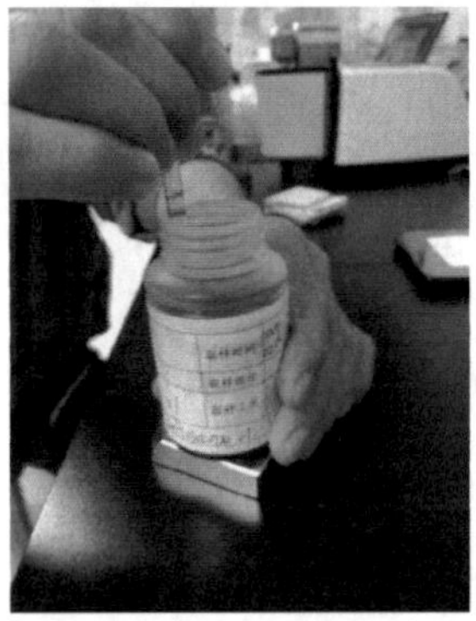

Figure 2. Draw oil samples using method 3

4.Analysis of the experimental results

When measuring two groups of oil samples using the method 1, it is found that the maximum value of the data standard deviation has reached 7.83, and the relative value compared to the average value of viscosity is as high as 7.79%, showing that it has no value to do using method 1 [3], as the method 1 can not effectively measure the viscosity of the lubricating oil and the viscosity data measured is shown in tab. 1.

Table. 1 The viscosity (40°C) measured through method 1

unit: cSt

Number Times	1	2	3	4	5	6	Average value	Standard deviation	Relative value of standard deviation
#1	103.8	114.8	108.9	99.1	104.9	98.6	105.0	6.12	5.83 %
#2	108.0	101.2	98.6	110.8	90.9	93.5	100.5	7.83	7.79 %

4.1. Gross error elimination

For methods 2 and 3, the relative values of the standard deviations of the viscosity values are less than 2% and have practical engineering value. (The values of viscosity measured by method 2 and method 3 are A and B respectively). The gross error of the data is eliminated. Because of the small sample value, the gross error is eliminated by using the Grabus quasi[4].

Taking the measurement data of # 1 sample as an example, the average value of the sample is: $\overline{X}$=40.40 cSt, the standard deviation is s = 0.626, and G_1 = 1.12, G_6 = 1.44, and we can get that the G_0

= 1.822 as n = 6 through the Grabus value table. Then, $G_1 < G_0$, $G_6 < G_0$, so there is no coarse error, the group of data does not need to be coarse error rejection.

Eliminate the gross errors of the remaining data in turn, and the average and standard deviations and the relative percentages of standard deviations are recalculated and showed in Tab. 2 and Tab. 3.

Table. 2 Viscosity values（40°C）of group A samples

unit: cSt

Numbers Viscosity Times	1	2	3	4	5	6	Average value $/\overline{X}$	Standard deviation/s	Relative value of standard deviation（%）
1	90.9	90.0	89.7	93.5	91.6	92.6	91.4	1.46	1.60
2	104.2	103.1	104.5	102.8	101.4	101.2	102.9	1.37	1.33
3	99.6	98.6	101.0	98.9	97.5	97.9	98.9	1.25	1.26
4	103.3	103.8	102.8	103.8	104.2	104.0	103.7	0.51	0.49
5	103.8	101.5	100.1	98.7	100.8	99.1	100.7	1.86	1.85
6	100.0	102.8	101.2	100.5	101.7	100.7	101.2	1.00	0.98
7	105.4	103.5	104.5	103.8	102.1	102.6	103.7	1.20	1.16
8	99.8		100.3	99.1	98.2	100.5	99.6	0.95	0.95
9	103.3	100.5	103.1	101.0	101.9	102.4	102.0	1.12	1.10
10	105.9	108.7	105.2	107.0	105.4	107.5	106.6	1.37	1.28

Table. 3 Viscosity values（40°C）of group B samples

unit: cSt

Numbers Viscosity Times	1	2	3	4	5	6	Average value $/\overline{X}$	Standard deviation/s	Relative value of standard deviation（%）
1	110.3	110.8	109.6	111.5	111.2	109.6	110.5	0.80	0.73
2	110.8	109.8	110.5	111.9	110.3	109.8	110.5	0.78	0.71
3	111.7	111.9	111.2	111.5	111.2	112.2	111.6	0.38	0.34
4	112.7	112.2	111.0	112.7	110.5	110.5	111.6	1.01	0.91
5	110.8	112.2	112.2	112.9	111.5	110.8	111.7	0.85	0.76
6	113.4	111.9	111.7	112.9	111.9	111.5	112.2	0.73	0.65
7	112.2	112.2		112.2	111.9	112.4	112.2	0.17	0.15
8	115.2	112.9	113.4	115.0	113.8	112.9	113.9	1.03	0.90
9	116.2	118.3	117.3	118.7	118.0	116.2	117.4	1.09	0.93
10		125.3	125.0	125.3	124.8	125.3	125.1	0.21	0.17

4.2. Data significance test

Since the average and standard deviations of each measurement data are known, the t distribution [5] is used to test the samples. Below, take the sample #1 and #2 in the A data set as an example and calculate.

From the t distribution test formula, we can be obtained that the degrees of freedom $v \approx 10$, then the experimental statistics is t = 34.45. Select the confidence of 95%, check t distribution table and we can get the critical value of 't' is that t = ± 1.812.

Since the obtained t value is within the range determined by the critical value, that is, at a confidence level of 95%, there is a significant difference between the two sets of data, so it can be determined that the viscometer based on the Hele-Shaw principle can clearly distinguish the the two oil samples.

The data of any two sets of data in group A and group B are analyzed statistically, and # 4 and # 7, # 5 and # 6 in group A data as well as # 1 and # 2 ,#3 and #4, #4 and #5, #6 and #7in group B data have no significant difference.

4.3. Data fitting test

The average value of the data measured by the two methods is tested by data fitting, and the curve is shown in fig. 3.

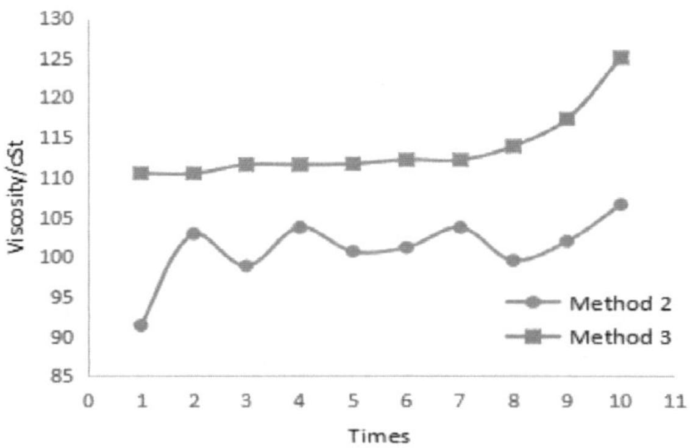

Figure. 3 The viscosity value fitting curve of group A group B

What we can see from the Fig. 3 are as follows.

① The viscosity of the method three shows an increasing trend. The viscosity of the sample measured by the method two shows great fluctuation.

② According to the data fitting, the value of viscosity fitting curve of the samples measured by method 3 is $y = 0.1205x^3 - 2.1043x^2 + 11.258x + 83.703$ and the $R^2 = 0.981$ when the #2, #4and#6 was removed as they can not to be distinguish through this viscositor , the $R^2 = 0.997$; The viscosity fitting curve of the samples measured by method 2 is $y = 0.1205x^3 - 2.1043x^2 + 11.258x + 83.703$ and $R^2 = 0.716$, when removed the #4 and #6, R^2=0.709.

As the sample to be measured in the experiment are got through heating the oil, and the heating time increase from #1 to #10, as the lubricating oil oxidation reaction occurs only, no fuel or water dilution that will decrease the oil viscosity . In the process of oxidation, lubricating oil will decompose and a series of chemical reactions such as polymerization will lead to the gradual deepening of the color of lubricating oil and the gradual increase of sediment, which will gradually increase the viscosity of lubricating oil [6-8]. Therefore, the viscosity of lubricating oil in theory should show a tendency of increase.

The percentage of the relative value of the standard deviation compared to the lubricating oil obtained by the method 2 is basically above 1%, while the percentage value of the method 3 is less than 1%. At the same time, it can be seen from the data fitting curve that the measured viscosity of the method 2 is very violent and does not accord with the theoretical situation, and it is not conducive to the study of laboratory data, too. On the contrary, the viscosity value of the method 3 has a significant increasing trend as the theoretical situation. And then the data fitting result of measuring viscosity value by method 3 is much higher than that of method 2.

4.4. Accuracy test

To test the accuracy of the two methods, the new oils and standard oils who have known viscosity were measured using methods 2 and 3 respectively, and the result are shown in tab. 4.

Table. 4. The viscosity of the new oil and standard oil through the two methods

unit: cSt

Samples Number		1	2	3	4	5	6	Average value	True value	Error value
Method 2	standard oil	83.4	86.2	82.5	85.5	85.8	84.4	84.6	88.1	4.1%
	new oil	103.1	107.5	105.4	107.0	104.5	103.5	105.2	115.4	8.8%
Method 3	standard oil	88.3	88.8	89.3	89.0	88.8	88.8	88.9	88.1	0.9%
	new oil	115.7	114.1	113.6	115.0	112.7	114.1	114.2	115.4	1.1%

Through the tab.4, we can see that when measuring the standard oil and new oil using the two methods, the error value of method 3 compared to the true value is much letter than the method 2.

To sum up, the method 3 is more accurate and reliable than the method 2.

5. Conclusion

1) Shake the sample before testing the sample, and the result has a greater error, because the bubble generated during the shake will affect the accuracy of measurement.
2) The lubricating oil sample to be measured is placed in the magnetic field formed by the magnet, and the oil sample is drawn from the same distance from the magnet, so that the viscosity value of the sample can be accurately measured.

Reference

[1] Larry A. Toms, Allison M. Toms. Machinery Oil Analysis: Methods, Automation and Benefits:A Guide for Maintenance Managers, Supervisors & Technicians[M]. Society of Tribologists and Lubrication Engineers, 2008.

[2] Y.W. Fei, F. Guo, J.T. Hao, et al. Analysis of thermal oxidation decay law of ester aviation lubricant base oil [J]. lubrication and sealing, 2015, 40 (10):74-79.

[3] ASTM Standard E2412-10, Standard Practice for Condition Monitoring of In- Service Lubricants by Trend Analysis Using Fourier Transform Infrared(FT-IR) Spectrometry[S]. New York: American Society for Testing and Materials, 2010.

[4] M.Y. Li , L. Li. Measurement technology foundation [M]. Xi'an: Xi'an Electronic and Science University press, 2007.

[5] C.Q. Zhuang , Y. He yuan. Basic of Applied Mathematical Statistics [M]. Guangzhou: South China University of Technology press, 2013.

[6] Y.G. Zhang , Y. P. Xiao, Y.L. Zhao , et al. GC/MS experimental study on high temperature oxidation of aeroengine lubricating oil [J]. lubrication and sealing, 2014, 39 (7):77-82.

[7] Q.M. Yang , X.P.Yan , S.Z. He.Practical technology of oil monitoring and analysis field [M]. Beijing: Mechanical Industry Press, 2006.

[8] Macian V, Tormos B, Gomez Y A, et al. Proposal of an FTIR Methodology to Monitor Oxidation Level in Used Engine Oils:Effects of Thermal Degradation and Fuel Dilution[J].Tribology Transactions, 2012, 55:872-882.

IOP Conf. Series: Materials Science and Engineering **280** (2017) 012018 doi:10.1088/1757-899X/280/1/012018

4D Biofabrication of Branching Multicellular Structures: A Morphogenesis Simulation Based on Turing's Reaction-Diffusion Dynamics

Xiaolu Zhu[1,2*] and Hao Yang[1]

[1]College of Mechanical and Electrical Engineering, Hohai University, Changzhou, Jiangsu 213022, China
[2]Changzhou Key Laboratory of digital Manufacture Technology, Hohai University, Changzhou, Jiangsu 213022, China

*Correspondence should be addressed to: Xiaolu Zhu (zhuxiaolu@hhu.edu.cn)

Abstract. The recently emerged four-dimensional (4D) biofabrication technique aims to create dynamic three-dimensional (3D) biological structures that can transform their shapes or functionalities with time when an external stimulus is imposed or when cell postprinting self-assembly occurs. The evolution of 3D pattern of branching geometry via self-assembly of cells is critical for 4D biofabrication of artificial organs or tissues with branched geometry. However, it is still unclear that how the formation and evolution of these branching pattern are biologically encoded. We study the 4D fabrication of lung branching structures utilizing a simulation model on the reaction-diffusion mechanism, which is established using partial differential equations of four variables, describing the reaction and diffusion process of morphogens with time during the development process of lung branching. The simulation results present the forming process of 3D branching pattern, and also interpret the behaviors of side branching and tip splitting as the stalk growing, through 3D visualization of numerical simulation.

1. Introduction

One of the main objectives of three-dimensional (3D) printing in health science is to mimic biological functions [1-3]; however, it is hard to control the shape and functions of the printed 3D bio-construct/materials after being printed because of the reorganization of the printed bio-construct via cellular self-organization. Thus, regulation the transformation or functionalities of 3D biological structures with time is critical for achieving the 3D bio-construct with expected shape and functions [4, 5]. To reach this goal, a four-dimensional (4D) biofabrication technique might be utilized, which aims to create dynamic 3D biological structures that can transform their shapes or functionalities with time when an external stimulus is imposed or when cell postprinting self-assembly occurs[6]. The evolution of 3D pattern with branching geometry via self-assembly of cells is critical for 4D biofabrication of artificial organs or tissues with branched geometry.

Recent research on organ morphogenesis has described an elaborate pattern of branching phenomena [7, 8]. In the lung, two primary forms of branching have been identified: side branching and tip splitting. The branching occur in sequence: first, side branching creates the primary stalks; then, there is a change of mode to tip splitting. The kidney developmental program encodes the intricate branching and organization of approximately 1 million functional units (nephrons)[9]. These phenomena have been hypothesized to be under genetic controls [7, 10, 11]. Since how genes could possibly act to produce these patterns is still not clear, the coding for branch patterning is possibly simplified by repeated use of a branching mechanism, as in Mandelbrot's fractal model and other elegant algorithms[12-15].

The reaction - diffusion (RD) model based on Turing-type mechanism has shown its potential power on simulating the dynamic process of molecules and tissue morphogenesis [16-18] with time. This is a suitable method to guide the 4D biofabrication via predicting the dynamic evolution of 3D self-organized microstructures. The RD model commonly involves a morphogen pair [19] (an activator- inhibitor pair) in which the inhibitor diffuse more rapidly than its activator. In recent study of lung branching morphogenesis based on reaction-diffusion model [20, 21], a four variable partial differential equation (PDE), due to Meinhardt[22], was utilized to describe the reaction and diffusion of morphogens creating branched lung development. Usually, the bone morphogenetic protein-4 (BMP4) can serves as activator, the matrix gamma-carboxyglutamic acid protein (MGP) can serve as inhibitor, and substrate S can be fibroblast growth factor 10 (FGF10). Here, we attempt to simulate the phenomena as the stalk growing, side branching and tip splitting by using a 3D branching model based on RD-based dynamics. The branching morphogenesis both in longitudinal (growth) direction and the transverse direction of the stalk is simulatively demonstrated.

2. Mathematical model

The mathematical model based on Turing's reaction-diffusion (RD) dynamics, due to Meinhardt[22], involves concentrations of activator, inhibitor, substrate and cell differentiation marker four variables. Here, the four variables are labeled as A, H, S and Y and the mathematical expressions are as follows:

$$
\begin{cases}
\dfrac{\partial A}{\partial t} = \dfrac{cA^2S}{H} - \mu A + D_A \nabla^2 A + \rho_A Y & (1) \\[2ex]
\dfrac{\partial H}{\partial t} = cA^2S - vH + D_H \nabla^2 H + \rho_H Y & (2) \\[2ex]
\dfrac{\partial S}{\partial t} = c_0 - \gamma S - \varepsilon YS + D_S \nabla^2 S & (3) \\[2ex]
\dfrac{\partial Y}{\partial t} = dA - eY + \dfrac{Y^2}{1 + fY^2} & (4)
\end{cases}
$$

The model postulates four spatially continuous variables, each of which is a spatio-temporal function. A, H, S, and Y are four concentration variables indicating concentrations of activator, inhibitor, substrate and cell differentiation marker. In the model, μ, v, γ and e represent the first-order degradation rate of A, H, S and Y respectively; D_A, D_H and D_S are the diffusion coefficient of activator, inhibitor and substrate; Differentiated Y cells secrete activator A and inhibitor H at the rates of ρ_A and ρ_H respectively; c is the positive rate of activator A in the autocatalytic reaction; c0 and ε are the production and consumption rates of Substrate S. cA^2S/H describes the autocatalytic generation of activator A under the interaction of substrate S and inhibitor H; cA^2S describes the generation of inhibitor H under the interaction of activator A and substrate S; dA describes cell differentiation marker generated by activator A; $D_A\nabla^2A$, $D_H\nabla^2H$ and $D_S\nabla^2S$ describe the diffusion of activator, inhibitor and substrate respectively. High concentrations of activator A produces cell commitment Y (the dA term in the equation (4)). When the concentration of activator A exceeds a certain critical value, cell differentiation (Y = 1 indicates that cells have been differentiated) is irreversibly activated (the sigmoidal term $Y^2/(1+fY^2)$ in the equation (4)). Here, all physical quantities are dimensionless. The mathematical model was solved by the finite difference method in the MATLAB platform and the parameters were traversed in a certain range, which was also implemented for solving other RD models[23, 24]. Three-dimensional data were displayed by 3D visual software (Voxler, trail version).

The simulation run on a computer with an 8-core CPU and at least 16G of memory. Usually one set of parameters needs about several hours to twenty hours.

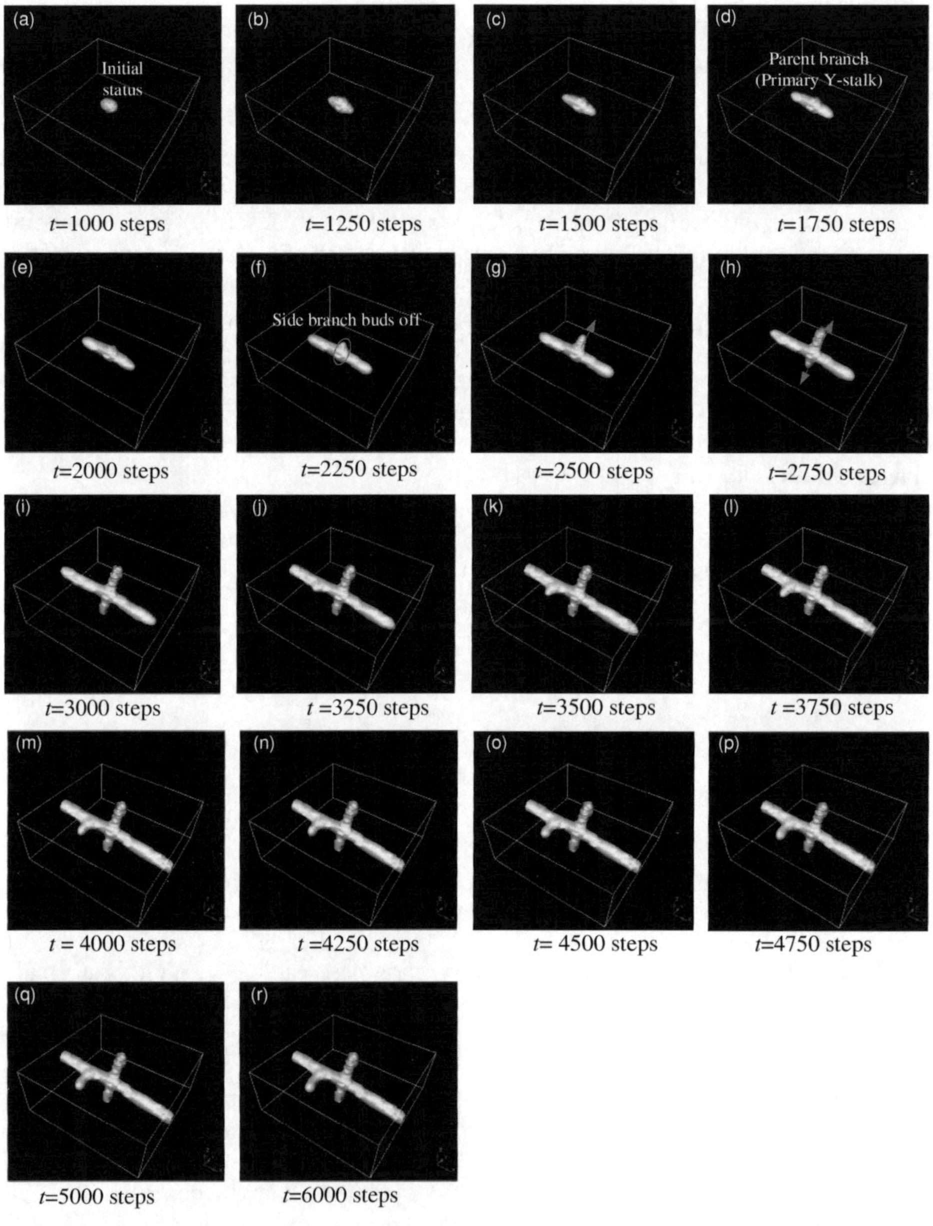

Figure 1. Simulation results for side-branching behaviors in 3D. The presented structure is the spatial distribution of the cellular differentiation marker Y. The number under image denotes the step number in simulation. Parameters: $c = 0.04$, $\mu = 0.3$, $v = 0.03$, $\rho_A = 0.03$, $\rho_H = 0.0001$, $c_0 = 0.02$, $\gamma = 0.02$, $\varepsilon = 0.042$, $d = 0.008$, $e = 0.1$, $f = 10$, $D_A = 0.1$, $D_H = 0.26$, $D_S = 0.06$.

3. The evolution of side branching

Lung development begins with side branches emerging in rows around the circumference of the parent/primary branch. The parent branch elongates and new side branches bud off in the typical domain branching. In this model, the primary branch (Y-stalk) grows through the positive feedback of the peaks of activator. Also, inhibitor H is produced is proportional to the activator distribution and H diffuses from high-concentration region to low-concentration region. The stalk elongates because the H that diffuses to the side of the primary branch serves as lateral inhibition which results in vimineous elongation of the Y stalk rather than isotropic expansion as shown in Figure 1. The elongation of activator peak continues to generate the differentiated cells Y in the local region. Thus, the elongated primary branch has been formed as shown in Figure 1a-d. However, Y cells consume S (the - εYS term in the equation (3)), and S concentration becomes lower at the location of Y-stalk, but relatively higher away from the Y-stalk. Therefore, the gradient of S forms, which is the main driver of activator migration because the activator A is always seeking the substrate S according to the term $+cA^2S/H$ in equation (1). Consequently, the newly formed activator peak will migrate in the direction from low S concentration to high S concentration as indicated by the arrows in Figure1g and h (the arrow directions are away from the present stalk). Along the direction perpendicular to the primary stalk, side branches emerge when the attraction of the substrate overcomes the inhibition (Figure 1 f-r). Each activator peak in 3D on the Y-stalk leads to a side branch as that activator peak migrates into regions of high-concentration S far away from the main Y-stalk, where Y cells have depleted S.

4. Tip bifurcation

In this simulation, the dynamic process leading to tip bifurcation begins with the lateral extending of activator A concentration profile. Since substrate S in the current position is exhausted, the activator no longer moves in the former direction and has to look for new substrate S perpendicularly to the former stalk extension direction. The specific dynamic process is shown in Figure 2.

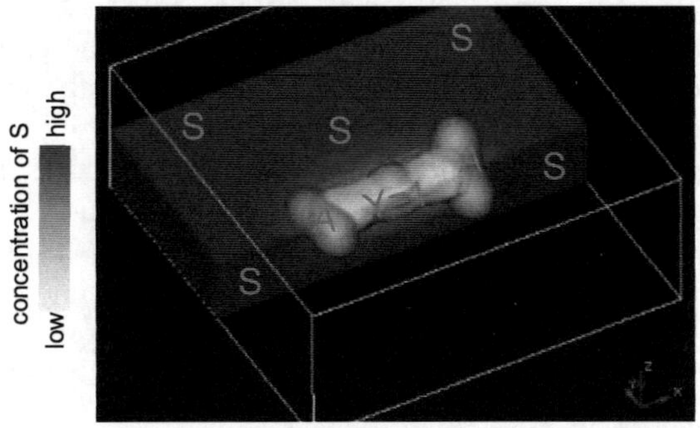

Figure 2. Schematic illustration of the profile of activator concentration along the direction transverse to stalk elongation.

There are two lateral extending behaviors of activator A concentration profiles in the horizontal and vertical planes respectively (figure 2). The white border represents the boundary of simulation domain; the red rectangle represents substrate S and substrate S distributes throughout the simulation domain space. Here only about 1/4 is taken for description and the deeper the color, the higher the substrate S concentration. The green part represents the differentiated cells (Y = 1). The orange ellipsoid indicates the isosurface of activator A. It can be seen from Fig. 1, concentration of substrate S is low around the differentiated cells (the term -εYS in equation (3) leads to the local concentration decrease of substrate S), whereas activator A extends to high concentration of substrate S direction (the term cA^2S/H in eq. (1) leads to large value of A in the region of high S). High concentration of substrate S exists in the domain perpendicular to original direction, which results in transversal expanding of activator A. And then concentration of activator A shows a flat long profile.

In order to study the distribution and relative time-dependent changes of activator A and inhibitor H in the transverse direction perpendicular to the stalk extension, the typical bifurcation (tip splitting) was simulated. Here, we select the simulation results of 2800, 3150 and 3600 steps and draw the corresponding 3D graphics of multicellular morphology for numerical analysis. Specific numerical values and variation tendency of activator A and inhibitor H along a certain spatial coordinate axis *k* are vividly shown in figure 3.

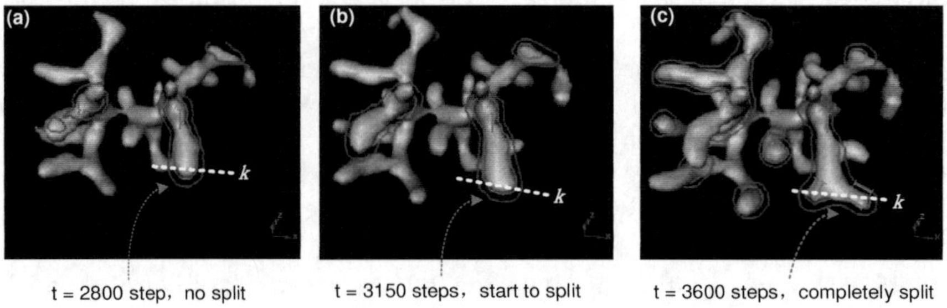

Figure 3. Simulation results for branching behaviors in the typical tip bifurcation. (a-c) Spatial distribution of the cellular differentiation marker Y; The tip bifurcation could be examined along the spatial axis *k* on the plane defined by the blue closed contours. Parameters: c=0.04, μ=0.8, ν=0.08, ρ_A=0.03, ρ_H=0.0001, c_0=0.02, γ=0.02, ε=0.042, d=0.008, e=0.1, f=10, D_A=0.1, D_H=0.26, D_S=0.06.

Figure 3 shows the dynamic process of tip bifurcation and concentration distribution relationships between activator A and inhibitor H. Figure 3a, b, c show 3D graphics of cell differentiation marker Y isosurface. Stalk I extends longitudinally along the original direction (figure 3a, steps 2800); Tip of stalk I becomes large and flat indicating the tip expands, perpendicularly, transverse to original direction (figure 3b, steps 3150); Tip of stalk I completely splits and stalk I is divided into two daughter branches, namely branch II and branch III (figure 3c, steps 3600). The symmetrical center plane of stalk I serves as tangent plane (the plane defined by blue and green contours, figure 3a, b, c). A line with the largest width through the tip of stalk I is made on the tangent plane (denoted by the blue closed contour) , which is denoted by *k* (the white dotted line, figure 3a, b, c). The concentration distribution change of inhibitor H lags behind that of activator A. The delayed inhibitor peak resembles a knife forcing the activator peak with lateral expansion into two (the inhibitor peak significantly reduces the concentration of activator at the corresponding region). After the tip splits completely, the inhibitor peak also divides into two peaks, and its distribution along the spatial axis *k* is almost the same as that of activator A (steps 3600). Further research work for the 4D biofabrication may rely on the 3D culture techniques [3, 24, 25] that are essentially constructing and regulating the dynamic structures composed of numeral cells.

5. Conclusion

Based on the Turing reaction-diffusion mechanism, the 3D simulation of the mathematical model describing bifurcation phenomena in biology, including concentrations of activator, inhibitor, substrate and cell differentiation marker were studied. This 3D simulation model can predict following patterns of 3D tissue morphology: side branches and tip bifurcation, and is able to demonstrate the evolution of side branching and tip bifurcation of the 3D multicellular structure. It lays a foundation for guiding 4D biofabrication of lung airway grafts via 3D cellular self-organization, which is expected to greatly reduce the complexity of future experimental research and number of trials.

Acknowledgments

This research was supported by grants from the National Natural Science Foundation of China (Grant No. 51505127), Natural Science Foundation of Jiangsu Province (Grant No. BK20161197), Changzhou Sci&Tech Program (Grant No. CE20165029), and the Fundamental Research Funds for the Central Universities of China (Grant No. 2015B04414).

References

[1] Chia H N and Wu B M. 2015 *Journal of Biological Engineering.* **9**: 4.

[2] Zhu X, Gojgini S, Chen T-H, Teng F, Fei P, Dong S, Segura T, and Ho C-M. 2016 *Biomedicine & Pharmacotherapy* **83**: 1203-1211.
[3] Zhu X, Gojgini S, Chen T-H, Fei P, Dong S, Ho C-M, and Segura T. 2017 *Journal of Biological Engineering* **11**(1): 12.
[4] Gao B, Yang Q, Zhao X, Jin G, Ma Y, and Xu F. 2016 *Trends in biotechnology* **34**(9): 746-756.
[5] Datta P, Ayan B, and Ozbolat I T. 2017 *Acta Biomaterialia*: doi: 10.1016/j.actbio.2017.01.035.
[6] Li Y C, Zhang Y S, Akpek A, Shin S R, and Khademhosseini A. 2017 *Biofabrication* **9**(1): 012001.
[7] Metzger R J, Klein O D, Martin G R, and Krasnow M A. 2008 *Nature* **453**(7196): 745-U1.
[8] Tzou D, Spurlin J W, Pavlovich A L, Stewart C R, Gleghorn J P, and Nelson C M. 2016 *Evodevo* **7**.
[9] Sampogna R V, Schneider L, and Al-Awqati Q. 2015 *Journal of the American Society of Nephrology* **26**(10): 2414-2422.
[10] Zhang Y, Yokoyama S, Herriges J C, Zhang Z, Young R E, Verheyden J M, and Sun X. 2016 *Proceedings of the National Academy of Sciences of the United States of America* **113**(27): 7557-7562.
[11] Morrisey E E and Hogan B L M. 2010 *Developmental Cell* **18**(1): 8-23.
[12] Mandelbrot B B, *The Fractal Geometry of Nature* [M]. 1983, New York: Freeman.
[13] Kitaoka H, Takaki R, and Suki B. 1999 *Journal of Applied Physiology* **87**(6): 2207-2217.
[14] Tebockhorst S, Lee D, Wexler A S, and Oldham M J. 2007 *Journal of Applied Physiology* **102**(1): 294-305.
[15] Kitaoka H, *Three-Dimensional Model of the Human Airway Tree Based on a Fractal Branching Algorithm*, in *Fractals in Biology and Medicine* [M], G.A. Losa, et al., Editors. 2002, Birkhäuser Basel: Basel. p. 39-46.
[16] Kondo S and Miura T. 2010 *Science* **329**(5999): 1616-1620.
[17] Menshykau D and Iber D. 2013 *Physical Biology* **10**(4): 046003.
[18] Menshykau D, Blanc P, Unal E, Sapin V, and Iber D. 2014 *Development* **141**(23): 4526-4536.
[19] Chen T H, Zhu X, Pan L, Zeng X, Garfinkel A, Tintut Y, Demer L L, Zhao X, and Ho C M. 2012 *Biomaterials* **33**(35): 9019-9026.
[20] Guo Y N, Sun M Z, Garfinkel A, and Zhao X. 2014 *Plos One* **9**(7).
[21] Guo Y N, Chen T H, Zeng X J, Warburton D, Bostrom K I, Ho C M, Zhao X, and Garfinkel A. 2014 *Journal of Physiology-London* **592**(2): 313-324.
[22] Meinhardt H. 1976 *Differentiation* **6**(2): 117-23.
[23] Zhu X and Yang H. In-Silico Constructing Three-Dimensional Hollow Structure via Self-Organization of Vascular Mesenchymal Cells [C]. in *The 16th International Conference on Nanotechnology (IEEE NANO 2016)*. 2016. Sendai, Japan: IEEE.
[24] Zhu X and Yang Y. Simulation for tubular and spherical structure formation via self-organization of vascular mesenchymal cells in three dimensions [C]. in *Image and Signal Processing, BioMedical Engineering and Informatics (CISP-BMEI), International Congress on*. 2016. IEEE.
[25] Zhu X L and Ding X T. 2017 *Slas Discovery* **22**(5): 626-634.

IOP Conf. Series: Materials Science and Engineering **280** (2017) 012019 doi:10.1088/1757-899X/280/1/012019

Three-dimensional Finite Element Modelling of Composite Slabs for High Speed Rails

Nhlanganiso Mlilo[1] and Sakdirat Kaewunruen[1,2]

[1] Department of Civil Engineering, The University of Birmingham, U.K.
[2] Birmingham Centre for Railway Research and Education, The University of Birmingham, U.K

nxm692@student.bham.ac.uk
KaewunrS@adf.bham.ac.uk

Abstract. Currently precast steel-concrete composite slabs are being considered on railway bridges as a viable alternative replacement for timber sleepers. However, due to their nature and the loading conditions, their behaviour is often complex. Present knowledge of the behaviour of precast steel-concrete composite slabs subjected to rail loading is limited. FEA is an important tool used to simulate real life behaviour and is widely accepted in many disciples of engineering as an alternative to experimental test methods, which are often costly and time consuming. This paper seeks to detail FEM of precast steel-concrete slabs subjected to standard in-service loading in high-speed rail with focus on the importance of accurately defining material properties, element type, mesh size, contacts, interactions and boundary conditions that will give results representative of real life behaviour. Initial finite element model show very good results, confirming the accuracy of the modelling procedure

1. Introduction

Composite steel-concrete beams due to their high stiffness, reduced deflections and high span/depth ratio are very popular in multistorey buildings [1]. In the say way, composite steel-concrete slabs connected to steel beams by means shear connectors usually shear studs represent an economic and efficient solution commonly used in building applications [2]. The steel sheeting acts as permanent formwork during construction and after hardening of the concrete acts as external tensile reinforcement. Considerable amount of research has been undertaken to date on the linear and nonlinear behaviour of composite floor beams using FE analysis [1]. Adoption of precast composite steel-concrete slab panels to resist rail loading was considered by Griffin [3] to address the problem of premature degradation associated with timber transoms on the Sydney Harbour Bridge. In this research a FE analysis software, ABAQUS was used to model and envisage the behaviour of theoretically determined composite steel-concrete panels when subjected to both standard in-service and derailment loading. However, the application of precast composite steel-concrete panels due to their nature and often complex loading conditions they are subjected to, the behaviour, interaction and response of the constituent components is often complex and requires a much thorough study. Present detailed knowledge of the behaviour of precast steel-concrete composite slabs subjected to rail loading is limited and there is therefore need to carry out thorough and detailed investigations in order to gain more understanding to inform future research.

IOP Conf. Series: Materials Science and Engineering **280** (2017) 012019 doi:10.1088/1757-899X/280/1/012019

In FE modelling, assigning reliable material properties, interactions and boundary conditions to the model that adequately represent the real situation or prototype is critical. A number of researchers have extensively and successfully used ABAQUS to analyse and investigate complex non-linear response of steel-concrete composite structures including Lam and Al-Lobody [4] who used ABAQUS to simulate the structural behaviour of headed shear studs in a steel-concrete composite beam and found that the FE results compared very well with experimental push-off test results while also accurately predicting the mode of failure of the headed shear studs. ABAQUS was also used to carry out an extensive parametric study to investigate the effect of changes in concrete strength and stud diameter on the capacity and behaviour of the shear connection [5]. In this research FE results were compared to those specified in design codes, namely; EC4 and AASTHO LRFD. A lot of other researchers have also developed FE models and successfully used ABAQUS.

2. Finite element modelling of composite steel-concrete slab

The accuracy of FE modelling greatly depends on the efficiency in simulating the non-linear behaviour of different materials within the composite structure when subjected to loading conditions.

2.1 Material Properties in ABAQUS

Steel-concrete composite slabs essentially consist of two structural materials i.e. concrete and steel which make up shear studs, profiled sheeting, reinforcing steel or prestressing steel and bridge stringer and it is important to carefully model the different components by accurately incorporating the stress-strain curves of these in the FE modelling. The accuracy of a FE analysis depends on constitutive laws used to define stress-strain characteristics of constituent materials [1].

2.1.1 Concrete The stress-strain curve of plain concrete in compression consists of two parts, the linear or elastic section which is defined by the modulus of elasticity, $E_{cm}$and the Poisson's ratio, v which is assumed to act up to a proportional limit of $0.4f_{cm}$. Beyond this point, the behavior is non-linear.

According to EC2 [6] the elastic deformation properties of plain concrete up to the characteristic compressive strength, f_{ck} of 90MPa can be accurately defined by the modulus of elasticity, E_{cm} in GPa, as:

$$E_{cm} = 22[(f_{cm})/10]^{0.3} \tag{1}$$

where

$$f_{cm} = (f_{ck} + 8)\ MPa \tag{2}$$

For this model plain concrete with a characteristic compressive cylinder strength, f_{ck} of 50MPa has been used and the modulus of elasticity has been calculated as 37GPa. The Poisson's ratio of concrete will be adopted as 0.2 for uncracked concrete and 0 for cracked concrete as detailed in EC2 [6].

There are several ways of modelling the plastic characteristics of concrete in ABAQUS including; plastic, cap plasticity, concrete damaged plasticity and concrete smeared cracking among many others. The non-linear compressive behaviour part of the stress-strain curve can be accurately found from the equation below [7].

$$\sigma_c = \frac{f_c' \gamma(\varepsilon_c/\varepsilon_c')}{\gamma - 1 + (\varepsilon_c/\varepsilon_c')} \tag{3}$$

where

$$\gamma = \left|\frac{f_c'}{32.4}\right|^3 + 1.55 \tag{4}$$

and

$$\varepsilon_c' = 0.002 \tag{5}$$

For concrete in tension, the stresses are assumed to increase linearly up to a point when the concrete cracks and after this they decrease linearly to zero [1].

2.1.2 Steel (Shear Studs, Profiled Steel Sheet, Reinforcing Steel and Bridge Stringer) The stress-strain behaviour of all the steel components i.e. shear studs, profiled sheet, bridge stringer, reinforcing steel is initially elastic after which yielding and strain hardening develops. Figure 1 below depicts the bi-linear stress-strain behaviour of steel sheeting and headed shear studs [3] while figure 2 depicts the expected strain-strain behaviour of bridge stringers and the reinforcing steel [1]. The stress-strain behaviour of all steel components is assumed to be the same in tension and compression.

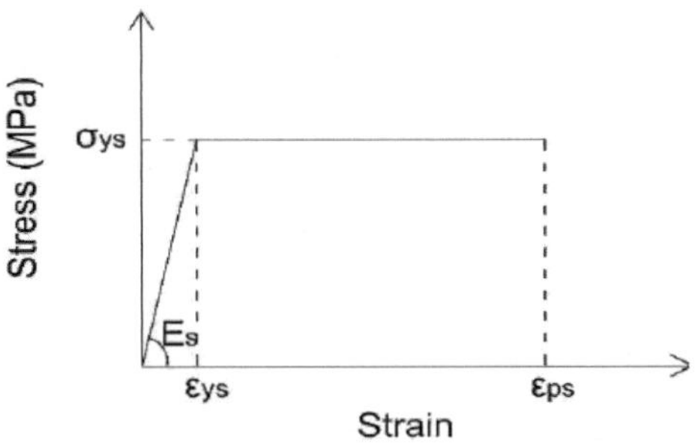

Figure 1. Bi-linear stress-strain curve

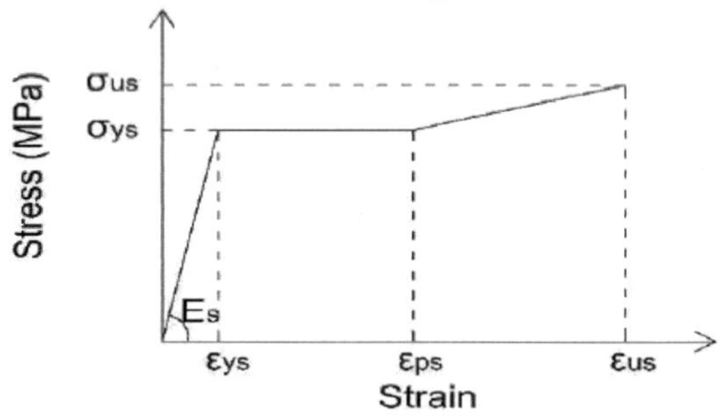

Figure 2. Tri-linear stress-strain curve

Table 1 below represents a summary of the component properties as well as the ratios for determining ultimate stress (σ_{us}), plastic strain (ε_{ps}) and ultimate strain (ε_{us}) [1].

Table 1. Summary of steel component properties

Component	*Young's Modulus E_s (GPa)*	*Poisson's Ratio, v*	*Yield Stress (f_y) MPa*	σ_{us} *(MPa)*	ε_{ps}	ε_{us}
Shear Studs	200	0.3	420	-	$25\varepsilon_{ys}$	-
Steel Sheeting	200	0.3	550	-	$20\varepsilon_{ys}$	-
Bridge Stringer	200	0.3	300	$1.28\varepsilon_{ys}$	$10\varepsilon_{ys}$	$30\varepsilon_{ys}$
Steel Reinforcement	200	0.3	500	$1.28\varepsilon_{ys}$	$9\varepsilon_{ys}$	$40\varepsilon_{ys}$

2.2 Finite Element Type and Mesh

In ABAQUS elements can be modelled as either two-dimensional or three-dimensional solid, shell, truss or beam elements. The stresses, strains as well as deformations are determined at the intersection of elements and because of this, to allow for an accurate analysis it is very important to select both the correct element type and element size. Solid, three-dimensional, eight node element with linear approximation of displacements, three translational degrees of freedom and reduced integration with hourglass control (C3D8R) can be used for linear and nonlinear models and is capable of handling large deformations, accommodate plasticity while accurately incorporating contact properties [8].

Mirza [1, 8] successfully used the C3D8R element to model the concrete slab and the steel beam and their use improved the rate of solution convergence.

To model thin walled sections, for example the profiled sheeting, it is recommended to use the S4R shell elements [1, 9 and 10]. These are four-node elements with 6 degrees of freedom per node and like the C3D8R incorporate reduced integration and when used by Mirza [1] to model the profiled steel sheeting provided accurate solutions and permitted quadratic deformation over the four nodal coordinates. The T3DR element which is a three-dimensional, two node truss element is used to model the steel reinforcement.

A good quality mesh is a very important in FE analysis as the shape and size of elements can significantly affect the results. A good balance needs to be achieved when the mesh size is being selected, the mesh needs to be fine enough to allow for good detail especially in the areas of most interest but at the same time not too fine as this can result in the analysis taking a lot of time to complete [1]. So in order to reduce the time for analysis, a fine mesh can be applied at regions where

IOP Conf. Series: Materials Science and Engineering **280** (2017) 012019 doi:10.1088/1757-899X/280/1/012019

there are interactions for example around the interface between the shear studs and concrete and around the interface between the shear studs and the steel beam/sheeting with a much coarser mesh applied everywhere else [5]. To ensure that a good model is produced which produces accurate results it is necessary to carry out a mesh sensitivity analysis. Table 2 below shows the adopted element types for each component as well as the optimum FE mesh sizes for the model which gave deflection results closest to the theoretically determined deflection.

Table 2. Summary of element type and mesh size

Component	*Element Type*	*Mesh Size*
Concrete	C3D8R	40
Shear Studs	C3D8R	40
Steel Sheeting	S4R	50
Bridge Stringer	C3D8R	40
Steel Reinforcement	T3DR	84.7

2.3 Contact and Interactions

Correct and accurate simulation of the interacting surfaces and parts within FE modelling is very important in order to obtain accurate results which reflect real life behaviour. The stability and adequacy of a FE model representing composite structures are influenced by the accurate definition of contact properties between the steel components and the concrete [2]. In structures like composite steel-concrete slabs where there are many interacting components, this is even more important. According to ABAQUS, there are many ways of defining interface interaction and the two most common are the constraint toolset and the interaction toolset. Surfaces are defined as master or slave with the stiffer material being assigned as a master and the other being assigned a slave. Reinforcing steel in concrete can sufficiently be modelled using the embedded technique. A summary of the surface designation and interface type is shown on the table 3 below.

Table 3. Surface and interface designation

Interacting Components	*Slave Surface*	*Master Surface*	*Interface Type*
Shear Stud - Concrete	Concrete	Shear Stud	Tie Constraint
Shear Stud – Steel Sheeting	Shear Stud	Steel Sheeting	Tie Constraint
Steel Sheeting – Stringer(weld)	Stringer	Steel Sheeting	Tie Constraint
Steel Sheeting - Stringer	Stringer	Steel Sheeting	Surface to Surface
Steel Reinforcement - Concrete	Concrete	Steel Reinforcement	Embedded
Concrete – Steel Sheeting	Concrete	Steel Sheeting	Surface to Surface

2.4 Boundary conditions and load application

Correct application of boundary conditions is crucial in FE modelling and the application should as much as possible accurately represent real-life situations otherwise, the model will not produce accurate results. In ABAQUS, boundary conditions can be applied to part edges, nodes or surfaces to define their translational or rotational response. If the models are symmetrical it is possible to either half or quarter them in order to reduce analysis time. When this is done the cut surfaces must have boundary conditions applied to them to represent the symmetry required. Several studies [1, 4, 5 and 10] investigating various aspects of composite slab behaviour produced accurate results when they used the above technique. The in-service model was considered symmetrical and is shown in figure 3.

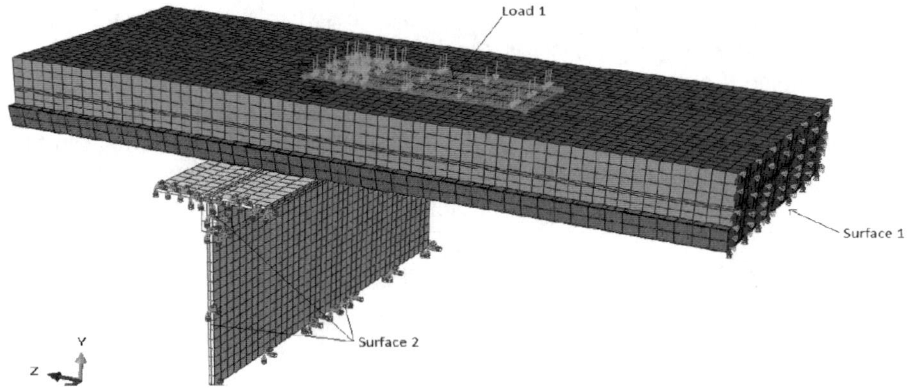

Figure 3. Load application and boundary conditions for the in-service half symmetrical model

Symmetric boundary conditions are applied to surface 1 and the nodes of the steel sheeting and concrete that lie on this surface are restricted from any translation in the z-direction. Surface 2 represents the cut edges of the bridge stringer and all the nodes on surface 2 are represented as encastre meaning that they are restrained from rotation and translation in all directions. The load combination which produced the worst loading case for bending moments was adopted for the simulation. The train load was applied as pressure load over the rail pads area and the factored dead load of the panel was also applied over the entire surface of the panel as the pressure load.

3. Results and conclusion

Figure 4 below represents the initial results from a symmetrical half in-service model developed following the above mentioned procedure.

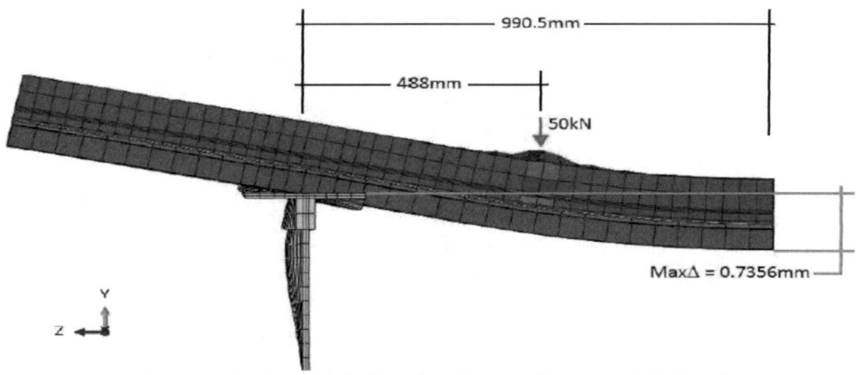

Figure 4. ABAQUS model showing in-service panel deflection

As shown on the figure above, the mid-span deflection which resulted from the ABAQUS model was 0.7356mm and comparing this to the theoretically determined deflection of 0.8137mm, it is clear that the results were comparable, i.e. 10.6% variation.

The maximum stresses developed in all the component parts compared to their corresponding yield stresses are summarised in the table 4 below.

Table 4. Limit strength summary of in-service panel

Component	*Material Yield Strength (MPa)*	*Maximum Stress developed (MPa)*	*Design ratio*
Concrete (in compression)	50	45	1.11
Concrete (in tension)	5.3	10.66	0.50*

Steel Sheeting	550	226	2.43
Stringer	300	133	2.26
Steel Reinforcement	500	32	15.63
Shear Studs	420	244	1.77

The design ratio is obtained by dividing the material yield strength by the corresponding maximum stress developed in the component. Looking at the design ratios, it is clear that all the components are adequate to resist the design loads apart from the concrete in tension which develops maximum tensile stress of 10.66MPa which is well over its maximum axial tensile strength, f_{ctk} of 5.3 MPa given in EC2 [6], implying that concrete does crack. However, upon cracking the steel sheeting continues to carry the tensile load. Figure 5 shows the concrete node (shown with arrow) which develops the maximum tensile stress.

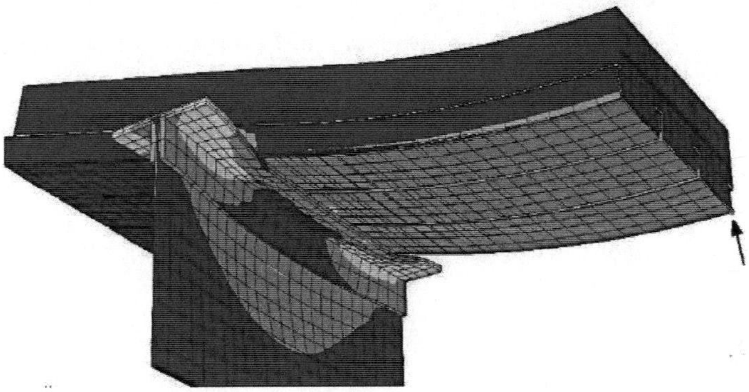

Figure 5. Critical node area in concrete

These results give us the confidence that the modelling procedure has been accurate and can be trusted as the theoretically designed in-service panel satisfies the ULS loading conditions.

References

[1] Mirza O 2008 *Behaviour and design of headed stud shear connectors in composite steel-concrete beams – PhD Thesis* (Sydney: University of Western Sydney)
[2] Tahmasebinia F, Ranzi G and Zona A 2012 *Journal of Constructional Steel Research* **80** 394-411
[3] Griffin D W P 2013 *Design of precast composite steel-concrete panels for track support: For use on the Sydney Harbour Bridge – BEng Thesis* (Sydney: University of Western Sydney)
[4] Lam D and El-Lobody 2001 *Structural Engineering, Mechanics and Computation* **1** 401-8
[5] Nguyen H T and Kim S E 2009 *Journal of Constructional Steel Research* **65** 1909-20
[6] British Standards Institute 2008 *BS EN 1992-1-1:2004 Eurocode 2: Design of concrete structures*
[7] Carreira D and Chu K 1985 *Journal of ACI Structures* **82** 797-804
[8] Mirza O and Uy B 2009 *Journal of Constructional Steel Research* **65** 662-74
[9] ABAQUS user's manual, version 6.14
[10] Shanmugam N E, Kumar G and Thevendran V 2002 *Finite Elements in Analysis and Design* **38** 579-99

IOP Conf. Series: Materials Science and Engineering **280** (2017) 012020 doi:10.1088/1757-899X/280/1/012020

Performance of Railway Sleepers with Holes under Impact Loading

Chie Hong Lim[1], Sakdirat Kaewunruen[1,2] and Nhlanganiso Mlilo[1]

[1] Department of Civil Engineering, The University of Birmingham, U.K.
[2] Birmingham Centre for Railway Research and Education, The University of Birmingham, U.K

CXL529@student.bham.ac.uk; KaewunrS@adf.bham.ac.uk; NXM692@student.bham.ac.uk

Abstract: Prestressed concrete sleepers are essential structural components of railway track structures, with the purpose of redistributing wheel loads from the rails to the ground. To facilitate cables and signalling equipment, holes are often generated in these prestressed concrete sleepers. However, the performance of these sleepers under impact loading may be a concern with the addition of these holes. Numerical modelling using finite element analysis (FEA) is an ideal tool that enables static and dynamic simulation and can perform analyses of basic/advanced linear and nonlinear problems, without incurring a huge cost in resources like standard experimental test methods would. This paper will utilize the three-dimensional FE modelling software ABAQUS to investigate the behaviour of the prestressed concrete sleepers with holes of varying sizes upon impact loading. To obtain the results that resemble real-life behaviour of the sleepers under impact loading, the material properties, element types, mesh sizes, contact and interactions and boundary conditions will be defined as accurately as possible. Both Concrete Damaged Plasticity (CDP) and Brittle Cracking models will be used in this study. With a better understanding of how the introduction of holes will influence the performance of prestressed sleepers under impact loading, track and railway engineers will be able to generate them in prestressed concrete sleepers without compromising the sleepers' performance during operation.

1. Introduction

Railway sleepers are a primary component in the railway track system, playing a significant role of transferring vehicle loads to the track structure below it. Other functions of the sleeper include maintaining the track gauge and providing insulation for the rails against electricity. While they may be constructed of materials such as steel, plastic composites, and timber, concrete sleepers are generally preferred for their better performance, flexibility for specific service requirements, and longer life-span.

Although concrete enjoys the listed merits, a main concern is that it is susceptible to deterioration issues. Cracks may occur in a sleeper which may expand and cause the latter to fail over time, and this could incur extra costs to replace them as they cannot be repaired after sustaining considerable damage.

A possible source of damage that may induce cracking in sleepers is the impact loads on track systems. It is discovered that impact loads may be resulted from the interaction between the vehicle

IOP Conf. Series: Materials Science and Engineering **280** (2017) 012020 doi:10.1088/1757-899X/280/1/012020

and the track, and the resonance produced among track components [1]. Despite the low cycle of impact loads, they may however cause severe damage to sleepers due to their high magnitude.

Given that holes are often drilled into sleepers for signalling equipment, this raises a concern that they may increase the risk of sleepers cracking and failing under impact loading. To make matters worse, there are limited studies pertaining to this concern. Hence, a better understanding of their behaviour under impact loading may be required to ensure that they have a life span which could in turn mean better returns on investment on the sleepers.

2. Finite element modelling

The commercial FE analysis software ABAQUS has been utilised for this study [2], where two methods are used for the FE sleeper models used in this research: The Concrete Damaged Plasticity models and the brittle cracking models.

The CDP models allow the resemblance of the behaviour of concrete by enabling strain hardening during compression and stiffness recovery, and the sensitivity to the straining rate can be adjusted accordingly. Heavy distortion may be avoided through the frequent re-meshing nature of the impacted zone during analysis, if adaptive meshing is adopted with CDP models. The CDP model is suitable for analysing models under cyclic loading conditions.

The brittle concrete models meanwhile, contain a failure criterion and the removal of elements during the analyses. This means that the crack propagation of the sleeper under impact loading can then be thoroughly examined, which is unachievable by using CDP models.

2.1 Elements and Mesh Sizes

There are four primary segments used for the model, which are the concrete sleeper, the prestressed tendons, the wheel, and the rail. The element sizes are 15, 35, 12 and 10 respectively. All components except the prestressed tendons are of C3D8R element type, while the prestressed tendons are of the C3D6 element type [3]. The element types and sizes were chosen to ensure accurate results while reducing the time for calculation and computational accuracy for contact analysis. Figure 1 shows the constructed mesh of the model setup.

Figure 1 - Constructed mesh of sleeper model

2.2 Contact definition

General contact has been defined for the entire model and a friction coefficient of 0.3 was applied for the interaction between the structural components.

2.3 Boundary Conditions

There are 3 main boundary conditions applied to the models, to replicate a real-life sleeper under impact loading. A vertical velocity of 1.94 m/s was applied at the center of the wheel and its DOF is constrained except the U2 direction [1,4,5] as shown in Figure 2, allowing it to resemble the motion of an impact loading. Roller boundary conditions are applied on the bottom of the sleeper as shown in Figure 3, acting as the supports of the sleeper.

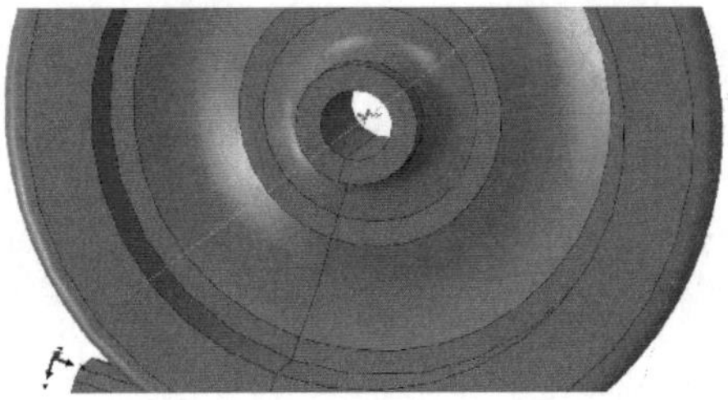

Figure 2. Boundary Condition at Wheel

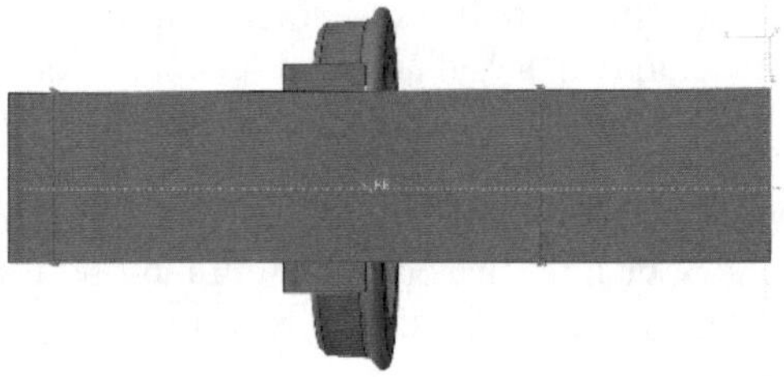

Figure 3. Boundary Condition under sleeper

2.4 Constraint definition

The following constraints have been defined in the models as shown in Table 1.

Table 1. Constraints definition

Component	Constraint
Wheel	Rigid
Surface between rail bottom and sleeper top	Tie
Prestressed tendons and concrete sleeper	Embedded Region

2.5 Material Properties

2.5.1 Concrete. The sleeper component is made of concrete and the typical properties of normal concrete that have been registered for the models, are listed in Table 2. The plasticity, tensile, and compressive properties used for the CDP models are listed in Tables 3, 4 and 5 respectively. Meanwhile the brittle behaviour and the brittle failure criteria are listed in Tables 6 and 7.

Table 2. Typical properties of normal concrete

Density ρ_c	Young's Modulus E_c	Poisson's Ratio v_c	Compressive Strength σ_{cc}	Tensile Strength σ_{ct}	Fracture Energy *GF*
2400 kg/m^3	36406 MPa	0.2	60 MPa	2.85 MPa	154 N/m

2.5.1.1 Concrete Damaged Plasticity (CDP)

Table 3. Plasticity input for CDP model

Dilation Angle	Eccentricity	Fb_0/fc_0	K	Viscosity Parameter
45	0.1	1.16	0.0067	0

Table 4. Tensile behaviour for CDP model

Yield Stress (MPa)	Fracture Energy
2.56	0.15

Table 5. Compressive behaviour for CDP model

Yield Stress (MPa)	Inelastic Strain
27.78650201	0
36.27619362	0.001
43.73984458	0.00125
49.88530231	0.0015
54.54914425	0.00175
57.70881121	0.002
59.46422457	0.00225
60	0.0025
59.54281442	0.00275
58.32516972	0.003
56.56074572	0.00325
54.43140618	0.0035
52.08318636	0.00375
49.62794701	0.004
32.10818842	0.006
21.33143134	0.008
15	0.01
11.07686332	0.012

2.5.1.2 Brittle Cracking Model [6]

Table 6. Brittle behaviour for brittle cracking model

Direct stress after cracking	Direct cracking strain	Field 1
3.17	0	0.5
0	0.0008	0.5
4.5	0	1.5
0	0.0008	1.5

Table 7. Brittle failure input for Brittle Cracking model

Failure Criteria	Direct cracking failure strain or displacement
Unidirectional	0.045

2.5.2 Steel and Prestressed steel tendon. The general properties of the steel used for the wheel, rail and tendons are assumed in Table 8 below.

Table 8. General properties of steel

Density ρ_s	Young's Modulus E_s	Poisson's Ratio v_s
7.8 g/cm^3	200 GPa	0.3

Table 9 summarises the plastic stress-strain relationship for the prestressed tendons.

Table 9. Plastic stress-strain property for prestressed steel tendon

Yield Stress (MPa)	Plastic Strain
1000	0
1703	0.0085
1750	0.0097
1797	0.01
1860	0.064

3. Validation

It is imperative to validate the FE models to ensure legitimate results. The developed models were calibrated to static loading conditions, for the validation against the results obtained in Erosha Gamage's research [7,8]. The ultimate bending moment at failure for the developed models was compared to the experimental and numerical results obtained in his research. The validation results can be seen in Tables 10, 11, 12 and 13, where positive correlations can be observed between the results for the models and the results from Erosha's research.

Table 10. CDP model validation against experimental results

Sleeper Case		Max Bending Moment at Rail Seat (kNm)		Difference (%)
		Experimental	FEA (CDP)	
No hole/web opening		67	66.3	1.04
Longitudinal hole	32mm	65	64	1.54
	42mm	61	59	3.28
Transverse Hole	32mm	57	59	3.51
	42mm	56	56	0
Vertical hole	32mm	65	65	0
	42mm	61	63	3.28

Table 11. CDP model validation against Erosha's FEA

Sleeper Case		Max Bending Moment at Rail Seat (kNm)		Difference (%)
		Erosha's FEA	FEA (CDP)	
No hole/web opening		65.7	66.3	0.91
Longitudinal hole	32mm	62.2	64	2.89
	42mm	59.4	59	0.67
Transverse Hole	32mm	55.5	59	6.31
	42mm	55	56	1.82
Vertical hole	32mm	63	65	3.17
	42mm	59	63	6.78

Table 12. Brittle Cracking model validation against experimental results

Sleeper Case		Max Bending Moment at Rail Seat (kNm)		Difference (%)
		Experimental	FEA (Brittle Cracking)	
No hole/web opening		67.0	66.3	1.82
Longitudinal hole	32mm	65.0	64.0	4.84
	42mm	61.0	59.0.	0.16
Transverse Hole	32mm	57.0	59.0	0.18

	42mm	56.0	56.0	1.41
Vertical hole	32mm	65.0	65.0	1.25
	42mm	61.0	63.0	2.71

Table 13. Brittle Cracking model validation against Erosha's FEA

Sleeper Case		Max Bending Moment at Rail Seat (kNm)		Difference (%)
		Erosha's FEA	FEA (Brittle Cracking)	
No hole/web opening		65.7	65.8	0.15
Longitudinal hole	32mm	62.2	62.0	0.32
	42mm	59.4	61.1	2.78
Transverse Hole	32mm	55.5	57.1	2.80
	42mm	55.0	56.8	3.17
Vertical hole	32mm	63.0	64.2	1.87
	42mm	59.0	62.7	5.90

4. Conclusion

This paper has examined the structural behaviour of railway sleepers with holes under impact loading by using finite element modelling. This includes the study of the load-deflection relationship, Von Mises Stress distribution and crack propagation of every case listed in the paper. It was then determined in both the CDP and Brittle Crack models that the sleeper with a 42mm transverse hole has the worst performance of all the cases under impact loading. This insight will probably help railway and track engineers to make critical decisions on the type of holes to generate in sleepers for signalling equipment without compromising the structural integrity of the sleeper.

References

[1] Kaewunruen S and Remennikov A M 2010 Dynamic crack propagations of prestressed concrete sleepers in railway track systems subjected to severe impact loads. *Journal of Structural Engineering* **136 (6)** 749–54.

[2] ABAQUS Benchmark manual: http://abaqus.software.polimi.it/v6.12/books/bmk/default.htm?startat=ch03s02ach172.html

[3] Ahmed A 2014 Modelling of a reinforced concrete beam subjected to impact vibration using ABAQUS. *International Journal of Civil and Structural Engineering* **4** 227-36

[4] Kaewunruen S and Remennikov A M 2007 Progressive failure of prestressed concrete sleepers under multiple high-intensity impact loads. *Engineering Structures* **31** 2460-73

[5] Kaewunruen S 2007 *Experimental and numerical studies for evaluating dynamic behaviour of prestressed concrete sleepers subject to severe impact loading. Ph.D. thesis* (Australia: School of Civil, Mining, and Environmental Engineering, University of Wollongong)

[6] Dassault Systèmes Simulia Corp. ABAQUS Analysis User's Manual

[7] Gamage E K, Kaewunruen S and Remennikov A M 2015 Design of holes and web openings in railway prestressed concrete sleepers. *Proceedings of the 13th International Railway Engineering Conference Railway Engineering Conference*. 28 June – 2 July 2015, Edinburgh

[8] Gamage E K, Kaewunruen S, Remennikov A M and Ishida T 2017 Toughness of railway concrete crossties with holes and web openings. *Infractructures* **2,3** 1 – 8

IOP Conf. Series: Materials Science and Engineering **280** (2017) 012021 doi:10.1088/1757-899X/280/1/012021

An Investigation and Prediction of Springback of Sheet Metals under Cold Forming Condition

A Elsayed[1], M Mohamed[1,2*], M Shazly[3] and A Hegazy[1]

[1] Department of Mechanical Engineering, Helwan University, Helwan, Egypt
[2] Impression Technologies Ltd, Coventry, UK
[3] Mechanical Engineering Department, The British University in Egypt, Cairo, Egypt

*Corresponding Author: m.mohamed@impression-technologies.com

Abstract. Low formability and springback especially at room temperature are known to be major obstacles to advancements in sheet metal forming industries. The integration of numerical simulation within the R&D activities of the automotive industries provides a significant development in overcoming these drawbacks. The aim of the present work is to model and predict the springback of a Galvanized low carbon steel automotive panel part. This part suffers from both positive and negative springback which physically measured using CMM. The objective is to determine the suitable forming process parameters that minimize and compensate the springback through robust FE model. The analysis of the springback was carried out following (Isotropic model and Yoshida – Uemori model) which are calibrated through cyclic stress strain curve. The material data of the Galvanized low carbon steel was implemented via lookup tables in the commercial finite element software Pam-Stamp$^{(TM)}$. Firstly, the FE model was validated using the deformed part which suffers from springback problem at the same forming condition. The FE results were compared with the measured experimental trails providing very good agreement. Secondly, the validated FE model was used to determine the suitable forming parameters which could minimise the springback of the deformed part.

1. Introduction

Sheet metal forming process is one of the most important manufacturing processes in the automotive sector. Achieving higher performance, higher productivity and lower cost is the main drive behind the development of the sheet metal forming technology in the automotive sector [1]. It is well known that the sheet forming processes, mainly cold forming, suffer from defects such as thinning, wrinkles, cracks and geometrical inaccuracies due to springback. Springback is the elastic recovery that causes change in the shape of the sheet metal part upon completion of the forming process and removal of the part from the tooling [2].

The springback phenomenon causes dimensional deviations and often resulting in rejection of the part due to failure in meeting dimensional tolerance requirements. Moreover, in many cases the springback may cause a delay in assembly time due to the additional post forming process to compensate the springback or additional machines with higher capacity to fit the parts correctly in the assembly [1].

A common countermeasure method against springback is to design the forming tool in a way that can compensate for the springback. However, calculating the required compensation accurately is a

IOP Conf. Series: Materials Science and Engineering **280** (2017) 012021 doi:10.1088/1757-899X/280/1/012021

challenging task even for highly experienced tool designers. Therefore, in practice, engineers mainly rely on trial and error approach to compensate for the springback [2].

Numerical simulations with an adequate material model with its parameters, tooling and part geometries is an effective alternative approach to the time consuming and expensive trial and error method. Finite element (FE) simulations provide an accurate prediction of springback which resulted in an optimized tool design that produces a sound part shape after forming at first attempt.

The current work is an attempt to investigate the springback of a case study which a part that experiences delays in assembly. The part, shown in figure 1a, is provided by INDE GROUP FOR ENGINEERING INDUSTRIES [3]. The main aim of the work is to develop an FE model coupled with an adequate with suitable material model to predict and compensate for the springback problem for the investigated part by testing different process parameters.

2. Analysis of the formed part

The Fog lamp part considered in this work is produced from 1.5mm low Galvanized low carbon steel. The part is produced in two stages; the first stage is a blanking process to produce the desired blank shape and the second stage is the forming process. The part is cold formed using 300 bar hydraulic press as shown in figure 1(b). The current forming parameters are a punch pressure of 130 bar and a cushion pressure (back pressure) of 70 bar with a punch speed of 15 m/s.

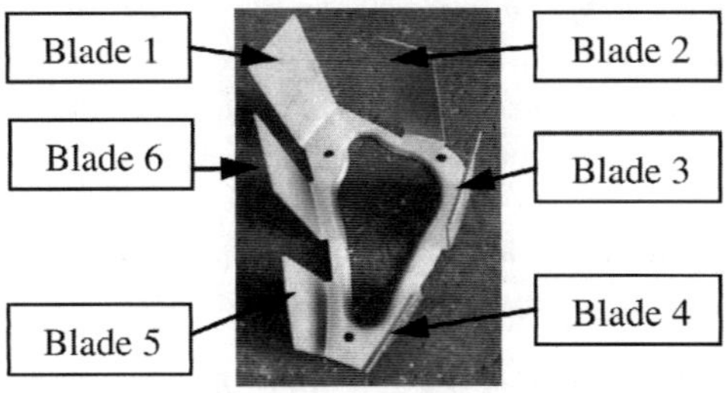

Figure 1. Fog lamp part investigated in the present work, INDE GROUP FOR ENGINEERING INDUSTRIES

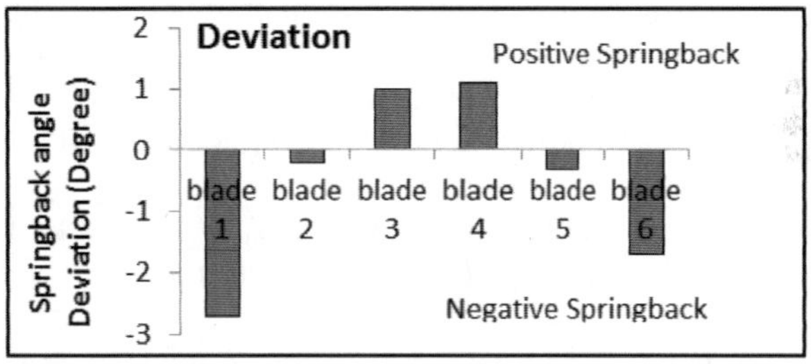

Figure 2. springback deviation from the CAD data

In order to characterize and analysis the springback behavior of the formed part, the physical part is scanned using Coordinate Measuring Machine (CMM). The required angle for all blades is 92°and by comparing the scanned and CAD parts, springback deviation was obtained as shown in figure 2, it is found that blade 1 has the largest deviation with 2.7° while the blades 6, 4 and 3 exhibits angle deviations of 1.7°, 1.1°and 1°respectively. Blades 2 and 5 show minimum deviation in the blades' angles.

The analysis of the formed part in figure 2 shows that the part is affected by both positive and negative springback. Blades 3 and 4 are affected by positive springback (i.e. the produced angles are more than the required angles). This type of springback occurs due elastic recovery during unloading. The existence of this type means poor tool design and/or improper springback compensation. On the other hand, blades 1 and 6 are affected by negative springback (i.e. the produced angles are less than the required angles). This type of springback usually occurs due to overloading and over stamping for the part during forming stage to compensate for the positive springback especially in closed bending dies [4].

The key factor in the forming process that directly affects the amount of springback in this part is the combination of the forming and cushion (back) pressures. Applying low pressures in forming and/or cushion pressures may cause insufficient forming and less pressure on the bent section leading to positive springback. On the contrary, applying high forming and/or cushion pressures means over stamping which leads to negative springback. In complex shapes, high pressures can cause thinning and cracking which must be taken into consideration during the design process. In the present work, there are three factors in addition that affect the springback and must be taken into account in the analysis process. First, blades have different orientations with respect to the parent sheet which is an important factor if the martial used has an anisotropic behavior. Second, the lengths of the blades are different, and as the blade length increases the springback sensitivity increases. Third, not all blades have straight rectangular cross section, some blade have curved cross section which affects the springback value [2].

3. Material Model

3.1. Material models and calibration

The accuracy of springback analysis is highly dependent on the predictions of stress levels at the final stage of stamping and during springback. The uneven stress distributions along the cross section of the component which occurs due to the stretch bending and subsequent unbending when drawn over a die corner are directly related to the material's hardening, Bauschinger effect, and cyclic hardening characteristics. As shown in figure 3, isotropic hardening plasticity models usually model an overrated cyclic hardening behavior under reverse loading. Moreover, it can't model the Bauschinger effect during cyclic deformation, the rapid change of material work hardening and early re-yielding. As a result, kinematic and combined hardening material models are introduced to overcome the limitations of isotropic models. These advanced material models have limited application in industrial models due to the increased number of material parameters that should be calibrated which require complex material testing and advanced mathematical techniques to determine these parameters. In addition, these complex material tests require high cost as compared to those used in the isotropic hardening models [4][5]. In order to predict the springback accurately, two different material models in Pam-Stamp$^{(TM)}$, isotropic hardening and kinematic hardening material model.

3.1.1. Isotropic hardening model. In the isotropic hardening model, the hardening part of the material behavior obeys Hollomon law, equation (1), while the plasticity law obeys Hill's plasticity [4].

$$\sigma = k\,\varepsilon^{n} \tag{1}$$

where $\boldsymbol{k}$ is the strength coefficient, $\boldsymbol{n}$ is the work hardening, $\acute{\varepsilon}_{\mathbf{p}}$ is the effective plastic strain and $\boldsymbol{\sigma_y}$ is the yield stress [5]. The input parameters for the model are given in Table 1.

3.1.2. Yoshida-Uomuri (Y-U) model. Y-U a model is a two-surface model that assumes the kinematic hardening of the yield surface $\boldsymbol{f_0}$ with backstress $\boldsymbol{\alpha}$ within the bounding surface $\boldsymbol{F}$ with back stress $\boldsymbol{\beta}$ of mixed isotropic–kinematic hardening. The centre of yield surface $\boldsymbol{f}$ is moving with the back stress $\boldsymbol{\alpha}$. The boundary of $\boldsymbol{F}$ is expanding due to the plastic strain hardening which is expressed by the (B+R) where $\boldsymbol{B}$ is the bounding surface $\boldsymbol{F}$ initial size and $\boldsymbol{R}$ it's isotropic hardening component[6].

$$\mathrm{f_0} = \varphi(\sigma - \alpha) - \mathrm{Y} = 0 \tag{2}$$

$$\mathrm{F} = \varphi(\sigma - \beta) - (\mathrm{B} + \mathrm{R}) = 0 \tag{3}$$

Where $\boldsymbol{\varphi}$ is the anisotropic yield function, $\boldsymbol{\sigma}$ is the Cauchy stress and $\boldsymbol{Y}$ is the yield strength. The isotropic hardening, which represents the expanding rate of boundary surface of $\boldsymbol{F}$, is a function of plastic strain rate $\boldsymbol{P}^{\circ}$ and is defined by

$$\mathrm{R}^{\circ} = \mathrm{m}\,(\mathrm{Rsat} - \mathrm{R})\mathrm{P}^{\circ} \tag{4}$$

while the Bauschinger effect is expressed in terms of the back stress $\boldsymbol{\alpha}$ which consists of two components $\boldsymbol{\beta}$ and $\boldsymbol{\alpha^*}$, defined by following equations:

$$\alpha^{\circ} = \beta^{\circ} + \alpha^{*\circ} \tag{5}$$

$$\alpha^{*\circ} = \mathrm{c}\left[\left(\frac{\mathrm{a}}{\mathrm{Y}}\right)(\sigma - \alpha) - \sqrt{\frac{\mathrm{a}}{\alpha*}}\,\alpha^{*}\right]\mathrm{P}^{\circ} \tag{6}$$

$$\beta^{\circ} = \mathrm{m}(\frac{2}{3}\mathrm{b}\,\varepsilon^{\mathrm{p}} - \beta\,\mathrm{P}^{\circ}) \tag{7}$$

$$\mathrm{a} = \mathrm{B} + \mathrm{R} - \mathrm{Y} \tag{8}$$

Where $\boldsymbol{Y}$, $\boldsymbol{C}$, $\boldsymbol{B}$, $\boldsymbol{m}$, $\boldsymbol{b}$ and $\boldsymbol{Rsat}$ are material parameters of Y-U model. Besides these 6 parameters, another parameter $\boldsymbol{h}$ is used to express hardening stagnation and considered as an adjusting parameter. The Young's modulus used in Y-U model varies with plastic strain which can be described by

$$E(\bar{\varepsilon}^{p}) = E_0 - (E_0 - E_a)\left[1 - \exp(-\xi\bar{\varepsilon}^{p})\right] \quad (9)$$

Where $\boldsymbol{E_0}$ and $\boldsymbol{E_a}$ are the Young's module for initial and infinitely large pre-strained materials, respectively. $\bar{\boldsymbol{\varepsilon}}^{\boldsymbol{p}}$ denotes effective plastic strain and $\boldsymbol{\xi}$ is a material constant [4,6].The two models (Isotropic hardening and Y-U model) are calibrated by fitting their mathematical (Numerical) expression with cyclic stress strain curve as shown in figure 3. The experimental cyclic stress strain curve was obtained from Yoshida and Uemori [6]. Tables 1 and 2 shows the material's constants for the calibrated material models; isotropic hardening and Y-U model, respectively.

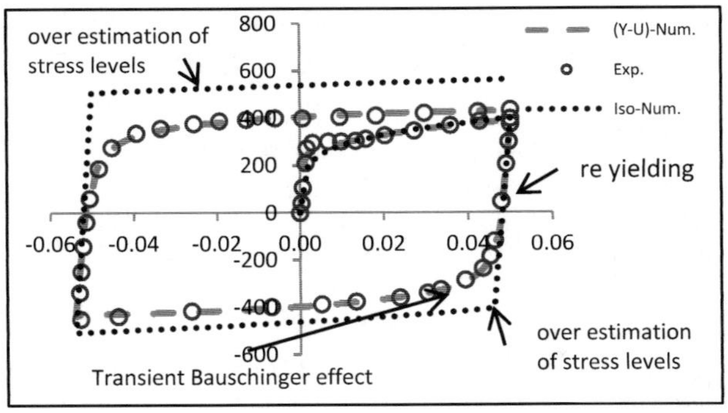

Figure 3. Calibration of the models by fitting the data with stress strain curve of the material.

Table 1. Material constants of the isotropic model for mild steel

Strength coefficient (k)	750MPa
Work hardening (n)	0.21
Young's modules(E)	207 GPa
Possion ratio (ν)	0.3
Density	7.85 g/cm^3

Table 2. Material constants of Y-U model for mild steel

Y	**C**	**B**	**m**	**b**
250	260	276	18	100
Rsat	**h**	**E0**	**Ea**	**ξ**
98	0.5	207000	160000	11

4. FE Model

4.1. Implementation of the material models

In the present work, Pam Stamp(TM) commercial finite element software is used to model and predict the stamping and springback stages of the investigated part. Both material models (Isotropic and Y-U model) are implemented in Pam Stamp(TM) via lookup tables. In order to validate and test the implementation process of the material models, one element cyclic test simulation is performed and the obtained cyclic stress strain curves are compared with their numerical curves as shown in figure 4. It should be noted here that the simulated stress strain curves of both isotropic and Y-U models are identical to their numerical curves.

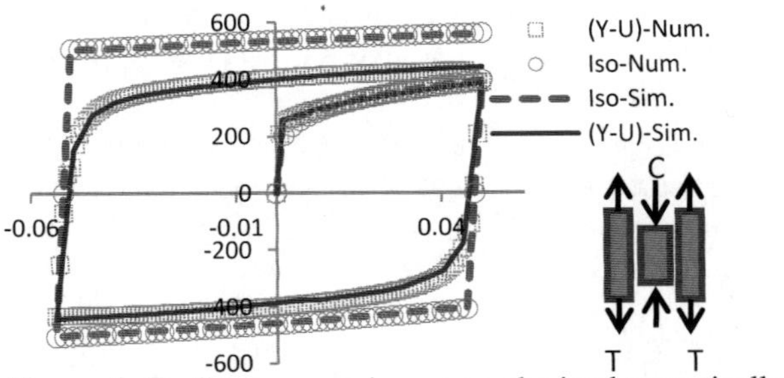

Figure 4. Cyclic stress strain curves obtained numerically and from one element cyclic test simulation T=Tension,

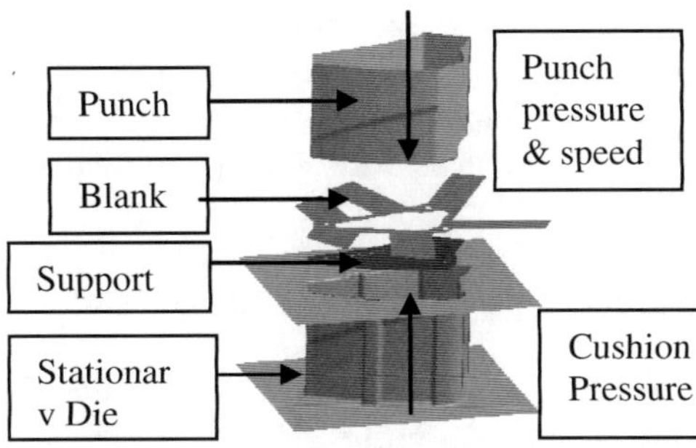

Figure 5. Tool setup with Boundary conditions the FE model.

4.2. Model setup

The blank of the part was modeled using shell elements with an element size of 3mm, thickness 1.5 mm using fully integrated analysis. The tool identification and the boundary conditions are applied as shown in figure 5. In the initial trails in the FE model, the blank is formed by a rigid punch with forming pressure of 130 bar and punch speed 15 mm/s. The support carries the blank with cushion pressure of 70 bar. The friction coefficient between all contacting surfaces is taken as 0.17.

5. Results and Discussions

5.1. Simulation of the Part with Punch Pressure of 130 Bar and Cushion Pressure of 70 Bar

Table 3 shows the predicted angles by the two selected models for the selected part. The deviations in the measurements of all angles between the formed and the simulated part are shown in figure 6. It can be seen that, the minimum deviation is obtained from Y-U model and hence is considered to be more reliable in predicting true material behavior.

Figure 7 shows a comparison between the FE model using the Y-U Model (F) and the produced part scanned by CMM (S) under the same forming conditions to validate the developed FE model for further investigations. As shown in figure 7, the FE model shows good agreement with the formed part. The maximum deviation measured is 0.7° indicating that the FE model could be adopted for further analyses.

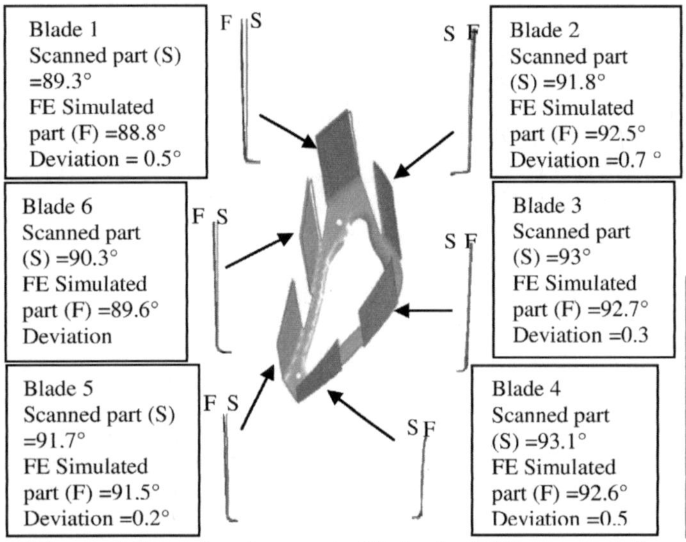

Figure 7. Comparison between the FE simulated (3D part- symbol (F)) and physical scanned part (wire frame – symbol (S)) formed at punch pressure of 130 bar and cushion pressure of 70 bar.

TABLE 3. Blades' angles Predicted by the two models.

	ANGELS					
	1	**2**	**3**	**4**	**5**	**6**
Y-U	**88.8**	**92.5**	**92.7**	**92.6**	**91.5**	**89.6**
Isotropic	**93**	**94.2**	**93.6**	**93.8**	**88.9**	**92.7**
Scanned Part	**89.3**	**91.8**	**93**	**93.1**	**91.7**	**90.3**

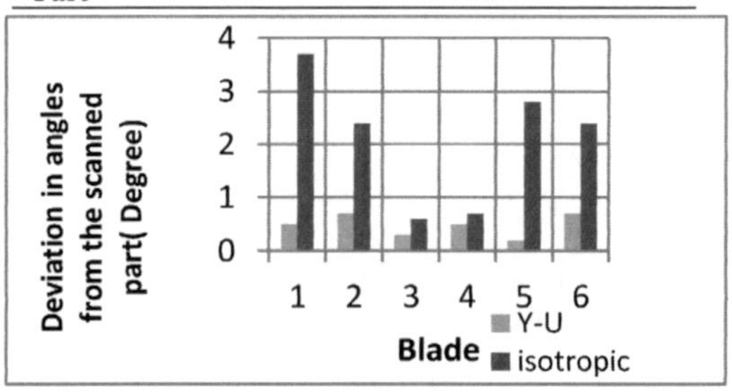

Figure 6. Deviation of the predicted angels by the material models from the formed part in degree.

FE model coupled with precise material model (Y-U model) is strongly able to predict the springback accurately. In the next step of the work, the FE model is used to determine suitable pressures combination which minimizes the springback. Several simulation trails with different forming conditions are carried out.

It is difficult to identify precisely the optimum pressure combination, but it is found that a punch pressure of 50 bar and a cushion pressure of 20 bar is the most suitable combination, which enabled the part to be formed with a minimum springback effect. To explore this further, the comparison between the FE model and scanned part at defined suitable condition is given below.

5.2. Prediction of the Part with a Forming Pressure of 50 Bar and a Cushion Pressure of 20 Bar

Figure 8 shows a comparison between the FE model and produced part for the proposed process parameters and the old ones. The FE model predicts the angles of blades 1, 3, 4, 5 and 6 successfully with very a maximum deviation 0.3° and a deviation of 0.8° for blade 2. In addition, figure 8 shows a comparison between the scanned physical part and the CAD data of the desired part (referred as (C)). The FE model results for blades 3 and 4 show angles identical to the design angle of 92°, blades 2 and 5 have deviations of 0.5° and 0.1°, respectively. However, deviations, between the FE model and the CAD data, in angle values can be seen for the blade 1 and 6. These deviations are summarized in figure 9.

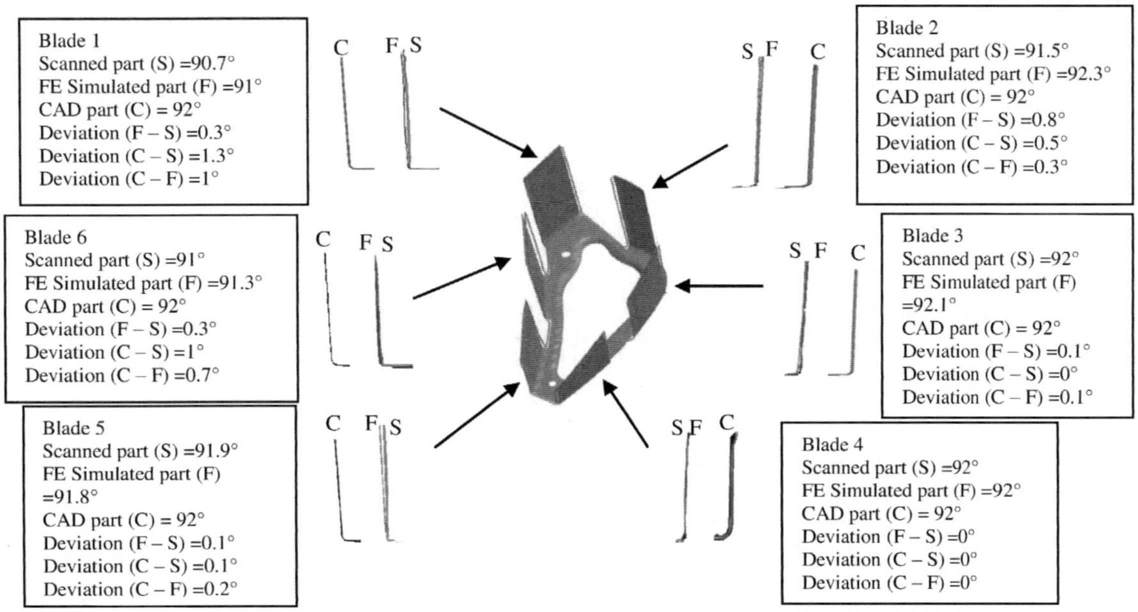

Figure 8. Comparison between the CAD Data (symbol(C)), FE simulated part (3D part-symbol (F)) and physical part (wire frame – symbol (S)) formed at punch pressure 50 bar and cushion pressure 20 bar.

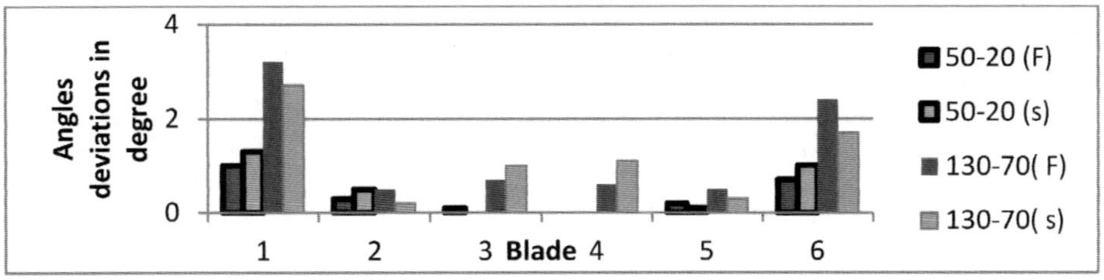

Figure 9. Deviation of the scanned (s) and FE simulated parts (F) at the two conditions (130-70) and (50-20) from the CAD part.

FE simulations coupled with advanced material models have a significant impact in developing the forming process and predicting the springback. As shown in figure 9, calibrating the FE model with the scanned formed part, fabricated with forming and cushion pressures of 130 and 70 bar (S), provide a robust model that was able to predict the springback under different forming conditions and by several simulation trails the suitable forming conditions of forming and cushion pressures of 50 and 20 bar (F) has been found to be the best pressure combinations and resulted in good agreements between the FE simulation and its experimental trail.

However, there is a little deviation between the FE simulations and the scanned parts from the actual forming process. These deviations can results from different sources of errors including inaccurate characterization of the anisotropy value of the material data that were obtained from the work of Yoshida [6] and/or errors in CMM measurements. However, despite of these errors and the resulted deviations, the results provide improvements in production line of the proposed part where the delay in assemble process caused by springback has been reduced.

6. Conclusions

In this paper, the forming process of an automotive part which affected by positive and negative springback was modeled using Pam Stamp$^{(TM)}$ and validated by comparisons against experimental trails. The accuracy of springback analysis depends on the predictions of stress levels at the final stage of stamping, and also at the springback stage. Two different material models (Isotropic and Y-U model) were used to predict the springback. It was found that Y-U model is considered to be more accurate material model in simulating bending processes and predicting springback providing maximum deviation measured (0.7^{o}) which is indicating that the FE model could be adopted for further analyses. By using the validated FE model coupled with advanced material models, the suitable pressure combination is determined which enabled the part to be formed with a minimum springback effect. The right compensation for the springback obtained by the FE simulation improves the productivity and minimize the delay caused by springback while assembling process.

Acknowledgments
The authors would like to thank INDE GROUP FOR ENGINEERING INDUSTRIES for providing opportunity to use the tool and the equipment.

References
[1] Ishikawa T and Mori K 2014 Springback analysis of high strength dual-phase steels 11th *Int. Conf. on Technology of Plasticity* Vol 81 (Japan : Congress Center) pp 975-980
[2] Yoshida T 2002 Material modeling for accuracy improvement of the springback prediction of high-strength steel sheets Nippon Steel Technical Report Vol 102
[3] INDE GROUP FOR ENGINEERING INDUSTRIES, 6OCT City, Cairo, Egypt. http://www.inde.com.eg/
[4] Pam Stamp user's guide manual 2017
[5] Ma N, UmezuY, Watanabe Y and OgawaT 2008 Springback prediction by yoshida-uemori model and compensation of tool surface using jstamp 7th *Int. Conf.and Workshop on Numerical Simulation of 3D Sheet Metal Forming Processes*
[6] Yoshida F and Uemori T 2002 A model of large-strain cyclic plasticity describing the bauschinger effect and work hardening stagnation *Int. Journal Of Plasticity* Vol 18 pp 661-686

IOP Conf. Series: Materials Science and Engineering **280** (2017) 012022 doi:10.1088/1757-899X/280/1/012022

Flow Field Characteristics and Lift Changing Mechanism for Half-Rotating Wing in Hovering Flight

Q Li, X Y Wang, H Qiu, C M Li and Z Z Qiu

School of Mechanical Engineering, Anhui University of Technology, Ma´anshan, 243002, P.R.China

liq_ahut@163.com, wangxy@ahut.edu.cn, qiuhan@ahut.edu.cn, licongmin_ahut@163.com, qzzhlx@ahut.edu.cn.

Abstract. Half-rotating wing (HRW) is a new similar-flapping wing system based on half-rotating mechanism which could perform rotating-type flapping instead of oscillating-type flapping. The characteristics of flow field and lift changing mechanism for HRW in hovering flight are important theoretical basis to improve the flight capability of HRW aircraft. The driving mechanism and work process of HRW were firstly introduced in this paper. Aerodynamic simulation model of HRW in hovering flight was established and solved using XFlow software, by which lift changing rule of HRW was drawn from the simulation solution. On the other hand, the development and shedding of the distal vortex throughout one stroke would lead to the changes of the lift force. Based on analyzing distribution characteristics of vorticity, velocity and pressure around wing blade, the main features of the flow field for HRW were further given. The distal attached vortex led to the increase of the lift force, which would gradually shed into the wake with a decline of lift in the later downstroke. The wake ring directed by the distal end of the blade would generate the downward accelerating airflow which produced the upward anti-impulse to HRW. The research results mentioned above illustrated that the behavior characteristics of vortex formed in flow field were main cause of lift changing for HRW.

1. Introduction

The movement forms of animal limbs are asymmetric swing motion in essence[1]. Swing is just a movement form that adapts to muscles for animal, whereas the asymmetry of motion is the basic condition to fly and walk. Flapping Wing Air Vehicle (FWAV) has been a topic of considerable interest for its wide application in military and civil fields[2-3]. Particularly, rich results had been got in FWAV aerodynamic research[4], such as theoretical modeling, lift experiments in different conditions and numerical simulation based on Computational Fluid Dynamics (CFD) theory. However, one insurmountable obstacle in the research for the development of FWAV in large size is the difficulty in overcoming the large inertial force generated when the flapping wing reciprocate and swing in a high speed[5]. Thence, our work team had developed a rotating mechanism named Half-Rotating Mechanism (HRM)[1], which could not only produce asymmetrical motion but avoid swing. HRW flight mechanism was reasonably developed to make HRW aircraft in large size possible.

The research on changing mechanism of the flight lift for Half-rotating Wing (HRW) is significant to determine the flight capability of HRW aircraft. The airflow on the airfoil of HRW is complicated, and the vortice formed in the flow field is the main source of lift changing for HRW. In this paper, the influence about the generation and variation of vortex of the blade on lift force for HRW was given by using the CFD software XFlow as the blade was rotating. According to the aerodynamic force of HRW and the airflow near the airfoil, deeper research on the lift characteristics of HRW was conducted to reveal its lift changing mechanism.

2. Driving mechanism and work principle of HRW

2.1. Driving mechanism of HRW

The driving mechanism of HRW was a kind of simplified HRM which was suitable for flight[6]. As shown in figure 1, point *C* was not only the center of shaft with groove but also the middle of blade of HRW. Point *H* denoted the center of positioning slider, which was the fixed point in HRM. HRW could produce continuous and certain motion when motor drove the crank to rotate. The blade of HRW would turn half round anticlockwise when crank turned one round anticlockwise, accordingly resulting in the asymmetric rotation.

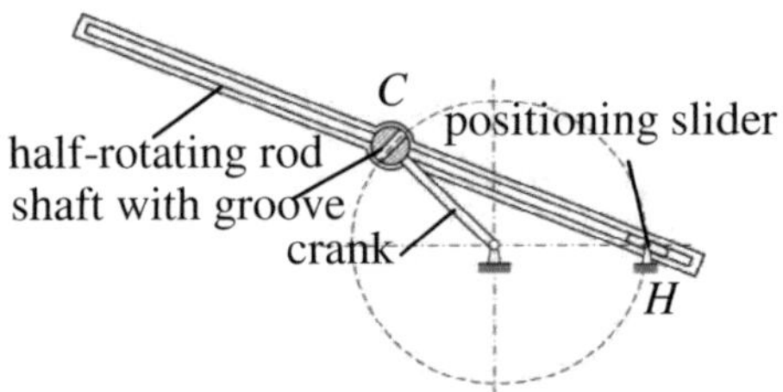

Figure 1. Driving mechanism of HRW

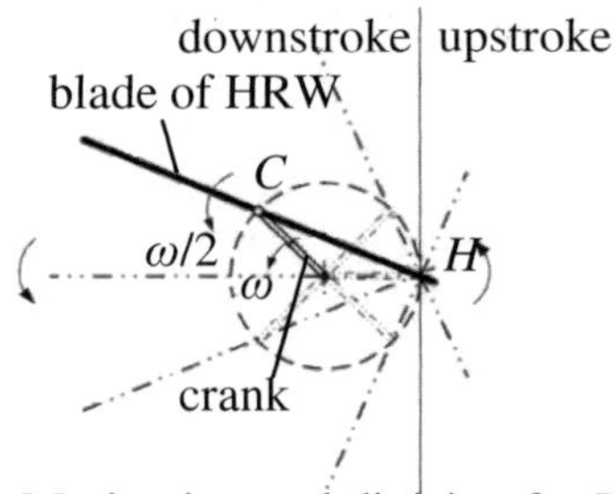

Figure 2. Motion interval division for HRW

As shown in figure 2, the blade of HRW was divided into two parts by fixed point *H* when HRW rotated counterclockwise. One part of HRW could be referred as downstroke on the left side of fixed point *H*, at which moment HRW moved downwards which was equivalent to the downstroke of insect wings. Similarly, the other part of HRW on the right side of point *H* could be called upstroke.

2.2. Work process of HRW

The blade of HRW, a working component driven by the crank, did the plane movement which could be decomposed into the translation following with shaft *C* and the rotation turning around point *C*. As shown in figure 3, *AB* represented the blade of HRW whose span and chord length were respectively $2a$ and h, while *OC* represented the crank whose length was R. The angular velocity of HRW was $\omega/2$ when the angular velocity of crank was ω. φ was indicative of the rotation angle of crank from original position where crank was in direction of horizontal-right while the rotation angle of HRW was $\varphi/2$. Our work team had developed a HRW aircraft prototype, which was symmetrically arranged with double blades as shown in figure 4. In this text, the simulation model was in reference of the prototype whose dimensions were R=0.06m, $2a$=0.28m, and h=0.21m.

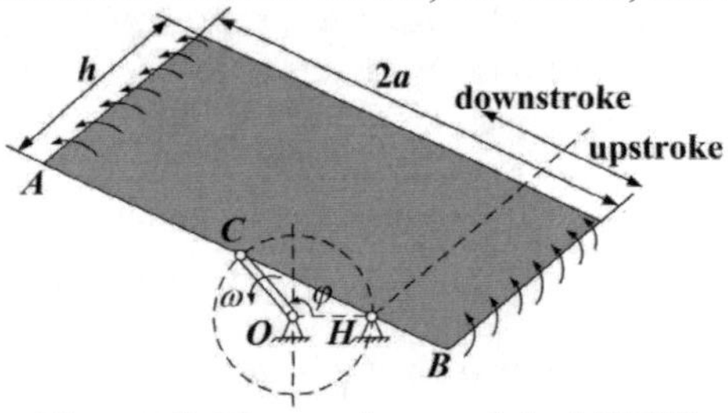

Figure 3. The motion model of HRW

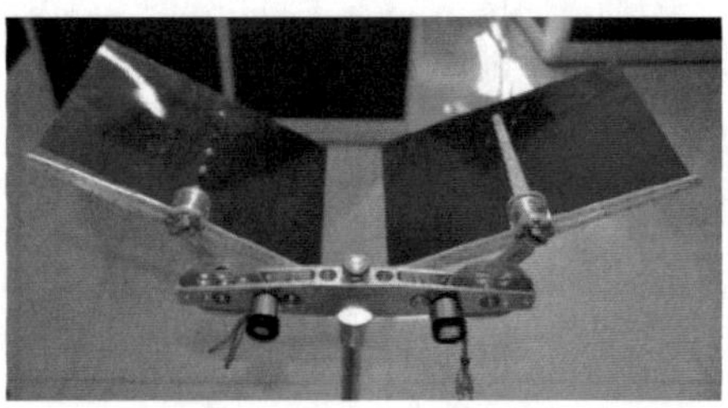

Figure 4. Aircraft prototype of HRW

3. Lift changing rules and flow field characteristics of HRW

3.1. Lift changing rules of HRW

The computing model of HRW was established based on software XFlow. The flight working condition was set as no incoming flow, namely it was in hovering flight, while ω was set as 20π rad/s. Figure 5 showed the time history of lift for HRW in 0.5s.

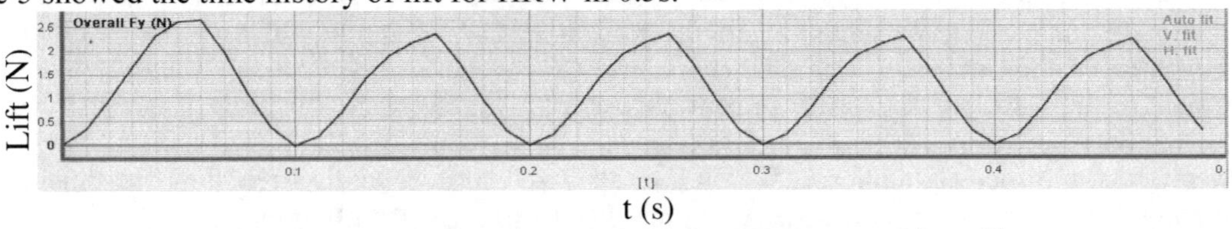

Figure 5. The time history of lift for HRW at ω=20π rad/s

Obviously, lift changed in cycle, and each stroke reached a force peak in the middle phase. At the onset of wing rotation, the lift began to rise slowly, and then reached a peak (to approximately 2.5-2.75N) followed by a relatively rapid decline. Moreover, the forces generated in the first stroke were similar to those generated during later strokes, with one distinct difference that the force peak in the first stroke was larger than later strokes.

3.2. Airflow changing rules of HRW in the flow field

Due to the different flapping mode of the blade, the lift curve of HRW was not consistent with the lift curve measured by Dickinson in Drosophila melanogaster experiment[7]. Therefore, the characteristics of the airflow near the blade were analyzed to research the relationship between the lift force and the changes of the flow field. The cross-section was created at midchord of the blade in the XFlow post-processing project tree. By gathering a complete time history of fluid velocity through five strokes, we could observe the growth of vorticity in select regions around the blade throughout each stroke. As shown in figure 6, HRW was started to rotate in the stationary flow field, and the plot of velocity at six moments in the first cycle was selected to analyze. The red point in the figure was the modeling center in XFlow, which was on the horizontal right side of the fixed point *H*, and the distance between them was 0.09m.

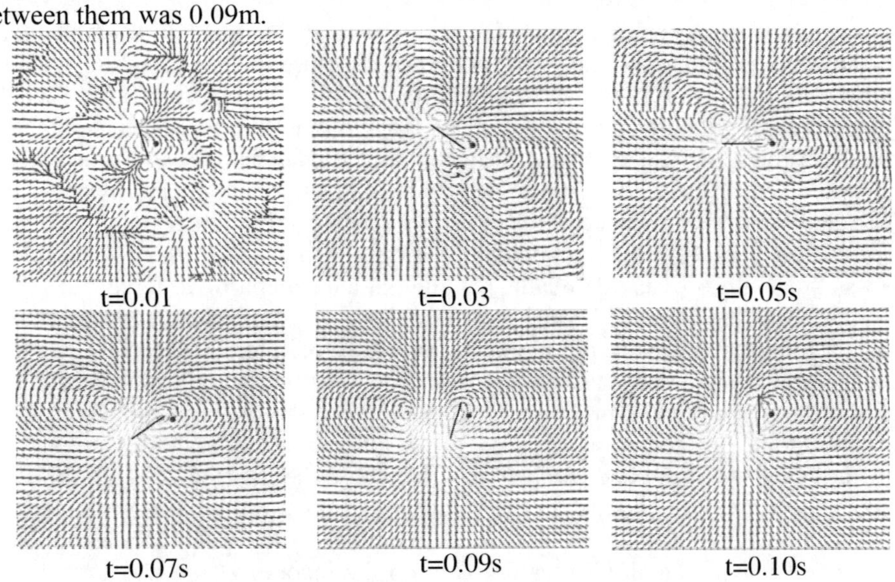

Figure 6. Velocity diagram of HRW during the first stroke

Vortex was mainly generated at the ends of the blade when HRW hovered in the air. In this paper, the end which was farther away from fixed point of the blade, namely the left end of the blade in figure 6, was called the distal end. Similarly, the right end of the blade could be called the close end. Hence, the vortex at the distal end of the blade was called the distal vortex(DV), and the other was called the close vortex(CV). Likewise, the vortex with the characteristic of attaching to the blade was called the attached vortex(AV), and the vortex which was staying at the blade was called the resident vortex(RV).

As shown in figure 6, at t=0.01s, the blade began to rotate in counter-clockwise direction, while the clockwise AV appeared at both ends of the back flow side of the blade with concomitant increase in force. Moreover, the fluid stationary state in the original equilibrium space was instantly broken by the movement of the blade and then a lift pulse was yielded. Therefore, the lift peak of the first cycle was higher than later few cycles as shown in figure 5.

At t=0.03s, the increase in force, which we attributed to acceleration-reaction movement of the distal end of the blade, occurred as the AV at the distal end significantly grew. Whilst the reverse AV at the close end was shed, leaving the RV rolled in the same direction as the blade. Accordingly, the rotational circulation rotated with the blade was produced by the RV, changing the direction of the near-wing airflow, whereas the angle between the airflow and the blade was increased. Thereby the speed which was parallel to the blade was reduced, reducing the the ability of pushing the DV out of the blade so that the DV would not shed so quickly. This prolonged attachment of the DV led to the increase in lift force.

With continued rotation, the DV was not shed until the blade was close to the horizontal position, namely it turned over 90° at t=0.05s. There were some flowing fan-shaped areas between the vortex and the rotational circulation that were consistent in orientation and size, which were called stream[8]. Such streams were often generated between vortices or vortex and surface of object. The changes of momentum of these streams would yield forces reverse to the movement of the streams on the surrounding related objects. These forces would produce moments that blocked the rotation of the blade and simultaneously yielded upward lift force against the streams. At this time the contact area of the blade and the streams was the largest, resulting in the lift peak.

From t=0.07s to t=0.09s, the DV was slowly shed while the RV was not shed during the whole stroke. At this time the contact area between the blade and the streams was small, so the lift value gradually decreased until the lift force was down to 0 by t=0.10s.

3.3. Distribution characteristics of vorticity, velocity and pressure for HRW in the flow field

The distribution characteristics of vorticity, velocity and pressure in the flow field would be further comprehended by the cloudatlas to analyze the flow field characteristics for HRW. Figure 7 showed the vorticity, velocity and pressure cloudatlas near the blade of HRW in the fourth stroke.

It was found that at t=0.01s, the AV at the distal end of the blade was formed with the wake directed by the close end generated in the prior two strokes underside of the blade in the vorticity cloudatla. The velocity of flow in the center of wake was considerably larger in the velocity cloudatla. The high pressure area was on the incoming side of the distal end of the blade, whereas the pressure of the back side was low in the pressure cloudatla, resulting in the generation in force.

At t=0.03s, the AV at the distal end was increased. The wake at the close end was elongated while the early wake was faded in the vorticity cloudatla. In the velocity cloudatla, the large flow velocity area at the distal end was increased, whereas the influence of the early wake was weakened. The high pressure area and the negative pressure were both enlarged in the pressure cloudatla, leading to the increase in lift force.

At t=0.05s, the AV at the distal end reached maximum, and the induced effect of the close end was getting weaker in the vorticity cloudatla. In the velocity cloudatla, the large flow velocity area of the AV at the distal end was continued to increase while the range of the wake area was enlarged but the flow velocity was decreased. The high pressure area on the incoming side was even bigger and the negative pressure area was also enlarged in the pressure cloudatla, leading to the increase in lift force.

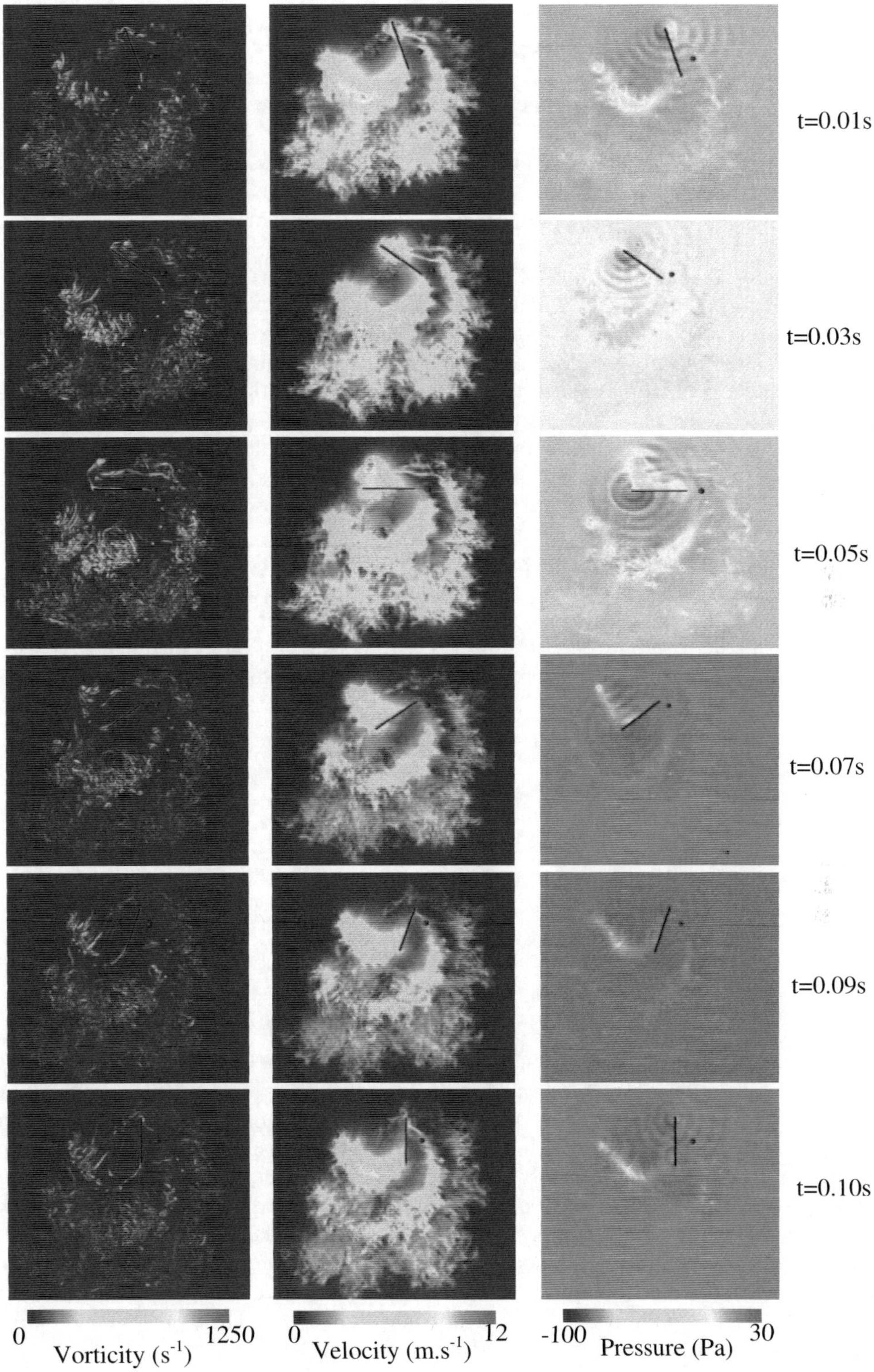

Figure 7. Vorticity, velocity and pressure cloudatlas for HRW in the fourth stroke

At t=0.09s, the AV at the distal end was shed and rolled up into a new wake, and the wake was detached from the close end in the vorticity cloudatla. In the velocity cloudatla, the range of the distal large velocity area was increased but the intensity was weakened. The high pressure area was big but the intensity was weakened, whereas the negative pressure area was smaller so that the lift force was lower in the pressure cloudatla.

At t=0.10s, the next stroke would begin and simultaneously the distal end and the close end would be swapped, resulting in a new distal AV and a large flow velocity area. The incoming side and the back side was swapped while the pressure direction was constant.

4. Lift changing mechanism for HRW

The elevated aerodynamic performance of insects has been attributed in part to the generation and maintenance of a stable region of vorticity[9]. The negative pressure on the surface of the blade would be improved if the blade was affected by the high-speed rotational vortex, leading to a great lift force. By analyzing the simulation results, the main features of the flow field for HRW were obtained. The change of velocity, pressure and lift was derived from the unique vortex, wherefore the lift changing mechanism for HRW was further analyzed on the basis of the characteristics of the vortex in this section.

4.1. Changes of the distal vortex

The vortex of HRW was first generated on the distal end of the blade, and then became gradually stronger and larger. Subsequently in the period of 0.05s-0.07s, the vortex was shed from the blade but still linked with it, forming the gradually expanding vortex ring directed by the distal end of the blade. Figure 8(a) and 8(b) were respectively the three-dimensional graphs of the vorticity in the flow field for HRW in the first and fourth strokes at t=0.07s, and the lower blade end was the distal end and another blade end was the close end.

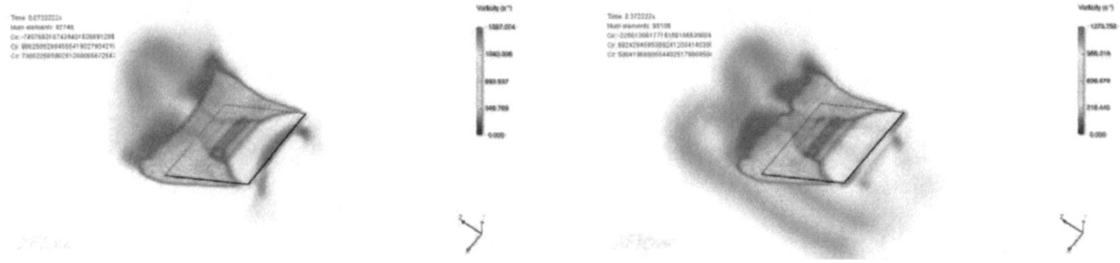

(a) (b)

Figure 8. Vortex ring at both ends of the blade

Figure 8(a) was corresponding to the moment t=0.07s in figure 6. It was obvious that the AV whose both sides were still connected with the blade after it shed from the distal end of the blade formed a small vortex ring which was gradually grew under the guidance of the end of the rotating wing. And then it formed the wake towards the underside of the blade in the later half stroke. The airflow in the vortex was upward outside and downward inside so that the downward accelerating airflow which was namely the airflow in the stream area in section 2.2 was formed in the middle of the vortex ring. This stream had a more obvious reflection in the fish parade[10]. The wake left behind the tail of swimmers was a staggered array generated as the caudal fin moved back and forth. A jet flow with alternating direction between the vortices would produce hydrodynamic force to push the fish forward. The structure of the wake was of a thrust-type, had a reversed rotational direction compared to the well-documented drag-producing Karman vortex street. Thus, HRW could still obtain upward lift by generating the downward accelerating air after the DV was shed.

Figure 8(b) was corresponding to the moment t=0.07s in figure 7. The long arc-shaped wake, which connected to the close end as shown in the bottom of the figure was formed below the blade after the distal AV of the prior stroke was shed.

4.2. Effect of the wake ring on airflow

The airflow in the vortex ring of the wake was upward outside and downward inside so that the downward accelerating airflow was formed in the middle of the vortex ring in figure 8(b). At the same time the incoming surface of the blade pushed the fluid forward, which was the driving force to accelerate the airflow. Accordingly, a directional airflow mass which flew rapidly downwards was formed in the flow field toward the bottom under the collective effect between the blade and the wake. Furthermore, the airflow could be pushed in all directions by the rotating blade but the wake was always located below the horizontal position of the blade. Consequently, most of the airflow pushed by the blade was moved downward through the wake ring.

The downward momentum of the airflow mass was added during the process of being driven by the blade and guided by the wake ring, so it was subjected to a downward impulse and produced the upward anti-impulse to the outside. Since the airflow was primarily pushed by the blade, the anti-impulse would affect on the blade. The high pressure area displayed on the incoming surface of the blade in figure 7 was the acting result of the anti-impulse, leading to the lift force of HRW.

4.3. Lift changing mechanism for HRW

It was indicated that the generation and attachment of the distal vortices on the blade caused the gradually increasing lift force by analyzing the aerodynamic and flow field characteristics for HRW. In the later downstroke, the distal vortex was still associated with the blade when it was shed, forming a vortex ring which was gradually enlarged under the guidance of the distal end. The lift force was still generated by the downward accelerating airflow in the middle of the vortex ring. However, the lift would gradually declined because of the decreasing contact area between the accelerating airflow and the blade.

After the start of the next stroke, the close end of the blade was changed to the distal end. As the continuous increase in the distal end velocity, the new AV was formed at the distal end with the increase in lift force in cycle. Moreover, the wake area was formed below the blade after the distal AV of the previous cycle was shed, and the upward anti-impulse which played a certain role on the increase of lift for HRW was produced by the downward accelerating airflow in the middle of the wake ring.

5. Conclusions

1) Aerodynamic solution showed that lift of HRW changed in cycle in hovering flight, and each stroke reached a force peak in the middle phase. Moreover, the changing rules of the lift force were consistent with the flow field characteristics of HRW.

2) The distal attached vortex produced by the rotation of the blade led to the increase of the lift force, while in later downstroke it gradually fell off from the blade with a decline of lift and then became the wake of the next stroke. The downward accelerating airflow generated in the middle of the wake ring produced the upward anti-impulse to HRW, which played a certain role on the increase in lift for HRW.

3) Distal vortex and wake ring formed by the shedding of distal vortex in the flow field were main source of the lift force for HRW during hovering flight.

Acknowledgments
This work is supported by the National Natural Science Foundation of China (Grant No. 51375014)

References

[1] Qiu Z Z 2011 *Half-Rotating Mechanism: Structure, Characteristics and Application* (Hefei: University of Science and Technology of China Press)

[2] Fearing R S, Chiang K H, Dickinson M H, Pick D L, Sitti M and Yan J 2000 *IEEE Int. Conf. on Robotics and Automation (San Francisco)* (Piscataway: Institute of Electrical and Electronics Engineers Inc) pp 1509-16

[3] Smith M 1996 *Aiaa J.* **34**(7) pp 1348-55
[4] Yang W Q, Song B F, Song W P and Chen L L 2015 *J. Exp. Fluid Mech.* **29**(3) pp 1-10
[5] Hu M L, Wei R X, Cui X F and Guo Q 2008 *J. Aero. Power* **23**(7) pp 1279-86
[6] Wang X Y, Wang H X, Qiu H, Chen F Q and Q Z Z 2017 *Mech. Sci. Technol. Aero. Eng.* **36**(7) pp 1-7
[7] Dickinson M H, Lehmann F O and Sane S P 1999 *Sci.* **284** pp 1954-60
[8] Wang B L 2013 *Aircraft Aerodynamics* (Beijing: Tsinghua University Press) pp 25-7
[9] Birch J M, Dickson W B and Dickinson M H 2004 *J. Exp. Biol.* **207**(7) pp 1063-72
[10] Sfakiotakis M, Lane D M and Davies J B C 1999 *IEEE J. Oceanic Eng.* **24**(2) pp 237-52

IOP Conf. Series: Materials Science and Engineering **280** (2017) 012023 doi:10.1088/1757-899X/280/1/012023

Study on Earthquake Response of High Voltage Electrical Equipment Coupling System with Flexible Busbar

Chuncheng Liu[1], Da Qu[1], Chongyang Wang[2], Chunlei Lv[1]and Guoqiang Li[3]

[1]Northeast Electric Power University Institute of Civil Engineering, Jilin, China
[2]Beijing Electric Power Transmission and Transformation Company, Beijing, China
[3]Guangdong Power Grid Co., Ltd.Aircraft Patrol Operation Center

E-mail: 1243306704@qq.com

Abstract. With the rapid development of technology and society, all walks of life in China are becoming more and more dependent on power systems. When earthquake occurs, the electrical equipment of substation is prone to damage because of its own structural features, top-heavy, and brittleness of main body. At the same time, due to the complex coupling of the soft electrical connection of substation electrical equipment, the negative impact can not be estimated. In this paper, the finite element model of the coupling system of the single unit of high voltage electrical equipment with the connecting soft bus is established and the seismic response is analysed. The results showed that there is a significant difference between the simple analysis for the seismic response of electrical equipment monomer and the analytical results of electrical equipment systems, and the impact on different electrical equipment is different. It lays a foundation for the future development of seismic performance analysis of extra high voltage electrical equipment.

1. Introduction

The previous seismic performance analysis of the substation electrical equipment often only studied on particular electrical equipment, and the same was true of traditional shaking table tests. However, in the actual structure, the electrical equipment constituted the substation system, and there was soft bus connection between the equipment. The simple analysis of seismic performance of the equipment monomer would ignore the coupling effect of the soft bus and had an effect on the analytical results. After wenchuan earthquake, the investigation of seismic damage showed that the complex coupling of soft busbar is one of the main reasons for the destruction of electrical equipment [1]. In 2004, Song J studied on the interaction between the equipment and devices connected by cable, and pointed out the dynamic amplification of the connecting wire to the electric equipment can not be ignored [2]. In the same year, domestic scholar Li Yaqi studied on the influence of conductor on seismic performance of electrical equipment, the results showed that wires significantly increase the seismic response of the equipment [3]. In April of 2013, Liu Min studied on the seismic performance of GIS high voltage electrical equipment, and the feasibility of test scheme was verified and its finite element model was established by comparing the shaking table tests with the finite element model [4]. However, these studies were only aimed at the single subject of electrical equipment, and the actual complexity was not completely reacted because of some limitations and lack of persuasion. This paper applied the structural system of practical substation design and studied on specific reaction of the structure for further study on the seismic performance.

2. Modelling of high voltage electrical equipment

This paper selects three groups of high voltage electrical equipment, including high voltage isolating switch, high voltage circuit breaker, high voltage current transformer, three sets of electrical equipment are connected with soft bus, which consists of the substation power transmission system. The finite element model was established by Ansys. Figure 1 gives the structural size of three groups of electrical equipment.

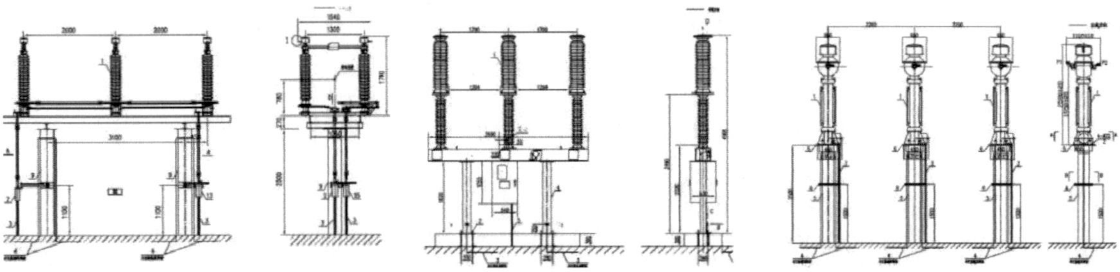

(a) Isolating switch (b) Circuit breaker (c) Mutual inductor

Figure 1. The structural size of high voltage electrical equipment

The high voltage isolator selects GW4-126DW, mainly including an insulation porcelain column, knife, aluminum alloy bracket, cement pillar, and the finite element model applies SOLID45 element; High voltage circuit breaker selects LW-126, mainly including a lower bracket, a control box, aluminum alloy flange, porcelain column, wiring board, wherein the control cabinet uses SHELL281 element, and remainder applies SOLID45 unit; Figure 2 gives the high voltage current transformer selects LVB-110, mainly including the lower cement column, bracket, ceramic column, flange and et al. All use SOLID45 unit.

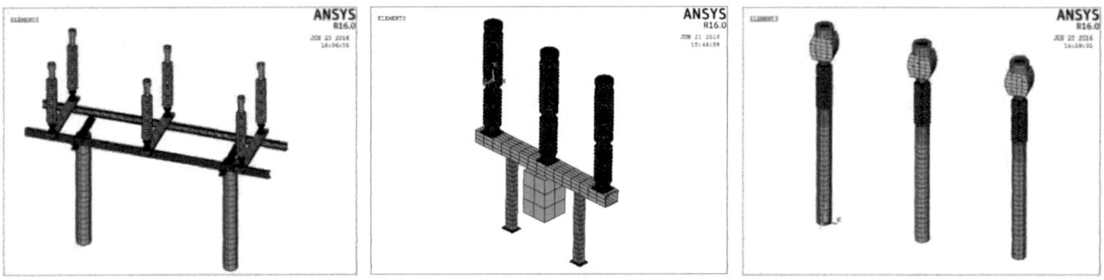

(a) Isolating switch (b) Circuit breaker (c) Mutual inductor

Figure 2. The finite element model of high voltage electrical equipment

3. Analysis of seismic response of high voltage electrical equipment monomer

3.1. Selection of earthquake waves

In this paper, the analysis of the seismic response of electrical equipment was conducted based on El-Centro and Taft two groups of seismic wave, each group of seismic wave contains two components, as shown in Figure 3 and Figure 4. El-Centro is the actual recorded seismic wave of the Imperial Valley earthquake in California of United States in 1940, which is suitable for the type of soil II; Taft is the actual recorded seismic wave in California of United States in 1952, which is suitable for the type of soil III.

3.2. Analysis of seismic response of electrical equipment monomer

The transient time history analysis of seismic response of high voltage electrical equipment is conducted based on Ansys. Porcelain column root of high voltage isolating switch, connection of flange, and the support joint of stress concentration are prone to damage, and the maximum

IOP Conf. Series: Materials Science and Engineering **280** (2017) 012023 doi:10.1088/1757-899X/280/1/012023

displacement is at the top, and the side pole displacement is intermediate pole column. Figure 5 and Figure 6 shows the time history curve of top displacement and acceleration of B pole under different seismic waves, the curves indicate that the equipment is relatively significant in the earthquake phase displacement and acceleration.

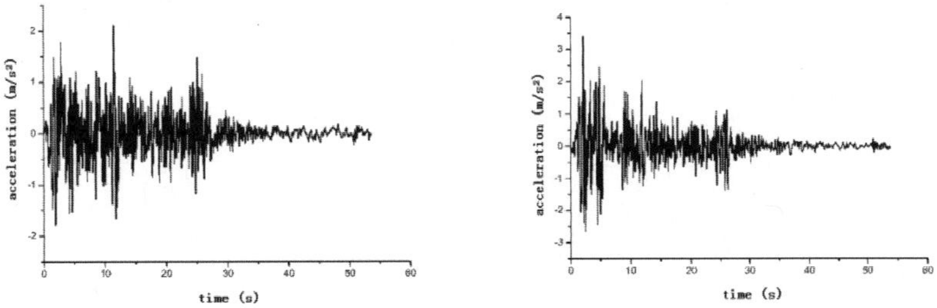

(a) E-W component (b) N-S component

Figure 3. Acceleration of EL-Centro seismic wave

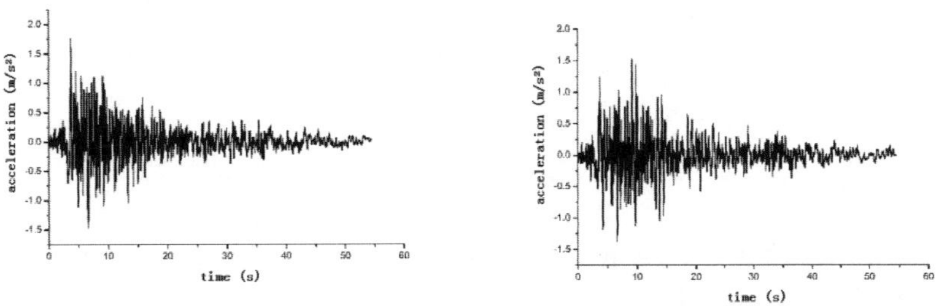

(a) E-W component (b) N-S component

Figure 4. Acceleration of Taft seismic wave

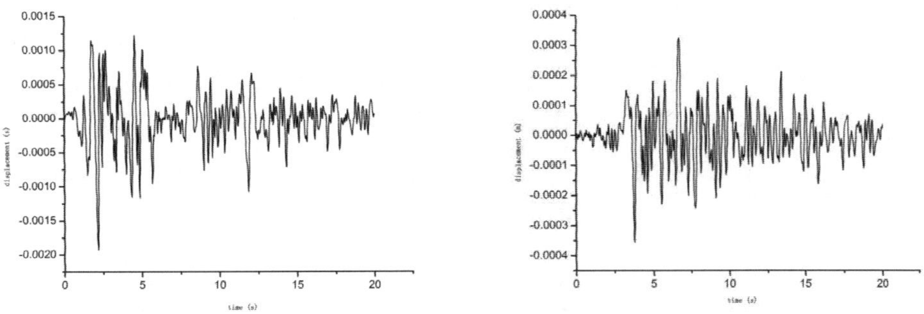

(a) EL-Centro wave (b) Taft wave

Figure 5. Displacement curve

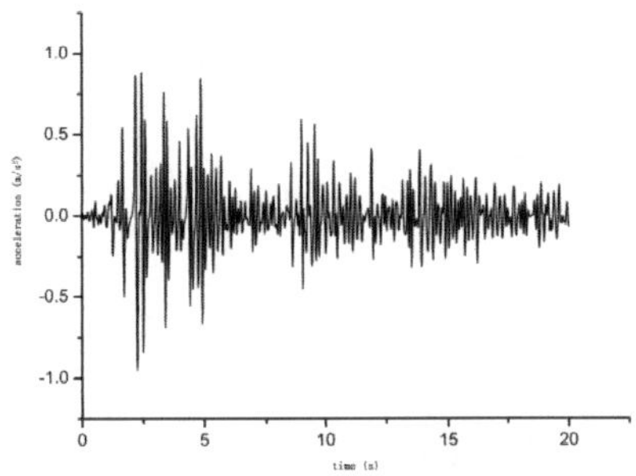

(a) EL-Centro wave

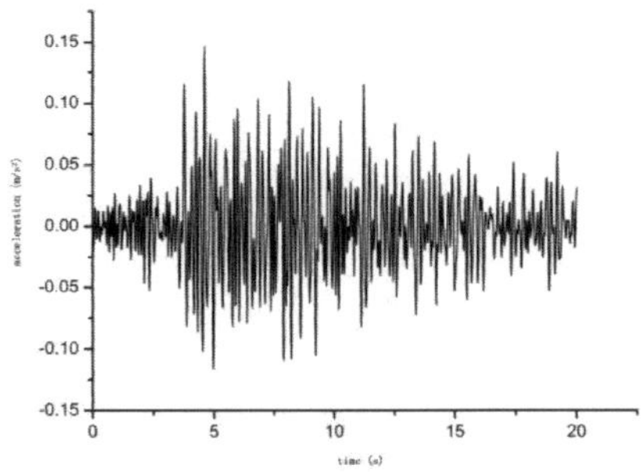

(b) Taft wave

Figure 6. Acceleration curve

Table 1. The stress of circuit breaker three-phase polar column under the seismic action/Pa

Terminal	Direction	EL Centro Seismic wave		Taft Seismic wave	
		Minimum stress	Maximum stress	Minimum stress	Maximum stress
A-pillar	X Direction	-816985	757606	-314145	286336
	Z Direction	-655480	613028	-252771	225969
B-pillar	X Direction	-229794	205720	-108192	82843.2
	Z Direction	-378321	312740	-163768	125569
C-pillar	X Direction	-147303	161028	-115583	86528.2
	Z Direction	-224396	195077	-88207.8	70657.5

The quality of high voltage circuit breaker has much greater mass than isolating. With the comparison of electrical equipment, the top displacement of pole is larger under the external seismic loading. Three poles of high voltage circuit breaker are asymmetric installed on the same bracket, the top displacement exists difference. Table 1 gives the root stress of poles. From table 2, we can see that the x direction displacement of three poles of high voltage circuit breaker is very small, the z direction displacement is larger, and the maximum displacement occurs in the column A under El-Centro earthquake wave. The x direction displacement is smaller, the z direction displacement is large, and the displacement of column A and C is larger and the values are similar under Taft earthquake wave. Therefore, under the earthquake load, three poles broken most easily occurred in column A and C, and column A more easily damaged in a strong earthquake [5].

Table 2. The top displacement of circuit breaker three-phase polar column/mm

Seismic wave	Direction	Displacement		
		A-pillar	B-pillar	C-pillar
EL-Centro Seismic wave	X Direction	0.696	0.587	0.505
	Z Direction	17.84	15.368	12.979
Taft Seismic wave	X Direction	0.69	0.43	0.7
	Z Direction	11.4	11.2	11.7

Figure 7 and Figure 8 gives the response of displacement and acceleration of the top of breaker pole under different seismic waves. We can see that the peak value of displacement and acceleration response is larger than isolating switch, main reason is that the whole structure of the circuit breaker is unevenly distributed in quality.

IOP Conf. Series: Materials Science and Engineering **280** (2017) 012023 doi:10.1088/1757-899X/280/1/012023

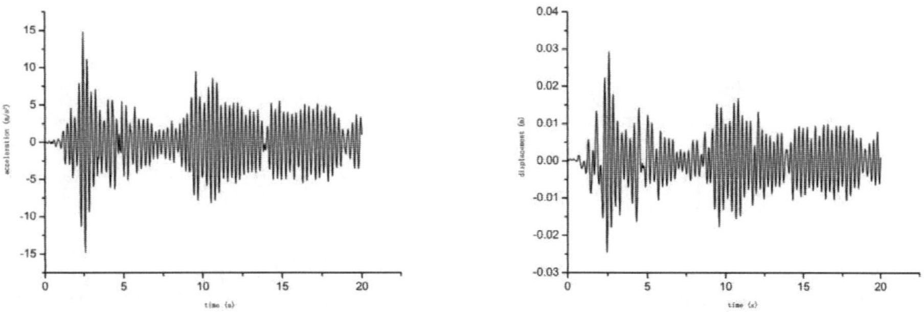

(a) EL-Centro Seismic wave (b) Taft Seismic wave

Figure 7. The displacement time history curve of the high voltage circuit breaker

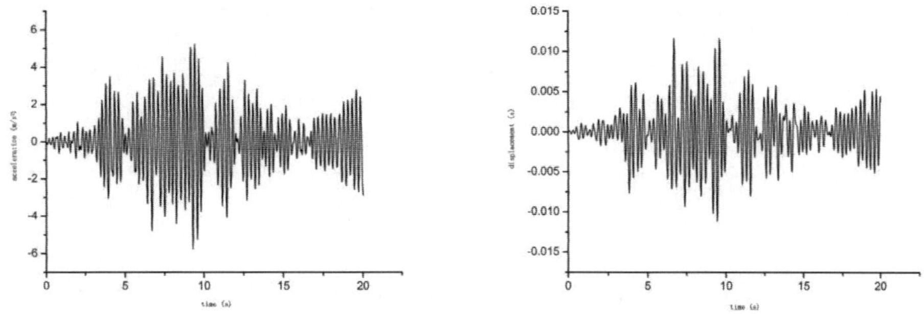

(a) EL-Centro Seismic wave (b) Taft Seismic wave

Figure 8. The acceleration and time curve of circuit breaker

The voltage transformer belongs to an independent structure. In the coupling system, three groups of transformers are required to work together, and each group of transformers is independent, and there is no interaction. Table 3 shows the maximum value of displacement and acceleration response of transformers in x and z direction under different seismic waves. Numerical analysis shows that the z displacement and acceleration response of the transformer are larger, and the x is relatively small.

Table 3. The response of voltage transformer under seismic wave

EL-Centro				Taft			
X Direction		Z Direction		X Direction		Z Direction	
Displacement (mm)	Acceleration (m/s^2)	Displacement (mm)	Acceleration (m/s^2)	Displacement (mm)	Acceleration (m/s^2)	Displacement (mm)	Acceleration (m/s^2)
2.47	0.686	4.65	1.763	1.106	0.383	1.08	0.5222

Table 4. The parameters of the conductor LGJ240/30

Calculated outside diameter (mm)	Direct-current resistance (Ω/km)	Quality (kg/km)	Elastic modulus (N/mm^2)	L-expansion coefficient (1/ ℃)	Breaking force (N)	Instantaneous failure stress (MPa)
21.60	0.1181	922.2	72569	19.6×10-6	75620	274.02

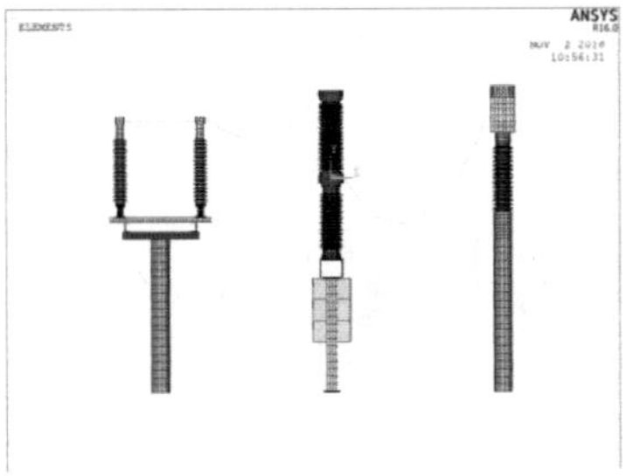

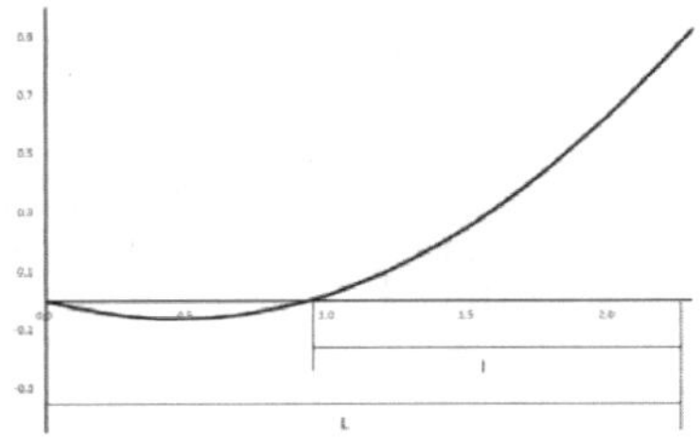

Figure 9. Finite element model of monomer flexible system **Figure 10.** Flexible busbar shape diagram

4. Analysis of seismic response of high voltage electrical equipment system connected with soft bus

4.1. Finite element model of soft bus coupling system

The soft bus is made of the conductor LGJ240/30, and its main parameters are shown in Table 4. According to its main parameters, the finite element modelling is carried out by using link180 unit. The finite element model of monomer flexible system is shown in Figure 9.

Table 5. Seismic response of different linear flexible busbars

C_1-C_2	Isolating switch		Circuit breaker		Mutual inductor	
	Displacement (m)	Acceleration (m/s^2)	Displacement (m)	Acceleration (m/s^2)	Displacement (m)	Acceleration (m/s^2)
0.2-0.2	0.01363	3.6022	0.02142	5.19285	0.01495	6.53148
0.4-0.2	0.02815	5.10595	0.01972	5.35657	0.01495	6.5323
0.6-0.2	0.03342	5.72878	0.02175	5.62671	0.01495	6.53176
0.8-0.2	0.03517	5.93835	0.02252	5.73394	0.01495	6.53162
1.0-0.2	0.03583	6.00856	0.02283	5.77684	0.01495	6.53158
1.0-0.4	0.03545	5.97743	0.02498	6.12407	0.01452	6.5294
1.0-0.6	0.03543	5.9717	0.02567	6.22499	0.01461	6.78012
1.0-0.8	0.03544	5.97061	0.02585	6.2456	0.01479	7.02982
1.0-1.0	0.03544	5.9702	0.02590	6.24445	0.01498	7.25598

4.2. Analysis of seismic response of coupled system

Figure 10 gives a soft bus shape diagram, where L is the distance between electrical equipment, and l is the distance between the horizontal intersection point on the right side of the soft bus and the electrical equipment on the right. C is defined as the ratio of l to L ($C = l / L$), C_1 is obtained from the disconnecting switch and the high voltage circuit breaker. C_2 is obtained from the voltage transformer and high voltage circuit breaker. The C values are taken from 0 to 1, and select one group every 0.2. Then further study on the seismic response of system under different C values.

The table 5 shows the response of displacement and acceleration of coupling system, isolating switch, circuit breaker and mutual inductor.It can be seen that the influence of different soft bus length connection on the electrical equipment is different, and there is certain regularity.

Figure 11 shows the regularities of displacement acceleration response of three kinds of electrical equipment under the coupling system.The analysis results show that the acceleration and displacement response of the high voltage isolation switch increases with the increase of C_1 value, and decreases with the increase of C_2 value, and the response is the biggest when C_1 and C_2 is 1.0-0.2;The acceleration and displacement response of the high voltage circuit breaker increases with the increase of the value of C_1 and C_2, and the response is the biggest when C_1 and C_2 is 1.0-1.0;The acceleration response of voltage transformer decreased gradually with the increase of C_1, increased with the increase of C_2, the displacement response decreased gradually with the increase of the C_1, decreased

IOP Conf. Series: Materials Science and Engineering **280** (2017) 012023 doi:10.1088/1757-899X/280/1/012023

slowly with the increase of C_2 at first then gradually increased, and the response is the biggest when C_1 and C_2 is 1.0-1.0.

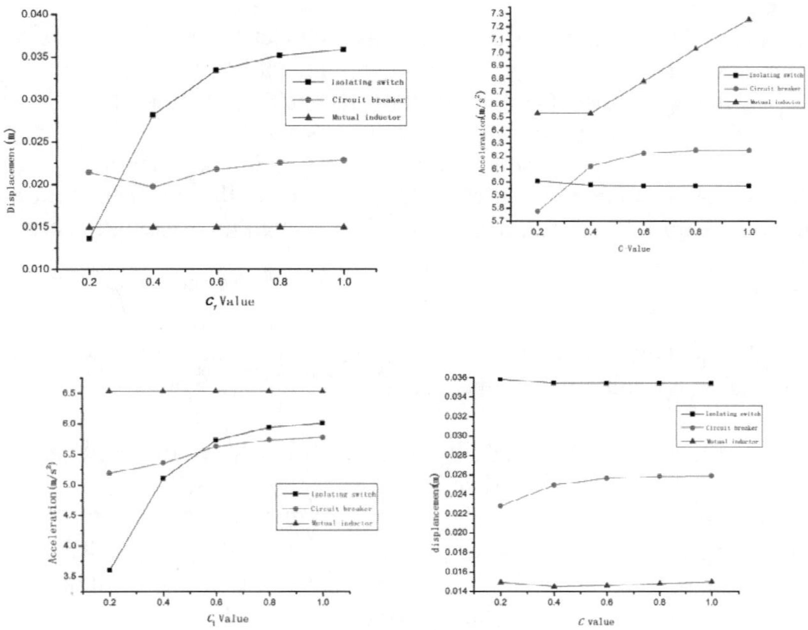

Figure 11. The regularities of electrical equipment response

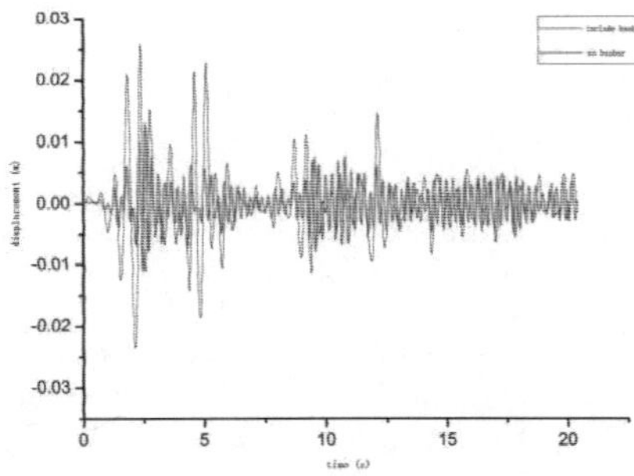

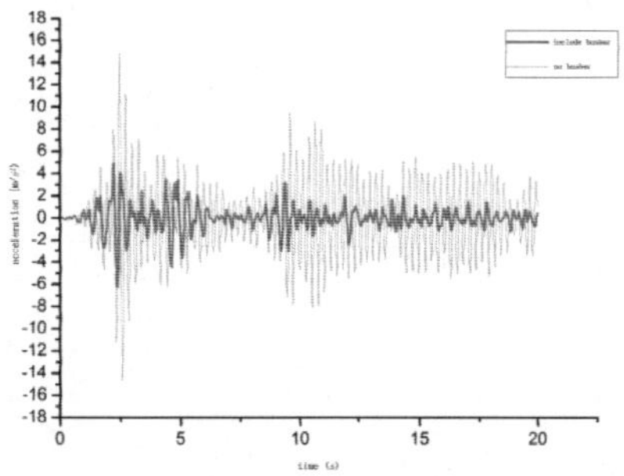

Figure 12. Comparison of displacement time history **Figure 13.** Comparison of acceleration time history

Figure 12 shows the displacement time history diagram of the high voltage circuit breaker connecting the soft bus and the unconnected soft bus. It can be seen that the displacement increases greatly during the period of seismic excellence when connects the soft bus; When the soft bus is connected, the acceleration of the coupling system is smaller than that of the single electric equipment, although the acceleration decreases, the excessive displacement will cause the soft bus to break the electric equipment, as shown in Figure 13.

5. Conclusions

In this paper, the finite element model of high voltage disconnecting switch, high voltage circuit breaker and voltage transformer is established. The main conclusions are as follows:

(1)The electric equipment belongs to large quality high-rise structure, and porcelain columns of electrical equipment are subject to greater outward tension under seismic loading because of the connection of soft bus coupling, which leads to brittle fracture at the joint between the flange and the porcelain column.

(2)The influence of the length of different soft bus connections on the main structure is different. The long soft bus has little influence on the top displacement of the main body, and the shorter bus will greatly increase the top displacement and reduce the peak acceleration at the same time.

(3)Through the time history analysis of soft bus, it is concluded that the connection of soft bus increases the peak value of displacement to some extent, while the vibration attenuation has little effect on the displacement. When the soft bus is not connected, the top acceleration of the equipment monomer is larger, and after the connection of the soft bus coupling system, the acceleration is smaller in the vibration period.

References

[1] Yongfeng Cheng, Quanjun Zhu, Zhicheng Lu. Current situation and development trend of seismic measures for substation power facilities [J]. Power System Technology, 2008, 32(22): 84-89.
[2] Song J and Kiureghian A D. Reliability of electrical substation equipment connected by rigid bus[J]. 2004.
[3] YaqiLi, Xiaojun Li. Effect of lead on the seismic response of electrical equipment [J]. Earthquake engineering and engineering vibration, 2004, 24(5): 127-130.
[4] MinLiu. Experimental study on seismic performance of GIS high voltage electrical equipment [D]. Xi'an University Of Architecture And Technology, 2013.
[5] Chuncheng Liu, Chongyang Wang. Seismic response and vibration control of high voltage electrical equipment based on TMD damper [J]. Science technology and engineering, 2016(35): 244-250.

Development of A New Automotive Active Suspension System

Eng. Yousef Abdulhammed[1,a] and Prof. Dr Eng. Hisham Elsherif[2,b]

[1]German University in Cairo, Mechatronics Engineering Department, Cairo, Egypt
[2]German University in Cairo, Head of Industrial Automation Department, Cairo, Egypt

[a]Yousef.abdulhammed@gmail.com; [b]Hisham.elsherif@guc.edu.eg

Abstract. The main objective was to develop a smart new vehicle suspension system that minimizes the road irregularities impact on the driver, also to increase performance and stability of the vehicle at high speeds. The central idea is based on modifying the normal passive suspension system into a computer controller hydraulic actuated active suspension system simply by adding a new component such as a hydraulic cylinder on a normal passive system. The new suspension system is economical to be wildly used in consumer's cars with low prices. The new added components was analytically tested and modeled according to different parameters. A new test rig was implemented to simulate a real quarter suspension system. The new suspension model was controlled by feedback controller according to the road conditions; the controller output controls the cylinder actuator to compensate the road oscillations and increases the vehicle stability for the passenger. Finally, to maximize the aerodynamics coefficients of the vehicle during high speeds by controlling the vehicle clearance level from the ground to achieve full stability, steering and fuel economy.

1. Introduction

Suspension system is the most important part which heavily affects the vehicle performance and ride quality. Traditional suspension systems use sets of springs and sets of dampers to decrease the vehicle vibration called passive suspension system as shown in figure 1. [1-2]

For the active suspension systems shown in figure 2 can automatically vary its height to maintain the vehicle steady have been an active area of research over the last two decades to achieve a better compromise during varying driving conditions. The active suspension system involves multiple complex technical problems, such as mechanology material science and electronics. [3]

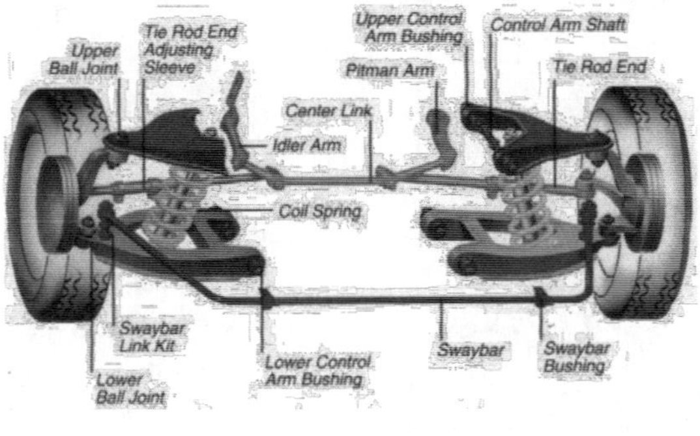

Figure 1. General Suspension System

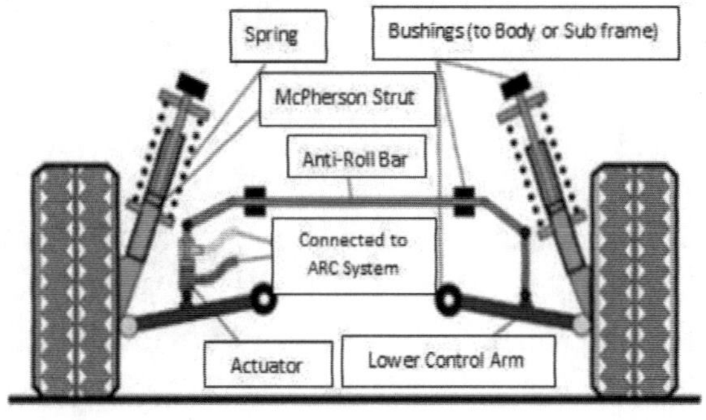

Figure 2. Active Suspension Systems

IOP Conf. Series: Materials Science and Engineering **280** (2017) 012024 doi:10.1088/1757-899X/280/1/012024

An active suspension more effectively reacts to the change in speed and while the car is cornering too, so that to make the ride to be more comfortable and steady, however it is more expensive than the passive suspension system [4].

This paper aims to develop a new suspension system to achieve passive and active suspension system advantages such as performance efficiency and cost.

2. The Design of the new Suspension System

Passive suspensions as illustrated in figure 3 achieves good ride comfort or good road holding since these two criteria conflict each other and require different spring and damper characteristics[5].

A significant improvement can be achieved by using of an active suspension system presented in figure 4 which supplied a higher power from an external source (Hydraulic Cylinder) to generate suspension forces to achieve the desired performance. The force is a function of several variables which can be measured or remotely sensed by various sensors, so the flexibility can be greatly increased.

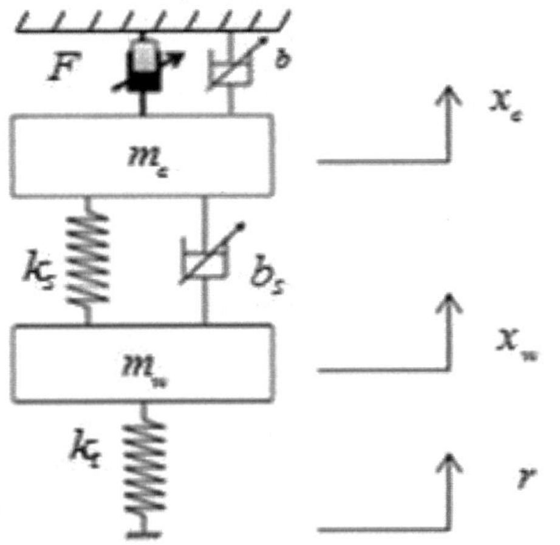

Figure 3. Passive suspension system

Figure 4. The New Designed Active Suspension System

To understand the new suspension system in a fundamental and quantitative way, a mathematical model shown in figure 5 must be obtained to represent the real world problem in mathematical form with some simplified assumptions. The mathematical model is also important to make sure that the new suspension system is controllable and observable.

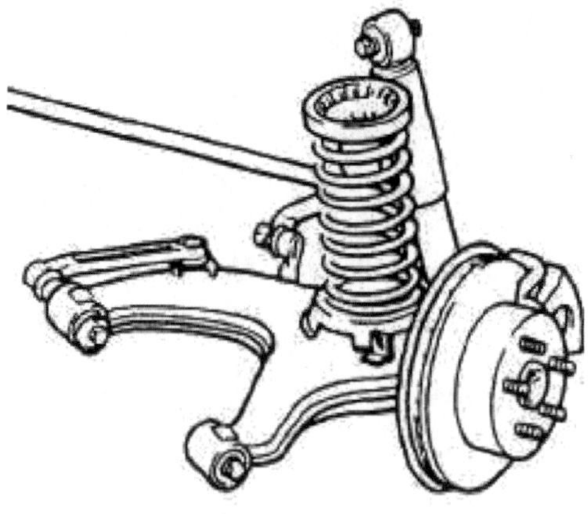

Figure 5. The New Designed Active Suspension System Mathematical Model

Automotive companies are challenging to make more developed cars, while comfort of passengers is an important demand and everyone expects from industries to improve it day by day. Therefore, in order to provide a smooth ride and satisfy passengers comfort while maintaining the high performance of a vehicle designing a modern suspension system is essential. A good and efficient suspension system must rapidly absorb road shocks and then return to its normal position, slowly, during low speeds however at high speeds the suspension system should increase the stability and performance by lowering the vehicle's clearance level from the ground [6-7].

3. Suspension System Control Mechanism

In order to fulfill the objective of designing a new active suspension system i.e. to increase the ride comfort, road handling, and performance, two control systems are developed.

The first is the disturbance road absorption control system that absorbs the oscillation on the vehicle's passenger from the road irregularities to increase the ride comfort. The other is the height adjustment control system that adjusts the clearance level from the ground of the vehicle according to its speed to increase the stability and the performance.

The Disturbance Absorption Cylinder control system consists of PID Controller, Ultrasonic Sensor, and an interface circuit.

The PID Controller input signals are the Ultrasonic Sensor and the Wheel Speed Sensor output signal. The error signal shown in figure 6 of the PID controller is the difference between the output signal of the Ultrasonic Sensor to the reference set point. Then according to the calculated error the PID Controller outputs a signal shown in figure 7 to the disturbance absorption cylinder's solenoid valve accordingly through the interface circuit, an Amplifier.

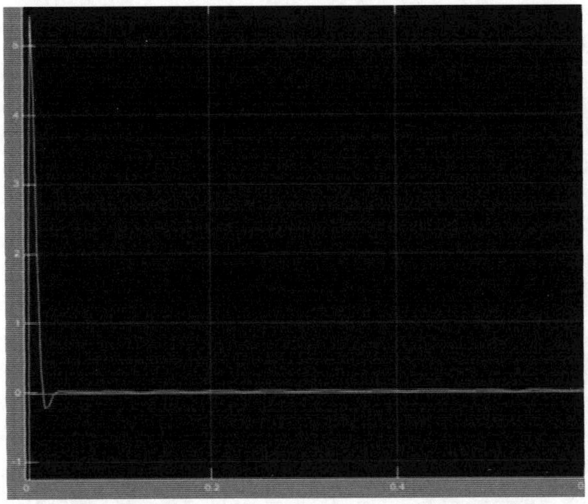

Figure 6. PID Controller error elimination: The blue line is the optimum error signal and the yellow line is tuned the error of the PID Controller

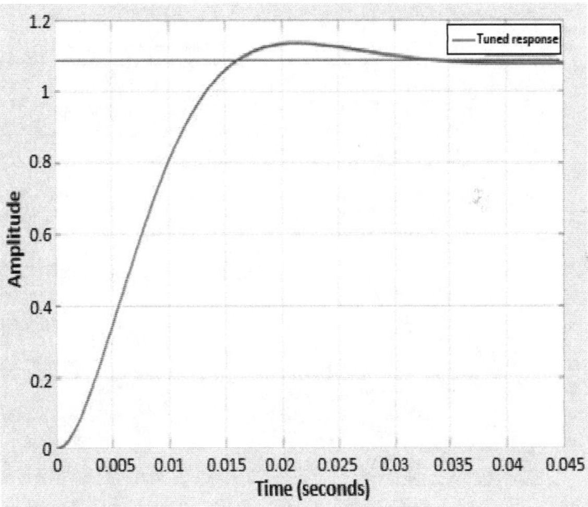

Figure 7. PID Controller Step Plot: Reference Tracking

The red line is the reference tracking constant meters and the blue line is the tuned response of the PID controller.

In the new active suspension system the ultrasonic sensor function is to measure the distance from the damper to the road disturbance absorption cylinder. The Ultrasonic Sensor readings output signal is shown in figure 9. The position of the ultrasonic sensor on the new suspension system is shown below in figure 8.

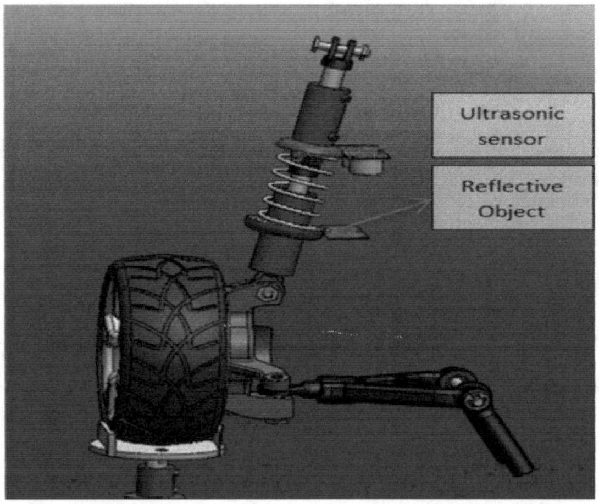

Figure 8. Passive Suspension System

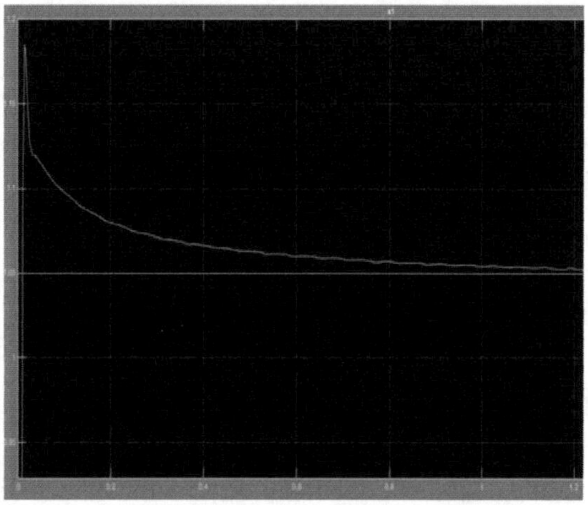

Figure 9. Displacement Position of m_c

In figure 9 the blue line is the optimum position of m_c , vehicle's displacement position which is 1.05 meters and the yellow line is the actual displacement position of the m_c , the Ultrasonic Sensor readings output.

For the Height Adjustment control system, to prove the fact that lowering the vehicle's height as the speed increases maximizes the vehicle's efficiency by decreasing the lift and drag coefficients; three Solid Works flow simulation tests are performed on a vehicle, shown in figure 10, designed according to the specifications of the minimum and maximum clearance level from the ground.

Figure 10. Solid Works Flow Simulation Test Vehicle

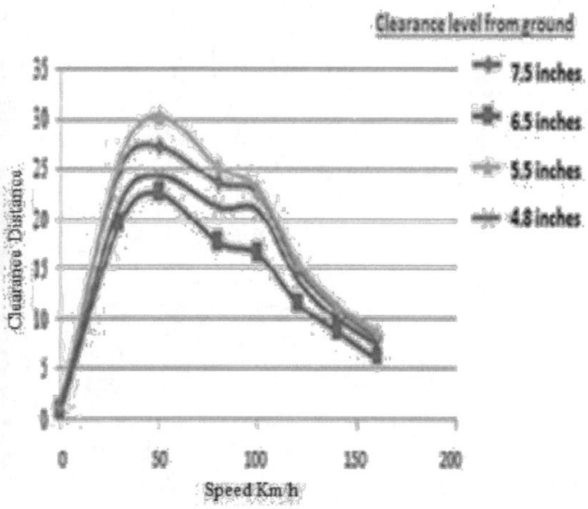

Figure 11. Solid Works Flow Simulation Test Vehicle

After applying the flow Solid Work simulation test on the vehicle at different clearance levels from the ground, comparisons of the lift and drag coefficients are shown in table 1 and in figure 11 to prove that as the clearance level from the ground decreases the lift and drag coefficients decrease thus increasing the vehicle's performance, stability, and fuel economy.

Table 1: Comparison of the Lift and Drag Coefficients for Normal, Mid, and Low Modes

Vehicle's clearance level from the ground	Lift	Drag	Comparison
Normal mode (clearance 6.5 inches)	**169.29**	**266.11**	**Low**
Mid mode (clearance 5.5 inches)	**132.94**	**262.69**	**Lower**
Low mode (clearance 4.8 inches)	**26.91**	**176.06**	**Lowest**

The height Adjustment Control System consists of an Arduino Controller, Wheel Speed Sensor, and an interface circuit.

The Arduino Controller input signal is the wheel speed sensor output signal. This signal is digital pulses with different frequencies according to the speed of the vehicle function. Then according to this output signal the Arduino Controller outputs a signal to the disturbance absorption cylinder's solenoid valve through the interface circuit, an Amplifier, and a Low pass filter.

In the height adjustment controller, the road disturbance absorption cylinder interface circuit connects the Arduino to the Directional proportional control valve. The interface circuit consists of an amplifier and a low pass filter.

4. Conclusion

A unique automotive active suspension system design, control, and implementation was the main thesis objective, to go through this task, the following research was covered through the thesis chapters.

Firstly different suspensions were analyzed from pervious researches. According to the variable gain for the car body velocity and acceleration, the combined suspension system with a hydraulic cylinder shows the best performance.

Then the PID controller was applied to let the suspension system with hydraulic actuator track the ideal force generated from the combined suspension system with hydraulic actuator. The performance of the four states including sprung mass position, velocity, unsprung mass position and velocity are worked on.

After that the electrical model control is applied to the suspension system with the hydraulic actuator. The control input is designed based on the model of PID control principle to minimize the tracking error between the actual force generated by hydraulic actuator and ideal force obtained by the combined suspension system with hydraulic cylinder.

Finally the practical implementation of the whole new suspension system was tested and compared to the theoretical results of the passive suspension system and new suspension.

References

[1] Abdolvahab Agharkali,Ghobad Shafiei Sabet and Armin Barouz, Simulation and Analysis of Passive and Active Suspension System Using Quarter Car Model for Different Road Profile, International Journal of Engineering Trends and Technology-Volume3Issue5-2012.

[2] Hongkun Zhang, Hermann Winner, and Wenjun Li, Comparison between Skyhook and Minimax Control Strategy for Semi-active Suspension System, World Academy of Science, Engineering and Technology 31 2009.

[3] J. De Jesus Lozoya-Santos, R. Morales-Menendez, J.C. Tudn-Martnez, O. Sename, L. Dugard, and R. Ramirez-Mendoza. Control strategies for an automotive suspension with an mr damper. volume 18, pages 1820{1825, 2011.

[4] B.L.J. Gysen, T.P.J. van der Sande, J.J.H. Paulides, and E.A. Lomonova. Eciency of a regenerative direct-drive electromagnetic active suspension. Vehicular Technology, IEEE Transactions on, 60(4):1384 {1393, may 2011.
[5] B. L. J. Gysen, J. L. G. Janssen, J. J. H. Paulides, and E. A. Lomonova, "Design aspects of an active electromagnetic suspension system for automotive applications," IEEE Trans. Ind. Appl., vol. 45, no. 5, pp. 1589–1597, Sep./Oct. 2009.
[6] W. D. Jones, "Easy ride: Bose Corp. uses speaker technology to give cars adaptive suspension," IEEE Spectr., vol. 42, no. 5, pp. 12–14, May 2005.
[7] O.Gendelman, Y. Starosvetsky, and M. Feldman Attractors of harmonically foced linear oscillator with attached nonlinear energy sink 1: Description of response regimes, Nonlinear Dynamics 51(2008). pp. 31-46.

IOP Conf. Series: Materials Science and Engineering **280** (2017) 012025 doi:10.1088/1757-899X/280/1/012025

Parameter Identification of Static Friction Based on An Optimal Exciting Trajectory

X Tu, P Zhao and Y F Zhou*

State Key Lab of Digital Manufacturing Equipment & Technology, Huazhong University of Science & Technology, Wuhan, Hubei, 430074, China

* corresponding author, E-mai: yfzhou@hust,edu.cn

Abstract. In this paper, we focus on how to improve the identification efficiency of friction parameters in a robot joint. First, the static friction model that has only linear dependencies with respect to their parameters is adopted so that the servomotor dynamics can be linearized. In this case, the traditional exciting trajectory based on Fourier series is modified by replacing the constant term with quintic polynomial to ensure the boundary continuity of speed and acceleration. Then, the Fourier-related parameters are optimized by genetic algorithm(GA) in which the condition number of regression matrix is set as the fitness function. At last, compared with the constant-velocity tracking experiment, the friction parameters from the exciting trajectory experiment has the similar result with the advantage of time reduction.

1. Introduction

Friction is one of the most relevant problems to realize a successful feedback control in an industry robot[1, 2]. It exists in all mechanisms to some extent and appears to the function of the physical interface between two surfaces and greases in contact[3-7]. The nonlinear phenomenon physically resulting from the contact geometry, properties of materials, relative velocity, wears, ambient temperature, etc, may cause steady-state errors, limit cycling and system vibration[3]. In order to compensate the effect of friction in motion control, it is necessary to describe the frictional behavior. For this reason, many static friction models have been put forward.

The classical static friction models contain Coulomb friction and linear viscous damping, in which the friction is assumed as a function of velocity[8]. However, the property of stiction can't be adequately described. In literature [9] , Morin introduced the situation where the starting friction is higher than friction at a nonzero velocity. Extra break-away forces are required to start the movement. In literature [10] , the Stribeck effect that the friction force decreases continuously as intermediate velocity increasing for a certain velocity regime was observed. It was described by the Stribeck model

$$F_f = \begin{cases} F(\dot{\varphi}) & (\dot{\varphi} \neq 0) \\ F_e & (\dot{\varphi} = 0, |F_e| < F_s) \\ F_s \cdot sign(F_e) & otherwise \end{cases} \tag{1}$$

Where $F(\dot{\varphi})$ for non-zero velocity describes the Stribeck, Coulomb, viscous friction as below,

$$F(\dot{\varphi}) = F_c + (F_s - F_c)e^{-\left|\frac{\dot{\varphi}}{v_s}\right|^{\xi}} + h(\dot{\varphi}), \quad h(\dot{\varphi}) = F_v \cdot \dot{\varphi} \tag{2}$$

Where F_c, F_s, F_v are the coefficients of Coulomb friction, static friction, viscous friction, respectively and v_s is called Stribeck velocity. The term $(F_s - F_c)\exp(-|\dot{\varphi}/v_s|^{\xi})$ describes the Stribeck phenomenon and tends to zero according to the speed v_s and exponent ξ. The term $h(\dot{\varphi})$ represents the viscous friction, which plays a heavy role at high speed. Obviously, the exponent term makes it impossible that the stribeck parameters are identified by linear predictors. In other words, friction has to be separated from other dynamic effects[11]. Indeed, an effective alternative is to use a third order polynomial function to represent both stribeck and coulomb phenomenons[2, 12]. It is also able to express the viscous effect that is not necessarily linear. Then the friction torque can be expressed as:

$$\tau_f(\dot{\varphi}) = [c_0 + c_1|\dot{\varphi}| + c_2|\dot{\varphi}|^2 + c_3|\dot{\varphi}|^3] \cdot sign(\dot{\varphi}) \tag{3}$$

Where $\dot{\varphi}$ is the joint velocity and a symmetric function is also be considered. Similarly, literature [13] proposed a simplified model that is still capturing the downward bends and possible asymmetries while remaining linear in the unknown parameters. The model is the following:

$$\tau_f(\varphi) = [\alpha_0 + \alpha_1|\dot{\varphi}|^{1/2} + \alpha_2|\dot{\varphi}|] \cdot sign(\dot{\varphi}) \tag{4}$$

It is noted that the parameters α_i no longer represent dry friction or viscous damping but the sum of all effects of the multistage gears into a mathematical description[11]. It can be estimated by minimizing a least-square estimation algorithm.

Generally, the static model is adequate enough to describe the effects of friction, and the relevant coefficients can be easily achieved from the selecting parts of the measurement with constant velocity[14]. If only one joint is moved, the influence of centrifugal and Coriolis forces can be eliminated. A forward trajectory from θ_1 to θ_2 with constant-velocity is usually designed to acquire the real motor torque $\tau_m{}^+$, and reverse $\tau_m{}^-$. As the gravity torque $\tau_g(\theta)$ is equal in both directions at the same joint angle θ, we can obtain that

$$\tau_f(\dot{\theta}) - \tau_g(\theta) = \tau_m^+, \quad \tau_f(-\dot{\theta}) - \tau_g(\theta) = \tau_m^- \tag{5}$$

Subtracting the equations above yields:

$$\tau_f(\dot{\theta}) - \tau_f(-\dot{\theta}) = \tau_m^+ - \tau_m^- \tag{6}$$

Since the friction forces in both directions are approximate, i.e. $\tau_f(\dot{\theta}) \approx -\tau_f(-\dot{\theta})$, then a direction-independent estimate of friction and gravity can be achieved as:

$$\tau_f(\dot{\theta}) = (\tau_m^+ - \tau_m^-)/2, \quad \tau_g(\theta) = -(\tau_m^+ + \tau_m^-)/2 \tag{7}$$

Obviously, in this way Multi- repetitive works, especially the friction samplings at an ultra low speed, will make the parameter identification so inefficiency. In order to evaluate the friction coefficients by moving the joint at one time and guarantee the friction accuracy, an optimal periodic trajectory is designed to minimize the condition numbers of observation matrix, which is from the linear dynamic model of the servomotor. The paper is organized as follows. Sect.2 presents the linear predictors of joint friction and an excitation trajectory based on periodic Fourier series with its boundary optimization. Sect. 3 contains the process of friction identification experiment with designed trajectory and the comparison with the constant-speed tracking experiment. Last, conclusions are presented in Sect.4.

2. Static friction model and parameter identification

2.1 the linearization of single joint system

For a robot, since each joint can be considered as a one degree-of-freedom(DOF) system, the dynamic of the joint system can be described as[2]:

$$J\beta = \tau_m - \tau_f + \tau_g \tag{8}$$

Where J is the nominal inertia(motor plus charge), β is the angular acceleration of the motor, τ_m is the motor torque, τ_f is the friction torque scalar and τ_g is the gravity torque of joint, which is the function of joint position and expressed as below:

$$\tau_g = p_x \cos(q) - p_y \sin(q) \tag{9}$$

Where p_x, p_y are called gravity observation coefficient, q is the joint angle. We select equation (4) as the static friction model. In this case, a standard least-squares method can be applied to estimate the friction coefficients. According to equation (4), equation (8) and equation (9), the motor torque in the motor side can be rewritten as:

$$\tau_m = m\phi \tag{10}$$

Where the vector

$$\phi = [J, \alpha_0, \alpha_1, \alpha_2, p_x, p_y]^T \tag{11}$$

The observation vector

$$m = [\beta, sign(\dot{q}), sign(\dot{q})|\dot{q}|^{0.5}, sign(\dot{q})|\dot{q}|, -\cos(q), \sin(q)] \tag{12}$$

and then vector m can be written in matrix form where all the samples are collected along a specific trajectory. Then,

$$\hat{\tau}_m = M\phi \tag{13}$$

where

$$\hat{\tau}_m = [\tau_1, \tau_2, \tau_3, \cdots \tau_k]^T \tag{14}$$

is the vector of measured motor torque at k sampling times. The observation matrix

$$M = \begin{bmatrix} \beta_1 & sign(\dot{q}_1) & sign(\dot{q}_1)|\dot{q}_1|^{0.5} & sign(\dot{q}_1)|\dot{q}_1| & -\cos(q_1) & \sin(q_1) \\ \beta_2 & sign(\dot{q}_2) & sign(\dot{q}_2)|\dot{q}_2|^{0.5} & sign(\dot{q}_2)|\dot{q}_2| & -\cos(q_2) & \sin(q_2) \\ \vdots & \vdots & \vdots & \vdots & \vdots & \vdots \\ \beta_k & sign(\dot{q}_k) & sign(\dot{q}_k)|\dot{q}_k|^{0.5} & sign(\dot{q}_k)|\dot{q}_k| & -\cos(q_k) & \sin(q_k) \end{bmatrix} \tag{15}$$

contains the motion data measured with an exciting trajectory. In this case, the coefficients are estimated by the use of the LS technique as

$$\phi = M^{+}\hat{\tau}_m \tag{16}$$

where M^{+} is the pseudo-inverse of the matrix M.

2.2 the design and optimization of exciting trajectory

The measurement noise and uncertainty of model can result in the deviation of parameter identification[15]. An excellent trajectory can greatly improve the identification precision and anti-noise capacity. Obviously, the trajectory needs to span all the joint velocities for many times in order to reduce the measurement noise and excite all the friction dynamics. If one or more parameters are not excited, the matrix M loses its full rank and equation (16) cannot be solved. Swevers [16, 17] etc. choose a finite Fourier series to be the exciting trajectory for each joint, which has the advantages of convenient data processing and insensitivity to noise. However, the Fourier series does not meet the boundary conditions of velocity and acceleration. In the beginning and end of the movement in a period, the mutations of velocity will lead to the joint vibration and reduce the parameter identification precision. Therefore， we add a quintic polynomial instead of the constant term in the Fourier series to ensure the continuity of the trajectory boundary. The improved Fourier series is written as

$$
\begin{aligned}
&q(t)=\sum_{l=1}^{N}\frac{a_l}{\omega_f l}\sin(\omega_f lt)-\frac{b_l}{\omega_f l}\cos(\omega_f lt)+\sum_{k=0}^{5}c_k(t-(j-1)t_f)^k\\
&t_f=2\pi/\omega_f\\
&j=\begin{cases}1 & if\ \ t=0\\ \mathrm{ceil}(t/t_f) & else\end{cases}\\
&\dot{q}(t)=\sum_{l=1}^{N}a_l\cos(\omega_f lt)+b_l\sin(\omega_f lt)+\sum_{k=1}^{5}c_k\cdot k\cdot(t-(j-1)t_f)^{k-1}\\
&\ddot{q}(t)=\sum_{l=1}^{N}-a_l\cdot(\omega_f l)\cdot\sin(\omega_f lt)+b_l\cdot(\omega_f l)\cdot\cos(\omega_f lt)+\sum_{k=2}^{5}c_k\cdot k\cdot(k-1)\cdot(t-(j-1)t_f)^{k-2}
\end{aligned}
\tag{17}
$$

Where t_f is the period of trajectory, ω_f is called the fundamental frequency, N means the number of the harmonic waves, $a_l, b_l (l=1,2,\cdots N)$, $c_k (k=1,2\cdots 5)$ are constant coefficients. In order to ensure that the robot can accurately track the exciting trajectory, we set the boundary condition as two part as below: the equality constraint,

$$q(0)=q_0,\ q(t_f)=q_0,\ \dot{q}(0)=0,\ \dot{q}(t_f)=0,\ \ddot{q}(0)=0,\ \ddot{q}(t_f)=0 \tag{18}$$

the inequality constraint,

$$q_{\min}\le q(t)\le q_{\max},\ \dot{q}_{\min}\le\dot{q}(t)\le\dot{q}_{\max},\ \ddot{q}_{\min}\le\ddot{q}(t)\le\ddot{q}_{\max} \tag{19}$$

where $q_{\min}, \dot{q}_{\min}, \ddot{q}_{\min}, q_{\max}, \dot{q}_{\max}, \ddot{q}_{\max}$ are the minimum or maximum values of joint angle, speed and acceleration, respectively. Combining the equations from (17) to (19), we can derive the matrix expression of boundary constraint as below,

$$
\begin{bmatrix} q_0\\ q_{tf}\\ \dot{q}_0\\ \dot{q}_{tf}\\ \ddot{q}_o\\ \ddot{q}_{tf}\end{bmatrix}=
\begin{bmatrix} 0 & 0 & 0 & 0 & 0 & 1\\ tf^5 & tf^4 & tf^3 & tf^2 & tf & 1\\ 0 & 0 & 0 & 0 & 1 & 0\\ 5tf^4 & 4tf^3 & 3tf^2 & 2tf & 1 & 0\\ 0 & 0 & 0 & 2 & 0 & 0\\ 20tf^3 & 12tf^2 & 6tf & 2 & 0 & 0\end{bmatrix}
\begin{bmatrix} c_5\\ c_4\\ c_3\\ c_2\\ c_1\\ c_0\end{bmatrix}+
\begin{bmatrix} s_0\\ s_1\\ s_2\\ s_3\\ s_4\\ s_5\end{bmatrix}
\tag{20}
$$

where

$$s_0=\sum_{l=1}^{N}-\frac{b_l}{w_f l},\ s_1=s_0,\ s_2=\sum_{l=1}^{N}a_l,\ s_3=s_2,\ s_4=\sum_{l=1}^{N}b_l w_f l,\ s_5=s_4 \tag{21}$$

and

$$c_0=q_0+\sum_{l=1}^{N}\frac{b_l}{\omega_f l},\ c_1=-\sum_{l=1}^{N}a_l,\ c_2=-\frac{1}{2}\sum_{l=1}^{N}b_l\omega_f l,\ c_3=-\frac{2c_2t_f+10c_1}{t_f^2},\ c_4=\frac{15c_1+c_2t_f}{t_f^3},\ c_5=\frac{-6c_1}{t_f^4} \tag{22}$$

Obviously, the multinomial coefficients $c_i (i=0,1,\cdots 5)$ can be seen as the function of a_l and b_l so that the number of Fourier parameters to be optimized has been reduced.

Generally, the random noises mixed in motion and torque samples have a bad effect on identification precision. In order to strengthen the anti-noise ability and robustness of the model (13), the Fourier parameters should be optimized. Since the condition number of the regression matrix M reflects the performance of the exciting trajectory, the optimization criteria can be expressed as,

$$\min cond(M)\big|_{a_l,b_l} \tag{23}$$

Combining with the limits for each joint position, speed, and acceleration (seen in equation (18),(19)), we can adopt genetic algorithm (GA) to solve the problem of nonlinear constrained optimization.

In the GA, the Fourier parameter N is set as 5, and the coefficients of exciting trajectory are set as genetic populations, i.e.,

$$x=[a_1,a_2,\cdots,a_5,b_1,b_2,\cdots b_5] \tag{24}$$

equation (23) is taken as the individual fitness function. The random sampling selection method is adopted for the best individual who is coded by decimal floating point coding method. The crossover probability is set as 0.9. For mutation operation, we select the Gauss mutation operator whose mutation probability is self-adjusted with the process of evolution. It can be expressed as,

$$p_m = (p_{mo} - (p_{mo} - p_{m\min}) \cdot \frac{g}{G} \tag{25}$$

where p_{mo} is the initial mutation probability, $p_{m\min}$ is the minimum probability, g is the current hereditary time, G is the maximum hereditary time and set as 100.

3. Experiment of friction parameter identification

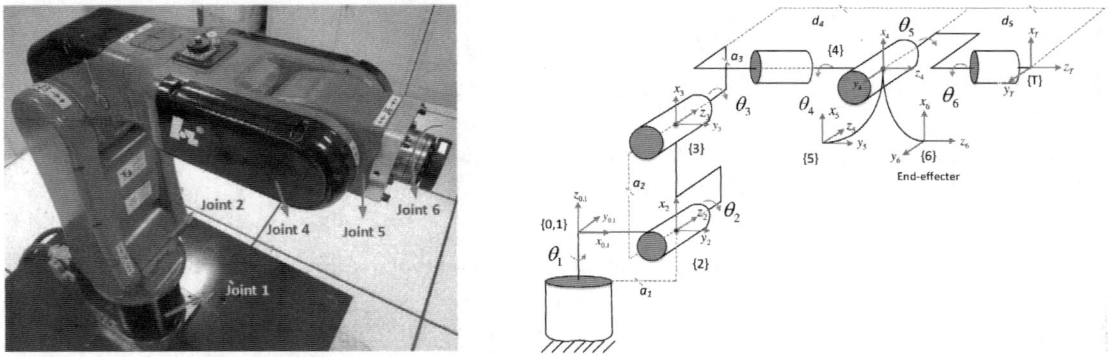

(a) **(b)**

Figure.1 The experiments were made in joint 4 of HSR robot. (a) the robot used in the experiments; (b) Modified D-H coordinate system.

In this paper, all the experiments are made in joint 4 of Robot HSR-JR605-C as seen in figure 1, and the kinematic parameters are listed in **Table** .

Table 1. Kinematic parameters of joint 4.

θ_4 (deg)	$\dot{\theta}_4$ (deg/s)	$\ddot{\theta}_4$ (deg/s^2)	*Jerk* (deg/s^3)	*(max)Gear Ratio* (η)
[-180,180]	[-40,40]	[-500,500]	[-45000,45000]	76.5:1
t_f (s)	ω_f	*Sampling rate(Hz)*	q_0 (degree)	*Rated current* (A)
3	$2\pi/3$	500	0	2.546

In this case, we derive the Fourier coefficients by GA method as $a_1 = 0.6340, a_2 = 0.0160, a_3 = -0.0032$, $a_4 = 0.0072, a_5 = 0.0072, b_1 = 0.4397, b_2 = -0.0194, b_3 = -0.14525, b_4 = -0.0697, b_5 = 0.0731$ .Based on the TwinCAT platform, an optimal trajectory tracking experiment(OTTE) is designed. we move the joint 4 along the optimal trajectory and sample the movement and motor torque for N(N=50) periods. At the same time, all the other axes keep the initial posture, i.e, $[\theta_1, \theta_2, \theta_3, \theta_5, \theta_6]$=[0,-$\pi$/2,0,0,0]. In order to improve the signal-to-noise ratio(SNR) of the sampling date, an average method in time domain is chosen as below,

$$\hat{q}(k) = \frac{1}{N}\sum_{i=1}^{N} q_{4i}(k),\ \hat{\dot{q}}(k) = \frac{1}{N}\sum_{i=1}^{N} \dot{q}_{4i}(k),\ \hat{\ddot{q}}(k) = \frac{1}{N}\sum_{i=1}^{N} \ddot{q}_{4i}(k),\ \hat{\tau}(k) = \frac{1}{N}\sum_{i=1}^{N} \tau_{4i}(k) \tag{26}$$

Where k represents the kth sampling in a cycle and $q_{4i}(k)$, $\dot{q}_{4i}(k)$, $\ddot{q}_{4i}(k)$, $\tau_{4i}(k)$ means the angle, speed, acceleration and motor torque of joint 4 in kth sampling and ith period, respectively. Moreover, to further improve the quality of measured signals, a high-speed nonlinear discrete tracking-differentiator(HSTD)[18] is adopted to clean the noise of the position and torque signals, and then its first and second order derivatives can be derived. The results are shown in figure 2:

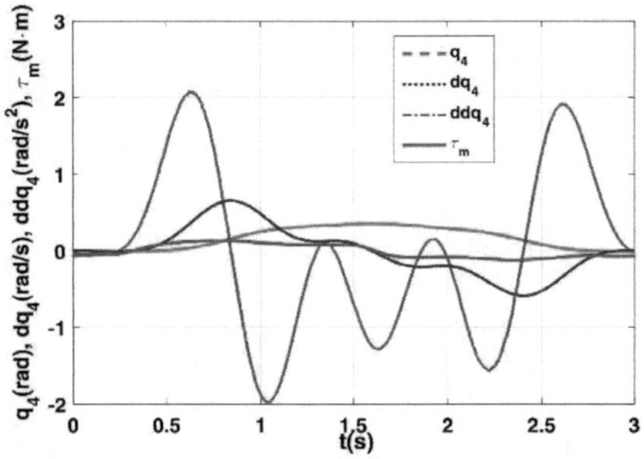

Figure. 2 Measured signals of position, speed, acceleration and motor torque of joint 4 along the exciting trajectory.

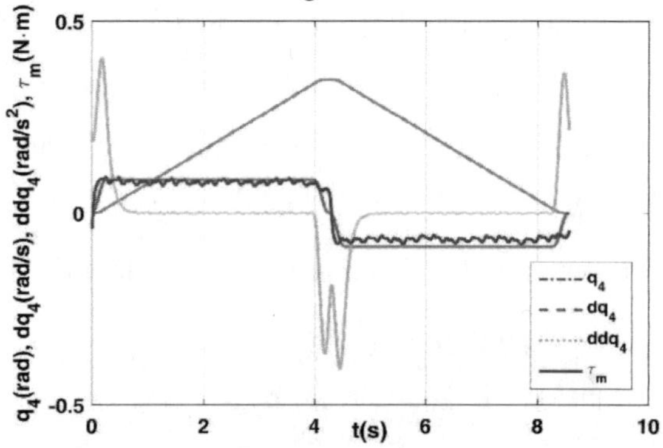

Figure. 3 Measured movement along a constant-speed trajectory

In this case, the speed is normalized and the parameter vector ϕ can be achieved according to equation (16) as ϕ=[0.0001887,0.087,-0.141,0.35,0.022087,0.132585] , then the friction parameters are obtained. In order to ascertain the validity of the identification results, the constant-velocity tracking experiments (CVTE) are made for comparison.

we select *S* wave (seen in figure 3) that traverses the constant-speed trajectory region [20:70]deg to acquire different joint angle mapping 29 sets of speed samples from 0 to 30deg/s in the joint side. Obviously, it takes a long time to move at an ultra-low speed. For reducing the impact of random noise, joint 3 is moved periodically at 50 times. In this case, based on equation (7), the results of the friction-velocity wave can be figured as below:

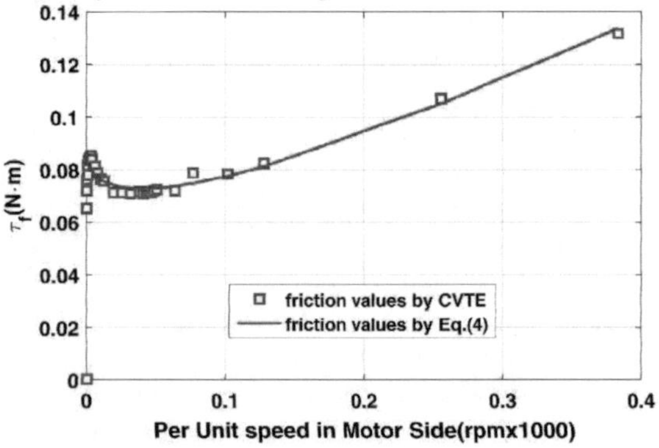

Figure. 4 Friction-velocity waves by CVTE and OTTE

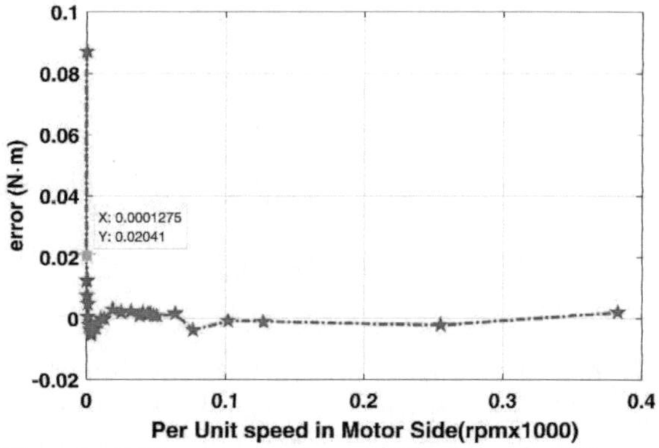

Figure. 5 The error values of friction to different velocities.

From the figure 4&5, we can derive that the results of those different experiments have a great similarity. Obviously, relatively large errors occur at low speeds, but fewer errors exist at higher speeds. In general applications, the friction parameter identification with an optimal trajectory can be adopted because of the good description of the measured characteristics.

4. Conclusion

In this paper, we take an exciting trajectory to identify the static friction parameters instead of the traditional constant-velocity trajectory. The Fourier series has been taken to build the period trajectory, in which the constant term is replaced by a quintic polynomial to meet the speed and acceleration's boundary condition of continuity. In this case, the GA is used to get the trajectory to satisfy the constraints and not sensitive to the measurement noise. On this condition, only one tracking trajectory is necessary to take the place of the multi-tracking experiments, thus the identification efficiency is greatly improved.

Acknowledgments
Research supported by National Natural Science Foundation of China (Project number: 51375194 and 51575411) and State Key Lab of Digital Manufacturing Equipment & Technology, School of Mechanical Science and Engineering, Huazhong University of Science & Technology.

Reference

[1] Guillo M, Dubourg L 2016 Impact & improvement of tool deviation in friction stir welding: Weld quality & real-time compensation on an industrial robot *Robotics&Computer-Integrated Manufacturing* **39** 22-31
[2] Simoni L, Beschi M, Legnani G, et al. 2015 Friction modeling with temperature effects for industrial robot manipulators *IROS. IEEE/RSJ International Conference on. IEEE* 3524-3529
[3] Bittencourt A C, Axelsson P 2014 Mmodeling and experiment design for identification of wear in a robot joint under load and temperature uncertainties based on friction data *IEEE-Asme Transactions On Mechatronics* **19**(5) 1694-1706
[4] Bittencourt A C, Gunnarsson S 2012 Static friction in a robot joint-modeling and identification of load and temperature effects *Journal of Dynamic Systems, Measurement, and Control* **134**(5) 51013
[5] Marton L, Lantos B. 2011 Control of robotic systems with unknown friction and payload *IEEE Transactions on Control Systems Technology* **19**(6) 1534-1539
[6] Bittencourt A C, Axelsson P, Jung Y, et al. 2011 Modeling and identification of wear in a robot joint under temperature uncertainties *IFAC Proceedings Volumes* **44**(1) 10293-10299
[7] Freidovich L, Robertsson A, Shiriaev A, et al. 2010 Lugre-model-based friction compensation *IEEE Transactions on Control Systems Technology* **18**(1) 194-200
[8] Olsson H, Åström K J, Canudas De Wit C, et al. 1998 Friction models and friction compensation *European Journal of Control* **4**(3) 176-195
[9] Morin A J 1833 New friction experiments carried out at metz in 1831 – 1833 *Proceedings of the French Royal Academy of Sciences* **4**(1) 128
[10] Stribeck R 1902 The key qualities of sliding and roller bearings *Zeitschrift des Vereines Seutscher Ingenieure* **46**(38) 39
[11] Grotjahn M, Daemi M, Heimann B 2001 Friction and rigid body identification of robot dynamics *International Journal of Solids and Structures* **38**(10) 1889-1902
[12] Visioli A, Legnani G 2002 On the trajectory tracking control of industrial scara robot manipulators *IEEE transactions on industrial electronics* **49**(1) 224-232
[13] Canudas De Wit C, Noel P, Aubin A, et al 1991 Adaptive friction compensation in robot manipulators: low velocities *The International Journal Of Robotics Research* **10**(3) 189-199
[14] Liu G, Goldenberg A A, Zhang Y 2004 Precise slow motion control of a direct-drive robot arm with velocity estimation and friction compensation *Mechatronics* **14**(7) 821-834
[15] Van Geffen V 2009 A study of friction models and friction compensation *DCT* **118** 24
[16] Swevers J, Ganseman C, De Schutter J, et al. 1996 Experimental robot identification using optimisced periodic trajectories *Mechanical Systems and Signal Processing* **10**(5) 561-577
[17] Swevers J, Ganseman C, Tukel D B, et al. 1997 Optimal robot excitation and identification *IEEE transactions on robotics and automation*1 **3**(5) 730-740
[18] Xie Y, Long Z 2009 A high-speed nonlinear discrete tracking-differentiator with high precision *Control Theory & Applications* **26**(2) 127-132

IOP Conf. Series: Materials Science and Engineering **280** (2017) 012026 doi:10.1088/1757-899X/280/1/012026

Modelling and Simulation on Multibody Dynamics for Vehicular Cold Launch Systems Based on Subsystem Synthesis Method

YAN Panyun[1,*], LIANG Guozhu[1], LU Yongzhi[2], QI Zhihui[2] and GAO Xingdou[2]

[1] School of Astronautics, Beijing University of Aeronautics and Astronautics, Beijing 100191, China.
[2] Beijing Institute of Space Launch Technology, Beijing 10076, China

[*] yanpanyun@163.com

Abstract. The fast simulation of the vehicular cold launch system (VCLS) in the launch process is an essential requirement for practical engineering applications. In particular, a general and fast simulation model of the VCLS will help the designer to obtain the optimum scheme in the initial design phase. For these purposes, a system-level fast simulation model was established for the VCLS based on the subsystem synthesis method. Moreover, a comparison of the load of a seven-axis VCLS on the rigid ground through both theoretical calculations and experiments was carried out. It was found that the error of the load of the rear left outrigger is less than 7.1%, and the error of the total load of all the outriggers is less than 2.8%. Moreover, time taken for completion of the simulation model is only 9.5 min, which is 5% of the time taken by conventional algorithms.

1. Introduction

The typical vehicular cold launch system (VCLS) is the SS-25 developed by Russia. In this system, the missile is ejected by compressed gas from an external gas generator, followed by ignition of the missile. In the existing VCLSs, the carrier vehicle is usually a multi-axis vehicle that has many independent pneumatic suspensions (IPS). For practical applications, it is essential to perform a fast simulation of the VCLS in the launch process. A system-level fast simulation model (SFSM) is especially suitable to help designers obtain the optimum scheme in the initial design phase. Therefore, establishing a feasible SFSM for VCLS is the focus of this paper.

The VCLSs are very complex and consist of many components interconnected by hinges and elastic elements. Thus, the VCLS is characterized by complex constraints and changing structure [1], and it is difficult to meet the requirements of high speed simulations for such a system by using the classical modelling approaches. However, the subsystem synthesis method [2-5], proposed by Kim et al., provides an efficient way to analyze multibody vehicle dynamics. The computational efficiency of this method is greatly improved with the increase in the number of subsystems and the subsystem degree of freedom.

Thus, in this paper, a SFSM based on the subsystem synthesis method will be established, where the absolute coordinates are chosen to define the system configuration, and the Lagrange's equation of the first kind is used to establish the governing equation. Taking the Euler parameters as the generalized coordinates, the subsystem synthesis method will be derived. Subsequently, a seven-axis

IOP Conf. Series: Materials Science and Engineering **280** (2017) 012026 doi:10.1088/1757-899X/280/1/012026

VCLS will be taken as an example to compare the theoretical results with the experimental results in order to verify the accuracy and efficiency of the SFSM.

2. Mathematical Modelling

2.1. Subsystem Structure of the VCLS

The VCLS consists of many components such as the carrier vehicle, the lifting device, the canister, the missile, IPS, and so on. These components are interconnected by hinges and elastic elements. The simplified schematic diagram of a typical VCLS is shown in figure 1.

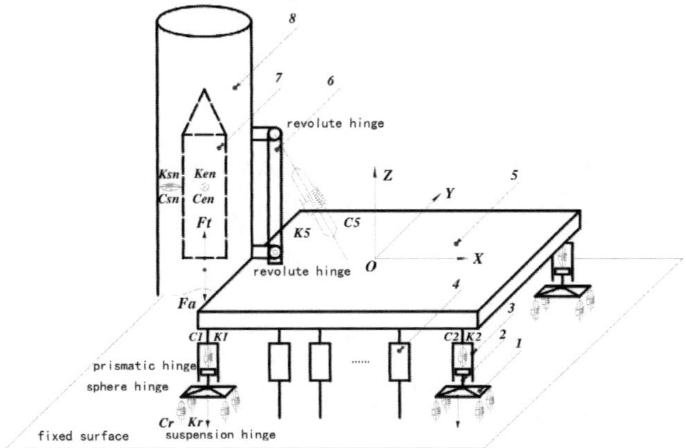

1-support plate, 2-inner cylinder of the hydraulic outrigger, 3-outer cylinder of the hydraulic outrigger, 4-hydraulic suspension, 5-chassis, 6-lifting auxiliary bracket, 7-missile, 8-canister

Figure 1. Schematic diagram of VCLS

The main assumptions are as follows:

- All hinge constraints are ideal and holonomic.
- The missile slides in the canister without jumping.
- Apart from the lifting auxiliary bracket, the deformations of other bodies need not be considered. The flexible bodies undergo small deformations, so the floating frame of reference formulation can be used to describe the deformations. Based on experience, the 20 lowest natural frequencies and corresponding modes of vibration of the flexible bodies are used.
- The interior ballistic pressure in the launch canister is simplified as the thrust F_t acting on the bottom of the missile, and the function of the rubber vesicle is modelled as the adjunctive force F_a acting on the canister.

The model of the VCLS based on the subsystem synthesis method is shown in figure 2. The IPS is simplified into subsystems, and the chassis is taken as a virtual base body which is the reference body to define kinematic relationships between bodies in the IPS.

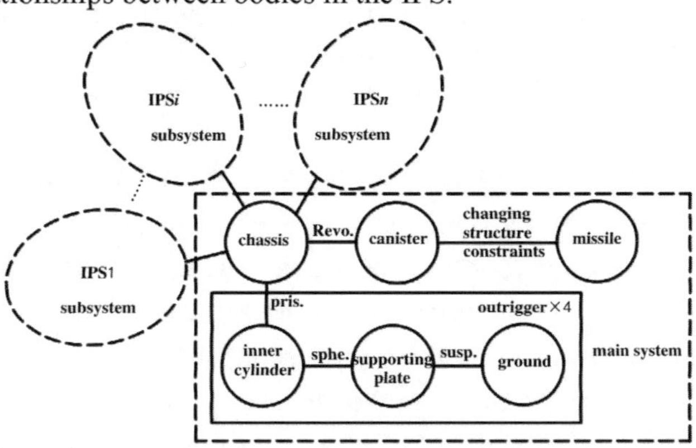

Figure 2. Subsystems structure of VCLS

IOP Conf. Series: Materials Science and Engineering **280** (2017) 012026 doi:10.1088/1757-899X/280/1/012026

2.2. Main System Equations of Motion

The rigid-flexible coupling dynamic equations of motion [6-9] are usually described as:

$$\begin{cases} M\ddot{q}+D\dot{q}+Kq+C_q^T\lambda=Q_e+Q_v \\ C(q,t)=0 \end{cases} \tag{1}$$

where M is the mass matrix, D is the proportional damping matrix, K is the stiffness matrix, Q_e is the vector of generalized forces, Q_v is a quadratic velocity vector, C_q is the Jacobian matrix associated with the constraint equations $C(q,t)=0$, and λ is the Lagrangian multiplier vector. With the constraints differentiated twice with respect to time, Eq. (1) is modified based on the Baumgarte's Constraint Violation Stabilization Method [10] (CVSM).

$$\begin{bmatrix} M & C_q^T \\ C_q & 0 \end{bmatrix}\begin{bmatrix} \ddot{q} \\ \lambda \end{bmatrix}=\begin{bmatrix} Q_e+Q_v-D\dot{q}-Kq \\ \gamma-\alpha\dot{C}-\beta C \end{bmatrix} \tag{2}$$

where α and β are the parameters of the CVSM, usually $1\le\alpha=\beta\le 50$. In this paper, the preferred parameters are $\alpha=\beta=2$. γ is the right term of the acceleration constraint equation.

2.3. Subsystem Synthesis Method

2.3.1. Relative Cartesian Coordinates Formulation. In order to obtain the effective mass matrices and force vectors of the IPS, motions of the bodies in the IPS must be represented in the body-fixed frame of the chassis. For this reason, relative Cartesian coordinates are used as shown in figure 3.

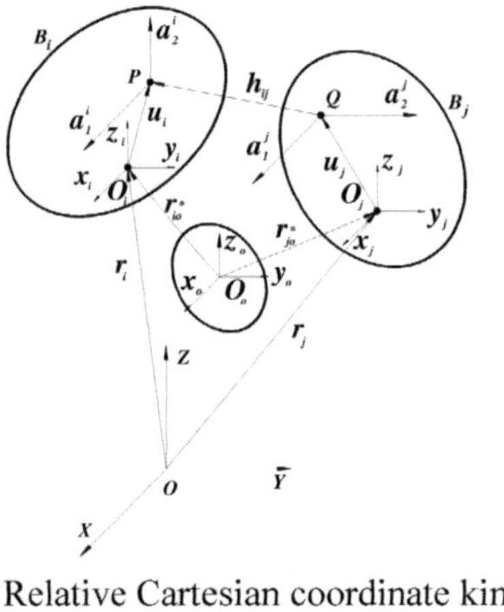

Figure 3. Relative Cartesian coordinate kinematics.

The generalized coordinates of the body i can be represented by the kinematic of the chassis and the relative kinematic of body i with respect to the chassis [3].

$$r_i=r_o+A_o r_{io}^* \tag{3}$$

$$\omega_i'=A_{io}^T\omega_o^*+\omega_{io}' \tag{4}$$

where r_i and r_o are the position vector of the body i and the chassis, respectively, A_o is the orientation matrix of the chassis, and r_{io}^* is the relative position vector of the body i with respect to the chassis, and it is expressed in the body-fixed frame of the chassis. The notation $(\cdot)^*$ indicates that the quantity $(\cdot)$ is expressed in the body-fixed frame of the chassis. ω_i' is the angular velocity vector of the body i, and the notation $(\cdot)'$ indicates that the quantity $(\cdot)$ is expressed in the body-fixed frame. A_{io} is the relative orientation of the body with respect to the chassis, ω_o^* is the angular velocity of the chassis, and ω_{io}' is the relative angular velocity of the body i with respect to the chassis.

When the Euler parameter Λ is used to describe the orientation, Eq. (4) is modified as:

$$\dot{\Lambda}_i = \boldsymbol{R}_i^T \boldsymbol{R}_o \dot{\Lambda}_o + \boldsymbol{L}_i^T \boldsymbol{L}_{io} \dot{\Lambda}_{io} \tag{5}$$

where the matrices $\boldsymbol{R}$ and $\boldsymbol{L}$ are two semi-transformations. Their product is the orientation matrix.

$$\boldsymbol{A} = \boldsymbol{R}\boldsymbol{L} \tag{6}$$

The velocity of the body i is obtained by differentiating Eq. (3) with respect to time as:

$$\dot{\boldsymbol{r}}_i = \dot{\boldsymbol{r}}_o + \boldsymbol{A}_o \tilde{\boldsymbol{\omega}}_o^* \boldsymbol{r}_{io}^* \tag{7}$$

where the notation $\tilde{(\cdot)}$ refers to the skew symmetric matrix made of the elements of vector $(\cdot)$. For convenience, the following compact velocity vectors are employed.

$$\boldsymbol{y}_i = \begin{bmatrix} \dot{\boldsymbol{r}}_i \\ \dot{\Lambda}_i \end{bmatrix}, \boldsymbol{y}_o = \begin{bmatrix} \dot{\boldsymbol{r}}_o \\ \dot{\Lambda}_o \end{bmatrix}, \boldsymbol{y}_{io} = \begin{bmatrix} \dot{\boldsymbol{r}}_{io}^* \\ \dot{\Lambda}_{io} \end{bmatrix} \tag{8}$$

Then, the following relative velocity relationship expressed by the Euler parameter is obtained.

$$\boldsymbol{y}_i = \boldsymbol{E}_{io} \boldsymbol{y}_o + \boldsymbol{G}_{io} \boldsymbol{y}_{io} \tag{9}$$

where

$$\boldsymbol{E}_{io} = \begin{bmatrix} \boldsymbol{I} & -2\boldsymbol{A}_o \tilde{\boldsymbol{r}}_{io}^* \boldsymbol{L}_o \\ \boldsymbol{0} & \boldsymbol{R}_i^T \boldsymbol{R}_o \end{bmatrix}, \boldsymbol{G}_{io} = \begin{bmatrix} \boldsymbol{A}_o & \boldsymbol{0} \\ \boldsymbol{0} & \boldsymbol{L}_i^T \boldsymbol{L}_{io} \end{bmatrix} \tag{10}$$

The relative acceleration between the body i and the chassis is obtained by differentiating Eq. (9) with respect to time as:

$$\dot{\boldsymbol{y}}_i = \boldsymbol{E}_{io} \dot{\boldsymbol{y}}_o + \boldsymbol{G}_{io} \dot{\boldsymbol{y}}_{io} + \boldsymbol{h}_{io} \tag{11}$$

where $\boldsymbol{h}_{io}$ is the velocity coupling term.

$$\boldsymbol{h}_{io} = \begin{bmatrix} \boldsymbol{A}_o \tilde{\boldsymbol{\omega}}_o^* \tilde{\boldsymbol{\omega}}_o^* \boldsymbol{r}_{io}^* + 2\boldsymbol{A}_o \tilde{\boldsymbol{\omega}}_o^* \dot{\boldsymbol{r}}_{io}^* \\ \dot{\boldsymbol{R}}_i^T \boldsymbol{R}_o \dot{\Lambda}_o + \dot{\boldsymbol{L}}_i^T \boldsymbol{L}_{io} \dot{\Lambda}_{io} \end{bmatrix} \tag{12}$$

2.3.2. Subsystem Equations of Motion. According to the Eq.(9), the acceleration relationship between all the bodies in the IPS and the chassis can be expressed in the following compact form as:

$$\dot{\bar{\boldsymbol{y}}} = \bar{\boldsymbol{E}}_{io} \dot{\bar{\boldsymbol{y}}}_o + \bar{\boldsymbol{G}}_{io} \dot{\bar{\boldsymbol{y}}}_{io} + \bar{\boldsymbol{h}}_{io} \tag{13}$$

Based on the subsystem synthesis method, the effective mass matrix and force vectors [3] can be expressed as:

$$\breve{\boldsymbol{M}}^e = \bar{\boldsymbol{M}}_{EE} - \boldsymbol{M}_{E\theta} \boldsymbol{M}_{\theta\theta}^{-1} \boldsymbol{M}_{E\theta}^T \tag{14}$$

$$\breve{\boldsymbol{g}}^e = \hat{\boldsymbol{g}}_E - \boldsymbol{M}_{E\theta} \boldsymbol{M}_{\theta\theta}^{-1} \boldsymbol{g}_\theta \tag{15}$$

where

$$\begin{aligned} &\bar{\boldsymbol{M}}_{EE} = \bar{\boldsymbol{E}}_{io}^T \boldsymbol{M} \bar{\boldsymbol{E}}_{io}, \bar{\boldsymbol{M}}_{EG} = \bar{\boldsymbol{E}}_{io}^T \boldsymbol{M} \bar{\boldsymbol{G}}_{io}, \bar{\boldsymbol{M}}_{GG} = \bar{\boldsymbol{G}}_{io}^T \boldsymbol{M} \bar{\boldsymbol{G}}_{io}, \bar{\boldsymbol{M}}_{EG} = \bar{\boldsymbol{M}}_{EG}^T, \bar{\boldsymbol{g}}_E = \bar{\boldsymbol{E}}_{io}^T \bar{\boldsymbol{h}}_{io}, \bar{\boldsymbol{g}}_G = \bar{\boldsymbol{G}}_{io}^T \bar{\boldsymbol{h}}_{io} \\ &\boldsymbol{M}_{E\theta} = \bar{\boldsymbol{M}}_{EG} \boldsymbol{N}, \boldsymbol{M}_{\theta\theta} = \boldsymbol{N}^T \bar{\boldsymbol{M}}_{GG} \boldsymbol{N}, \hat{\boldsymbol{g}}_E = \bar{\boldsymbol{g}}_E - \bar{\boldsymbol{M}}_{EG} \boldsymbol{p}, \boldsymbol{g}_\theta = \boldsymbol{N}^T (\bar{\boldsymbol{g}}_G - \bar{\boldsymbol{M}}_{GG} \boldsymbol{p}) \end{aligned} \tag{16}$$

$\boldsymbol{M}$ is the constant mass and inertia matrix of the subsystem, and $\bar{\boldsymbol{g}}$ is the generalized composite force vector acting on all the bodies in the subsystem. It should be noted that the matrix $\boldsymbol{M}$ and vector $\bar{\boldsymbol{g}}$ are calculated in the inertial frame.

In order to obtain the expressions for the transform matrix $\boldsymbol{N}$ and the matrix $\boldsymbol{p}$, the acceleration in relative Cartesian coordinates is expressed in terms of acceleration of the independent generalized coordinates $\ddot{\boldsymbol{\theta}}$ as:

$$\dot{\bar{\boldsymbol{y}}}_{io} = \boldsymbol{N}\ddot{\boldsymbol{\theta}} + \boldsymbol{p} \tag{17}$$

The hinge constraint equations in the IPS can be expressed as:

$$\boldsymbol{C}(\bar{\boldsymbol{z}}_{io}) = \boldsymbol{0} \tag{18}$$

where $\bar{\boldsymbol{z}}_{io}$ is the displacement vector of the composite relative Cartesian coordinates of all the bodies in the IPS.

The acceleration constraint equations can be obtained by differentiating Eq.(18) with respected to time twice as:

$$C_{\bar{z}_{io}} \ddot{\bar{y}}_{io} = \gamma \tag{19}$$

where $C_{\bar{z}_{io}}$ is the associated Jacobian matrix of the subsystem constraint equations.

The L-U factorization is applied to the Jacobian matrix. Assuming that the constraints are independent, the Jacobian matrix has full row rank. Therefore, there is always at least one non-singular submatrix by applying column transformation to the Jacobian matrix.

$$C_{\bar{z}_{io}} P_2 = [C_u \quad C_v] \tag{20}$$

where P_2 is the column transform matrix.

Now, the transform matrix N is obtained when the subsystem constraint equations are holonomic.

$$N = P_2 \begin{bmatrix} -C_u^{-1} C_v \\ I \end{bmatrix} \tag{21}$$

The matrix p in Eq. (16) can be expressed as:

$$p = P_2 \begin{bmatrix} -C_u^{-1}(\gamma - \alpha \dot{C} - \beta C) \\ 0 \end{bmatrix} \tag{22}$$

As all the bodies are rigid in the subsystem, the preferred parameters of the CVSM are: $\alpha = \beta = 40$.

With the effective mass matrix and force from each of the subsystems, the following equation of motion for the chassis is obtained to solve for accelerations.

$$(M^o + \sum_{i=1}^{n} \breve{M}^{ei}) \ddot{q}^o = Q_e^o + Q_v^o + \sum_{i=1}^{n} \breve{g}^{ei} \tag{23}$$

where M^o is the inertia matrix, Q_e^o and Q_v^o are the vectors of generalized force and quadratic velocity, respectively, and q^o refers to the generalized coordinates, where the superscript ^{o} represents the chassis.

3. Program Implementation

3.1. Static Analysis

The launch dynamics simulation generally begins with a static equilibrium state. In order to obtain the static equilibrium state, dynamics analysis is carried out by adding additional damping force to the system. Thus, the generalized external force is corrected to the following formula.

$$F^a = F^a - M\dot{q}, if \ |\dot{q}| > preci \tag{24}$$

where *preci* is a given precision.

3.2. Launch Dynamics Simulation

FORTRAN is used for the programming language. There are five main data structures defined in the program including body, hinge, mark, force and force element. These data structures are linked by a linked list for easy deletion and addition. Eq. (2) represents the stiff differential algebraic equations (SDAEs), and the L_1 condition number is approximately 5.0×10^{18}. The variable-step Gear implicit integral method is effective for handling the SDAEs, and the step size $\Delta t \in [10^{-3}, 10^{-6}]$ is used in this work. The calculation flow chart of the VCLS is given in figure 4.

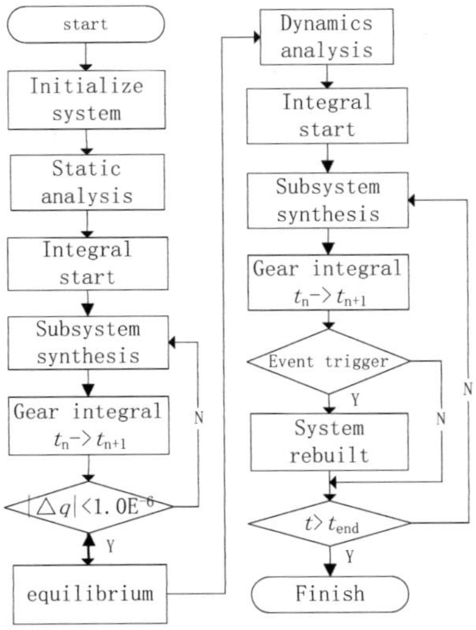

Figure 4. Calculation flow chart of the VCLS

4. Simulation and Verification

In order to verify the above general algorithm for the VCLS shown in figure 1, a seven-axis VCLS is taken as an example. The seven-axis VCLS has 72 bodies including the flexible lifting auxiliary bracket, which can be divided into a main system comprised of 16 bodies and 14 subsystems. Each subsystem is comprised of 4 bodies. The seven-axis VCLS is supported by tires together with outriggers on the rigid ground, where the stiffness K_r of the rigid ground is 3×10^8 N/m, and the damping C_r of the rigid ground is 1×10^6 Ns/m.

For the convenient display of the results, the non-dimensional time $\bar{t}$ is defined as:

$$\bar{t} = \frac{t}{t_e} \tag{25}$$

where t is the current time, and t_e is the length of the simulation time.

The non-dimensional force $\bar{F}$ is defined as:

$$\bar{F} = \frac{F}{M_0 g} \tag{26}$$

where F is the current force, M_0 is the total mass of the VCLS, and g is the local gravitational acceleration (9.8 m/s^2).

A comparison of the theoretical and experimental results for non-dimensional force $\bar{F}_1$ of the rear left outrigger in the model is carried out and the results are shown in figure 5. The comparison of the total non-dimensional force $\bar{F}_{\text{total}}$ of all the outriggers is shown in figure 6. The error of $\bar{F}_1$ is less than 7.1%, and the error of $\bar{F}_{\text{total}}$ is less than 2.8%. Moreover, the time taken for completion of the SFSM is only 9.5 min, which is 5% of the completion time of the HHT-SI2 method [11] which takes approximately 180 min for the seven-axis VCLS. The above results illustrate that the mathematical model of VCLS is feasible and can be used for practical engineering applications.

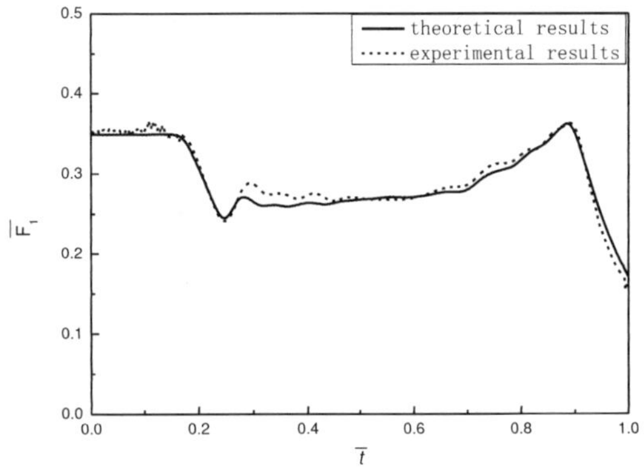

Figure 5. Simulation results of the load of the rear left outrigger.

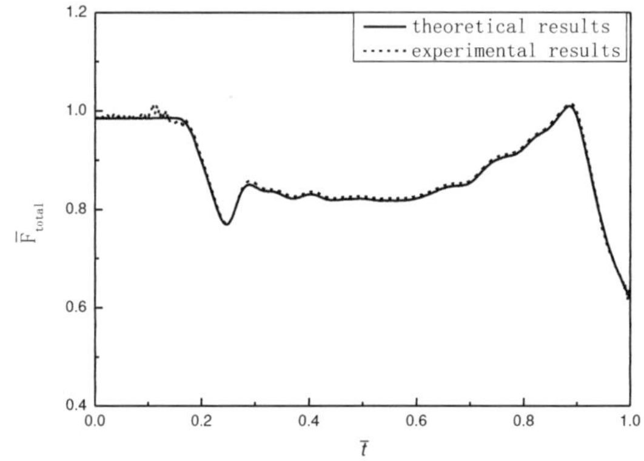

Figure 6. Simulation results of the total load of all outriggers.

5. Conclusion

In this paper, a SFSM was established based on the subsystem synthesis method, and a comparison of the load of a seven-axis VCLS on the rigid ground in theory and experiment has been carried out. It was found that the error of the load of the rear left outrigger is less than 7.1%, and the error of the total load of the outriggers is less than 2.8%. Moreover, the completion time is only 9.5 min, which is just 5% of the completion time of the HHT-SI2 method. Therefore, the SFSM is able to meet the requirements for practical engineering applications.

References

[1] Yan Panyun, Liang Guozhu, Lu Yongzhi, et al. Fast simulation on rigid multibody dynamics for vehicular cold launch systems[J]. Journal of Beijing University of Aeronautics and Astronautics, 2017, 43(1): 71-78.

[2] Kim Sung-Soo, Lee Chang Ho. A recursive subsystem synthesis method for repeated closed loop structure in multibody dynamics[J]. Journal of Mechanical Science and Technology, 2009, 23(4): 946-949.

[3] Kim Sung-Soo, Wang Ji-Hyeun. Subsystem synthesis methods with independent coordinates for real-time multibody dynamics[J]. Journal of Mechanical Science and Technology, 2005, 19(1): 312-319.

[4] Kim Sung-Soo. A Subsystem Synthesis Method for Efficient Vehicle Multibody Dynamics[J]. Multibody System Dynamics, 2002, 7(2): 189-207.

[5] Kim Sung-Soo, Oh Young-Seok.A Real-Time Multibody Vehicle Dynamics Model Using a Subsystem Synthesis Method:18th Biennial Conference on Mechanical Vibration and Noise[Z].New York, N.Y.:American Society of Mechanical Engineers,2001.

[6] Shabana Ahmed A. Dynamics of Multibody Systems[M]. Third Edition ed. New York: Cambridge University Press, 2005: 223-225.

[7] Yoo Wan S., Haug Edward J. Dynamics of Articulated Structures. Part I. Theory[J]. Journal of Structural Mechanics, 1986, 14(1): 105-126.

[8] Shabana A. A. Substructure synthesis methods for dynamic analysis of multi-body systems[J]. Computers & Structures, 1985, 20(4): 737-744.

[9] Hinton E., Rock T., Zienkiewicz O. C. A note on mass lumping and related processes in the finite element method[J]. Earthquake Engineering & Structural Dynamics, 1976, 4(3): 245-249.

[10] Baumgarte J. Stabilization of constraints and integrals of motion in dynamical systems[J]. Computer Methods in Applied Mechanics and Engineering, 1972, 1(1): 1-16.

[11] Negrut Dan, Jay Laurent O., Khude Naresh. A Discussion of Low-Order Numerical Integration Formulas for Rigid and Flexible Multibody Dynamics[J]. Journal of Computational and Nonlinear Dynamics, 2009, 4(2): 21008.

IOP Conf. Series: Materials Science and Engineering **280** (2017) 012027 doi:10.1088/1757-899X/280/1/012027

The Innovative Design and Prototype Verification of Wheelchair with One Degree of Freedom to Perform Lifting and Standing Functions

Long-Chang Hsieh[1] and Tzu-Hsia Chen[2,*]

[1]Department of Power Mechanical Engineering, National Formosa University, Huwei, Yunlin, Taiwan.
[2]Department of Mechanical Engineering, Minghsin University of Science and Technology, Taiwan.

E-mail: lochsieh@nfu.edu.tw; *son.summer@msa.hinet.net

Abstract. Traditionally, the mechanism of wheelchair with lifting and standing functions has 2 degrees of freedom, and used 2 power sources to perform these 2 motion function. The purpose of this paper is to invent new wheelchair with 1 degree of freedom to perform these 2 motion functions. Hence, we can use only 1 power source to drive the mechanism to achieve lifting and standing motion functions. The new design has the advantages of simple operation, more stability, and more safety. For traditional standing wheelchair, its' centre of gravity moves forward when standing up and it needs 2 auxiliary wheels to prevent dumping. In this paper, by using the checklist method of Osborn, the wheelchair with 1 *DOF* is invented to perform lifting and standing functions. The centre of gravity of this new wheelchair after standing up still located between the front and rear wheels, no auxiliary wheels needed. Finally, the prototype is manufactured to verify the theoretical results.

1. Introduction

Wheelchair is an important assisted vehicle for the disabled person for body moving and participating social activities. With the progress of sociality, the demands of wheelchair are more and more diversified. Hence, the standing up [1-3], lifting up [4-6], lying down [7-9], tilting down [10-12], and multi-function [13-15] wheelchairs are invented to satisfy the disabled person's need. Figs. 1(a) ~ 1(d) show the wheelchairs with standing, lifting, lying, and tilting functions. For a wheelchair with more than two functions is called multi-function wheelchair. This paper focuses on the innovative design of wheelchair with lifting and standing functions.

(a) Standing wheelchair (b) Lifting wheelchair (c) Lying wheelchair (d) Tilting wheelchair

Figure 1. Functional wheelchairs

IOP Conf. Series: Materials Science and Engineering **280** (2017) 012027 doi:10.1088/1757-899X/280/1/012027

Currently, the mechanism of wheelchair with lifting and standing functions has 2 degrees of freedom, and used 2 power sources to perform these 2 motion function. The 1st purpose of this paper is to invent mechanism with 1 degree of freedom (DOF) to perform lifting and standing functions. Hence, we can use only 1 power source to drive the mechanism to achieve lifting and standing motion functions. According to Figure 1(a), for the traditional standing wheelchair, its' centre of gravity moves forward when standing up and it needs 2 auxiliary wheels to prevent dumping. The 2nd purpose of this paper is to improve the above problem.

By referring to such related patents [1-15], the studies on kinematics of mechanisms [16-18], and the studies on the creative design [19-24], we synthesize a mechanism with 1 DOF to perform the lifting and standing functions. The centre of gravity of this new wheelchair after standing up and tilting down still located between the front and rear wheels, no auxiliary wheels needed. Hence, the new design has the advantages of simple operation, more stability, and more safety. The results of this research will enhance R&D capability in the multifunctional wheelchair to improve industrial competitiveness.

2. Multifunctional wheelchair

Before executing the innovative design, we have to collect and analyse the structure of multi-function wheelchairs including the mechanism from academic paper, catalogue, and technical report. Figure 2 shows a wheelchair, which is the product of AIDC, with standing and lying functions. Figure 3 shows a wheelchair, which is the product of LEVO, with standing and tilting functions. Figure 4 shows a wheelchair, which is the product of Sunrise Medical, with lifting and tilting functions.

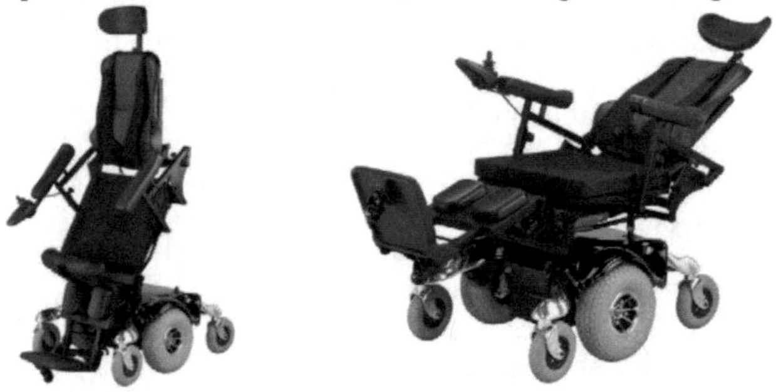

(a) Standing up position (b) Lying down position

Figure 2. Wheelchair with standing and lying functions (AIDC)

(a) Standing up position (b) Tilting down position

Figure 3. Wheelchair with standing and tilting functions (LEVO)

(a) Seat position

(b) Tilting down position

Figure 4. Wheelchair with lifting and tilting down functions (Sunrise Medical)

3. Osborn's Check-List Method

Alex Faickney Osborn (May 24, 1888 – May 5, 1966) was an advertising executive and the author of the creativity technique named brainstorming. In 1953, Osborn published ***Applied Imagination: Principles and Procedures of Creative Problem Solving,*** in which Osborn presented the technique of *Check-List*. In the book "Applied Imagination", Osborn listed 83 questions that could be used to check for creativity. Professor Dior of Harvard University summarized Osborn's 83 questions to 9 items. The common 9 items are: 1. Reverse, 2. Transfer, 3. Combine, 4. Change, 5. Extend, 6. Enlarge, 7. Reduce, 8. Substitute, 9. Rearrange. When we use of Osborn check-list method to create new ideas and designs, we have to make sure that the creations could make advantages instead of just for invention. Therefore, we must make the innovation which can bring better performance than current status.

4. Innovative Design

Currently, the mechanism of wheelchair with lifting and standing functions has 2 degrees of freedom, and used 2 power sources to perform these 2 motion functions. The 1st purpose of this paper is to invent mechanism with 1 degree of freedom (DOF) to perform lifting and standing functions. According to figure 1(a), for the traditional standing wheelchair, its' centre of gravity moves forward when standing up and it needs 2 auxiliary wheels to prevent dumping. The 2nd purpose of this paper is to improve the above problems. Here, we adopt the transfer, combine, and reduce methods of Osborn check list to design the wheelchair with lifting and standing functions.

4.1. Design Requirement

The requirements of the mechanism for wheelchair with standing and tilting functions are:

1. The mechanism of wheelchair has 1 degree of freedom.
2. The lifting and standing functions can be achieved by using only 1 power source.
3. The center of gravity of this new wheelchair after standing up still located between the front and rear wheels, no auxiliary wheels needed.

4.2. Transfer Method

The "transfer method" is to apply same ideas from other fields. It could be similar or totally different task. There are some old proverbs described the experience transfer, such as "The advice from others are better than 10-years studies." and "The experience of others could be used as the reference." "Transfer method" is not just to duplicate. It should not only absorb the essence from the other situation but also create a better idea and built a better result. Based on the transfer method, the brand new and practical invention could be realized.

Figure 5 shows a mechanism with 6 links and 7 joints (J_1=7), its' degrees of freedom can be written as:

$$F = \lambda(N - J - 1) + \sum f_i = 3(6 - 7 - 1) + (7) = +1 \tag{1}$$

IOP Conf. Series: Materials Science and Engineering **280** (2017) 012027 doi:10.1088/1757-899X/280/1/012027

Where, λ is mobility number of mechanism (for planar mechanism, λ=3), N is number of links, J is Number of joints, and Σf_i is Total degrees of freedom of joints. The mechanism, shown in figure 5, has 1 degree of freedom and can achieve vertical lifting function.

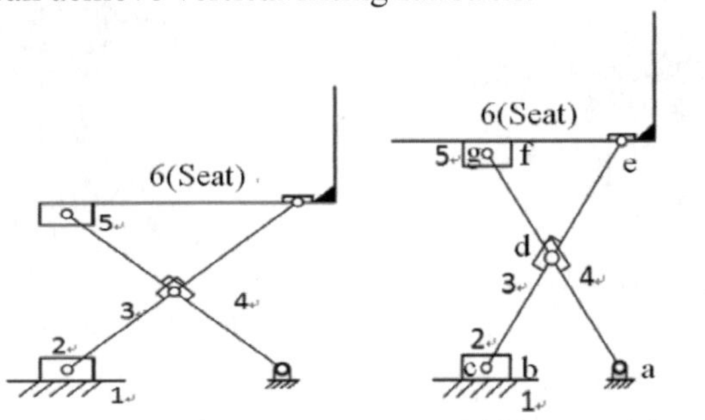

(a) Seat position (b) Lifting position

Figure 5. A mechanism with vertical lifting function

4.3. Combine Method (1)

Many creative thinking processes involve the combination of different concepts. The "combine method" of Osborn check-list is to combine the different concepts for having the new creativity. The "combine method" has the following characteristics: 1. the effects of combining many individual concepts are greater than the sum of individual effects. 2. The cost of the combining many individual concepts is less than the sum of individual cost. Due to the reason that the follower of cam can move the prescribed motion, we combine the vertical lifting mechanism, shown in Fig. 5, and can pair to for designing wheelchair with lifting and standing functions. Figure 6 shows the corresponding combining process.

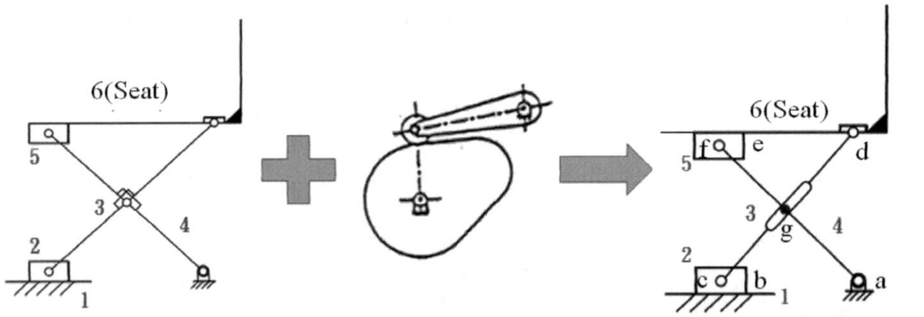

(a) Lifting mechanism (b) Cam pair (c) Temporary mechanism (I)

Figure 6. Temporary mechanism (I) obtained by "combine method (I)"

The temporary mechanism (I), shown in figure 6(c), has 6 links and 7 joints (including 1 cam pair g), i.e. J_1=6 and J_2=1, its' degrees of freedom can be written as:

$$F = \lambda(N - J - 1) + \sum f_i = 3(6 - 7 - 1) + (6 + 2) = +2 \tag{2}$$

This temporary mechanism (I), shown in figure 6(c), has 2 degrees of freedom. The temporary mechanism (I) does not meet the design requirements. So, the temporary mechanism (I) must be modified to have 1 degree of freedom.

4.4. Reduce Method

The "reduce method" of Osborn check-list is to lessen the fact, the appearance, or the significance of an existing design for having the new creativity. At this step, we use reduce method to reduce the degrees of freedom of temporary mechanism (I). In the "reduce method", slider 5 is eliminated and link 3 is adjacent to link 5 (seat 5) by revolute pair, figure 7 shows the corresponding reducing process.

The temporary mechanism (II), shown in figure 7(c), has 5 links and 6 joints (including 1 cam pair f), i.e. J_1=5 and J_2=1, its' degrees of freedom can be written as:

$$F = \lambda(N - J - 1) + \sum f_i = 3(5-6-1) + (5+2) = +1 \tag{3}$$

The temporary mechanism (II), shown in figure 7(c), has 1 degree of freedom.

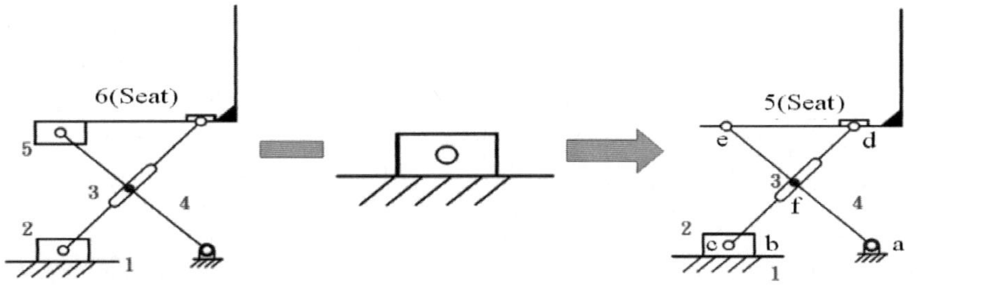

(a) Temporary mechanism (I) (b) Slider (c) Temporary mechanism (II)

Figure 7. Temporary mechanism (II) obtained by "reduce method"

4.5. Combine Method (II)

Due to the reason that temporary mechanism (II), shown in Fig. 7(c), does not have back link and foot link, we combine the temporary mechanism (II) and parallel four-bar linkage to obtain temporary mechanism (III). Figure 8 shows the corresponding combining process. The temporary mechanism (III), shown in figure 8(c), has 8 links and 10 joints (including 1 cam pair j), i.e. J_1=9 and J_2=1, its' degrees of freedom can be written as:

$$F = \lambda(N - J - 1) + \sum f_i = 3(8-10-1) + (9+2) = +2 \tag{4}$$

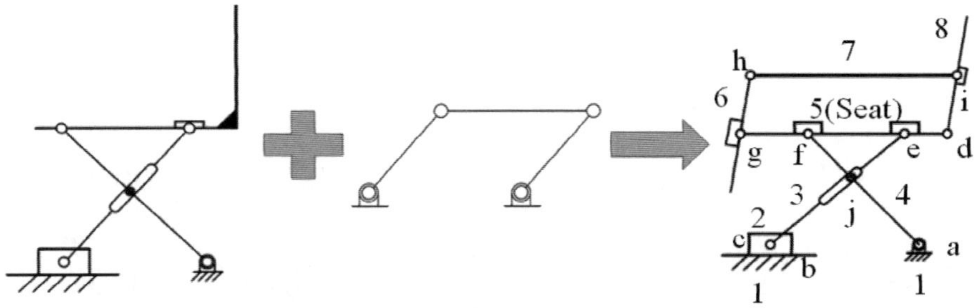

(a) Temporary mechanism (II) (b) Parallel four-bar linkage (c) Temporary mechanism (III)

Figure 8. Temporary mechanism (III) obtained by "combine method (II)"

This temporary mechanism (III), shown in figure 8(c), has 2 degrees of freedom. The temporary mechanism (III) does not meet the design requirements

4.6. Enlarge Method

The "enlarge method" of Osborn check-list is to subjoin the fact, the appearance, or the significance of an existing design for having the new creativity. Due to the reason that the temporary mechanism (III) has 2 degrees of freedom, we must reduce its' degrees of freedom to 1. At this step, we use"enlarge method" to reduce the degrees of freedom of temporary mechanism (III). In the"enlarge method", link 9 is subjoined into the temporary mechanism (III) and adjacent to link 4 and link 8 by 2 revolute pairs (joints *k* and *l*). Figure 9 shows the corresponding enlarging process. The new mechanism (IV) will

IOP Conf. Series: Materials Science and Engineering **280** (2017) 012027 doi:10.1088/1757-899X/280/1/012027

have 9 links and 12 joints (including 1 cam pair j), i.e. J_1=11 and J_2=1, its' degrees of freedom can be written as:

$$F = \lambda(N - J - 1) + \sum f_i = 3(9 - 12 - 1) + (11 + 2) = +1 \quad (5)$$

The new mechanism (IV), shown in Figure 9(c), has 1 degree of freedom.

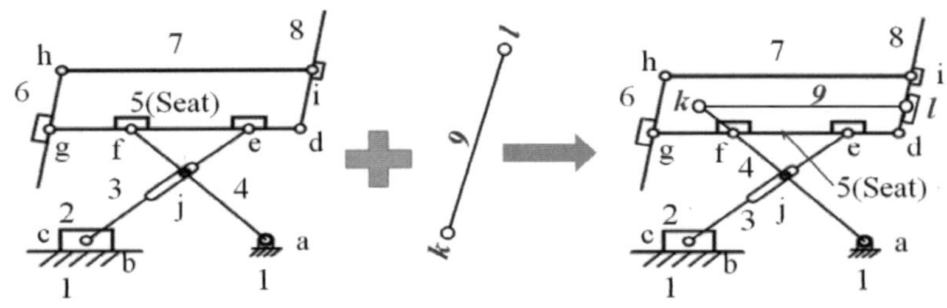

(a) Temporary mechanism (III) (b) Binary link (c) New mechanism (IV)

Figure 9. New mechanism (IV) obtained by "enlarge method"

4.7. Combine Method (3)

Due to the reason that the new mechanism, shown in figure 9(c), has 1 degree of freedom, we can arrange 1 power source to drive the mechanism to have the lifting and standing functions. At this last step, we use the "combine method" to combine the new mechanism and power source together. Figure 10 shows the corresponding combining process. The final mechanism will have 11 links and 15 joints (including 1 cam pair j), i.e. J_1=11 and J_2=1, its' degrees of freedom can be written as:

$$F = \lambda(N - J - 1) + \sum f_i = 3(11 - 15 - 1) + (14 + 2) = +1 \quad (6)$$

The final mechanism, shown in figure 10(c), has 1 degree of freedom.

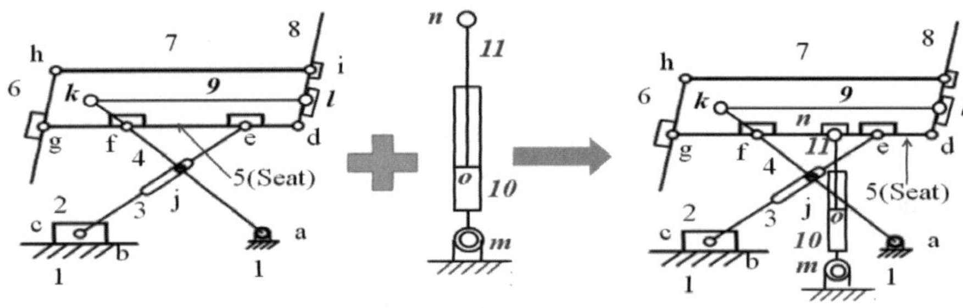

(a) New mechanism (IV) (b) Power source (c) Final mechanism

Figure 10. Final mechanism obtained by "combine method"

5. Prototype Design and Manufacture

According to Figure 10(c), the engineering drawings of the wheelchair are completed and its corresponding prototype is manufactured. Figures 11(a), 11(b), and 11(c) show the corresponding diagrams of multifunctional wheelchair at lower seat position, high seat position, and standing position. Figure 12(a), 12(b), and 12(c) shows the three situations of prototype of wheelchair with lifting and standing functions.

IOP Conf. Series: Materials Science and Engineering **280** (2017) 012027 doi:10.1088/1757-899X/280/1/012027

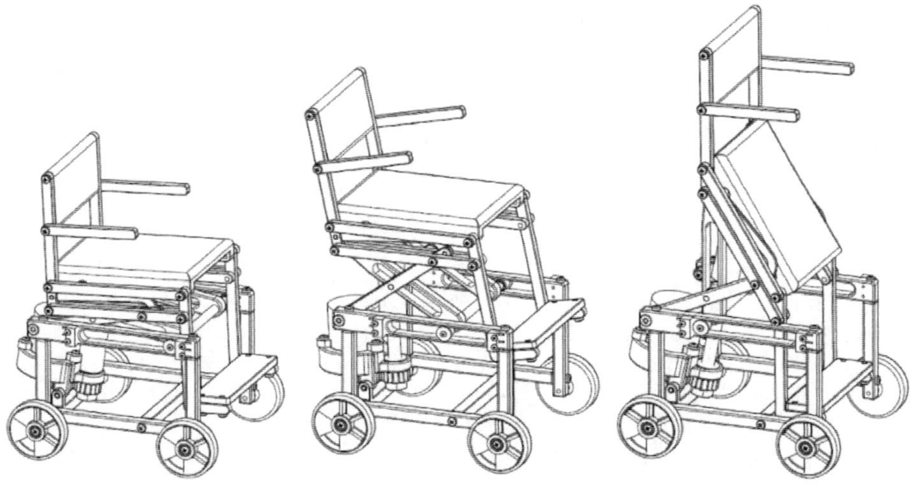

(a) Low seated position (b) High seated position (c) Standing position

Figure 11. Wheelchair with lifting and standing functions

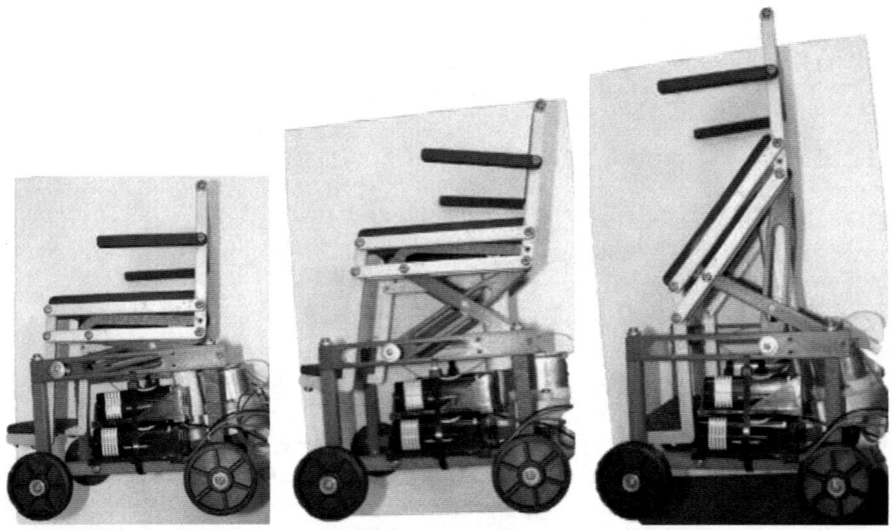

(a) Low seated position (b) High seated position (c) Standing position

Figure 12. The prototype of the wheelchair with lifting and standing functions

6. Conclusions

In this paper, we adopt the transfer, combine, and reduce methods of Osborn check list to invent "Wheelchair with One Degrees of Freedom to Perform Lifting and Standing Functions". The results of this paper can be used as a reference for engineering to design the functional wheelchair and will enhance the industrial competitiveness. We conclude that this invention "Wheelchair with One Degrees of Freedom to Perform Lifting and Standing Functions" has following characteristics:

1. The mechanism of wheelchair is a mechanism with 1 degrees of freedom, it is driven by only 1 power source to drive the mechanism to achieve lifting and standing functions.
2. The center of gravity of this new wheelchair after standing up (or lifting up) still located between the front and rear wheels, no auxiliary wheels needed.

3. This invention has the advantages of simple operation, more stability, and more safety.

Acknowledgements
The authors are grateful to the Ministry of Science and Technology of the Republic of China for the support of this research under MOST 104-2221-E-150-055.

References

[1] Perk H. and Kretschmer O.2009 *US Patent* No 5,772,226.
[2] Hunziker K 2010 *U.S. Patent* No. 0207354 A1.
[3] Hunziker K. and Steffisburg 2010 *U.S. patent* No 20100207354 .
[4] Bobichon G.1998 *US Patent* No 5,772,226.
[5] Chen C.C. and Su F.C. 2004 *Taiwan patent* I577,321.
[6] Liu P.P., Cheng L.H., Hsieh W.T., and Zheng K.W. 2006 *Taiwan patent* M 294,932.
[7] Huang S.F. and Zhu S.Y. *Taiwan patent* M 310,704 (2007).
[8] Yu Y.C. 2015 *Taiwan patent* M 492,731.
[9] Wang T.H., Chen S.Y., Chang Y.Y., and Liao Y.Y. 2015 *Taiwan patent* M 513,020.
[10] Huang C.C. 2012 *Euro. Patent* EP 2471499 A1
[11] Bergman H. 2012 *U.S patent* No 8419130.
[12] Mascari N.A. and Juchem C.M2013 *Euro. Patent* EP 2039334 B1.
[13] Shyu J.H., Liu D.Y., and Huang X.C 2006 *Taiwan patent* I250013.
[14] Hsieh L.C. and Wei S.J. 2013 *Taiwan patent* I440,451.
[15] Hsieh L.C. and Huang J.B. 2015 *Taiwan patent* I495,461.
[16] Hsieh L.C., Liu J.Y., and Hsu M.H., 1996 *Transaction of Canadian Society for Mechanical Engineering* Vol. 20 (4) p.421-435.
[17] Hsu M.H., Chen Y.G., Liu J.Y., and, Hsieh L.C., and, Chen 2007 *Proceeding of 12th World Congress on the Theory of Machines and Mechanisms* No 945.
[18] Hsieh L.C. and Chen T.H. 2012 *Journal of Advanced Science Letters* Vol. 9 p318-324.
[19] Osborn A.F. 1953 Applied Imagination: Principles and Procedures of Creative Problem Solving *New York: Charles Scribner's Sons.*
[20] Altshuller G. 1998 40 Principles– RIZ Keys to Technical Innovation Technical Innovation Center *Inc., Worcester, MA.*
[21] Yan H.S. 1998 Creative Design of Mechanical Device *Springer-Verlag, Singapore.*
[22] Hsieh L.C. and Lee H. S.2004 *The TRIZ Journal* 2004/07 p 1-30.
[23] Hsieh L.C. and Chen T.H. 2012 *Journal of Advanced Science Letters* Vol 9 p 24-29.
[24] Hsieh L.C. and Tang H.C. 2013 *Transactions of the Canadian Society for Mechanical Engineering* Vol.37 (3) p 741-753.

IOP Conf. Series: Materials Science and Engineering **280** (2017) 012028 doi:10.1088/1757-899X/280/1/012028

Passive Control of Vortex Shedding via Screen Shroud

A M Azmi[1] and T. Zhou[2]

[1] Fakulti Kejuruteraan Mekanikal, Universiti Teknologi MARA Selangor, 40450 Shah Alam Selangor, Malaysia .
[2]School of Civil, Environmental and Mining Engineering, The University of Western Australia, 35 Stirling Highway, Crawley, WA 6009, Australia.

corresponding author: azlinma@gmail.com

Abstract. The turbulent wake of screen-shrouded cylinders were experimentally investigated using flow visualization. Screen cylinders made from screen mesh of various porosity (37%, 49%, 61% and 67%) were used as the shrouds. The main purpose of the study is to examine the effect of screen porosity, β and screen diameter ratio, d_w/D (wire diameter to cylinder diameter ratio) on the vortex development behind the shrouded cylinders, particularly in supressing the vortex shedding from a circular cylinder. The diameter ratio between the screen shroud and the plain cylinder, D/d was 2.0. The flow Reynolds number based on the shroud diameter, Re_D was about 1000. Results showed that the inclusion of the screen shrouds has significant impact on the wake of the circular cylinder. With larger value of the non-dimensional parameter $\beta d_w/D$, vortex was impaired and the formation length was longer in the shrouded cylinder wake. The vortex generation mechanism was also discussed.

1. Introduction

Suppression of vortex shedding and vortex-induced vibration (VIV) has been one of the most active topics of research and patenting in fluid dynamics for many decades due to its significance in engineering applications. Previous investigations on control devices to suppress VIV have contributed significantly to the fields of buildings, bridges and marine. VIV can be suppressed by active or passive methods. Active methods involve applying energy from an external force while passive methods involve modification of the flow or structure. Each method works by disrupting the vortex shedding, either by diminishing the strength of the shear layers on either side of the cylinder or by disrupting the interaction of the shear layers. Extensive reviews of these devices can be found in available literatures [1].

Shrouds are one of the passive solutions that are less frequently employed. The used of shrouds in mitigating the impact of vortex formation have long been introduced since the work of Price in 1956. Price pioneered the work in circular holes shroud postulating that it would break up the flow into large number of small vortices [2]. The use of a screen mesh as shroud for vortex shedding suppression was first explored by Zdravkovich and Volk [3]. Zdravkovich and Vork [3] conducted experiments at high reduced velocities for three types of shroud; the square holed, circular holed and a fine-mesh gauze all having 36% porosity. They found that the shroud made of fine-mesh gauze was the most effective one. The gauze prevented separation and affected the pressure distribution more significantly than the other two shrouds and the uniform part of the pressure distribution curve started at 120° instead of 90°. The study of screen mesh as controller of a bluff body wake has been revisited in recent years. Azmi et al.

[4, 5] found that the vibration amplitude of a bare cylinder was suppressed by about 50% and 78% using a screen shroud of 72% and 67% porosity at a shroud-to-cylinder diameter ratio of 1.25 and 2.0, respectively.

In studying the wake characteristics of a single screen body, Takeuchi et al. [6] measured the airflows around permeable cylinders made of wire mesh of 54% and 71% porosities. The flow just behind the permeable cylinders had forward velocity due to the permeability, which relaxed the near field excitation of Kármán vortex. As a result, the reattachment point behind the cylinder was delayed several times of that a solid cylinder wake. This result is consistent with Azmi et al. [7], who showed that that there exists a critical location that differentiates the screen cylinder wake into two regions where the formation of the fully formed large-scale vortices was delayed until this critical location. Gansel et al. [8] tested cylinders made from metal mesh of various porosities. The results indicated that the wake characteristics changed toward the wake characteristics of a solid cylinder at a porosity just below 75%. [9].

The various studies on perforated bodies proved that such flow cannot be easily predicted based on porosities alone, since there are multiple length scales involved such as apparent diameter, wire diameter and aperture (hole size) that govern the flow [10]. However, very limited studies have been conducted so far in understanding the effect of these parameters. The present study aims to examine the effect of a non-dimensional parameter (which is based on porosities and screen diameter ratio) on the vortex development behind a screen shrouded cylinder, paying attention to the vortex formation length. The study would certainly complement the relatively limited literature on this topic.

2. Experimental Details

The experiments were conducted in a small open circuit type wind tunnel with a test section of 380 mm (width) × 255 mm (height) and 1.8 m (long). The free stream velocity in the test section was uniform to 0.2%, and the longitudinal turbulence intensity was less than 5%. Experiments were conducted in the wakes of both the plain cylinder and shrouded cylinder. The solid cylinder was a plain and smooth cylinder made of polished aluminium of diameter, d of 12.7 mm. The screen cylinder was of 25 mm diameter, D, rolled from a stainless steel wire mesh of a square-cross-section of various porosity, β, aperture, a and wire diameter, d_w The porosity of the screen mesh β is defined as the ratio of the opening area to the total area, i.e. $a^2/(a+d_w)^2$. The properties of the screen cylinders are given in Table 1. The shrouded cylinder had an outer to inner diameter gap ratio of about 2.0. The plain cylinder and the shrouded cylinder had an aspect ratio of $l/d = 29.9$ and 15.2, respectively. Figure 1 shows the shrouded cylinder set up and one of the screen cylinders used in the study.

Table 1. Screen mesh properties

Screen	Hole Size, a (mm)	Wire Diameter, d_w (mm)	Porosity, β (%)	Dimensionless parameter, $\beta\, d_w/D$	Formation length, L_f/D
1	0.25	0.16	37	0.237	8
2	0.71	0.32	48	0.614	8
3	1.6	0.45	61	1.098	14
4	2	0.45	67	1.206	18

Smoke wire flow visualizations were carried out using a high speed camera to obtain a qualitative understanding on the evolution of vortices in the wake of the screen shrouded cylinder. Smoke is produced from a thin (0.1 mm) steel wire coated with oil. The wire is connected as a resistor to an electric circuit. When the circuit is on, the wire heats causing the oil to generate smoke which was carried downstream by the flow to reveal the flow motion. The visualizations were conducted at Reynolds numbers of around 1000.

IOP Conf. Series: Materials Science and Engineering **280** (2017) 012028 doi:10.1088/1757-899X/280/1/012028

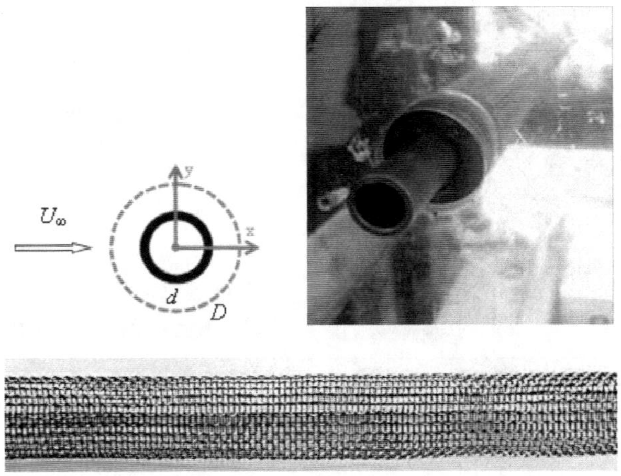

Figure 1. Shrouded cylinder set up and one of the screen cylinders used in the experiment.

3. Results

For the solid cylinder wake (Figure. 2), the large-scale organized vortices can be seen clearly with two rows of counter-rotating vortex structures connected by the streamwise rib-like structures. The rib-like structures are stretched due to the counter-rotating vortices. The two rows of vortices also interact with each other when evolve downstream by penetrating to the opposite side across the wake centreline. The formation length of the Kármán vortices is about $2d$ which is in good agreement with [11].

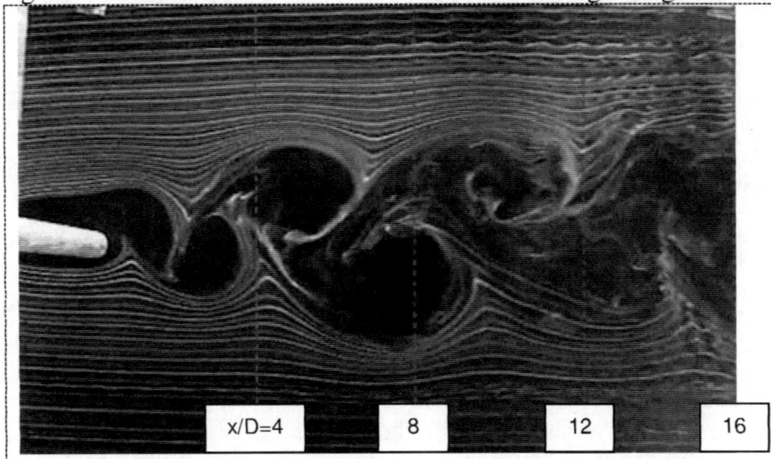

Figure 2. Flow visualization in the wake of a solid cylinder.

The screen-shrouded cylinder wake using screen 1 and screen 2 behaves quite similarly where the onset of shear layer interaction occur in the wake at about $8D$ as shown in Figure 3 and 4, respectively. There is an onset of transition in the shear layers and the development of turbulence in the wake further downstream. The near wake is surrounded by a laminar shear layer. An undulation of the shear layer that looks like a transition waves is evident, followed by the formation of large-scale structures which become irregular further downstream once the turbulence sets in. It can be seen in Figure 4 that the shear layer rolls up into small-scale vortices at x/D= 4. The formation length, L_f for both flow configurations using screen 1 and 2 was found to be at about $8D$ from the cylinder.

IOP Conf. Series: Materials Science and Engineering **280** (2017) 012028 doi:10.1088/1757-899X/280/1/012028

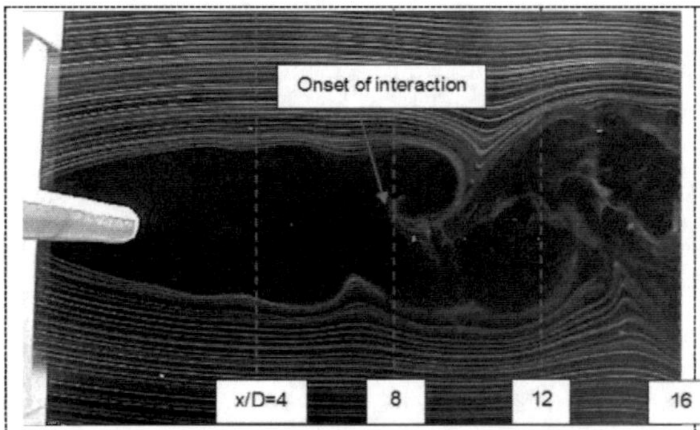

Figure 3. Flow visualization in the wake of a screen-shrouded cylinder (screen 1).

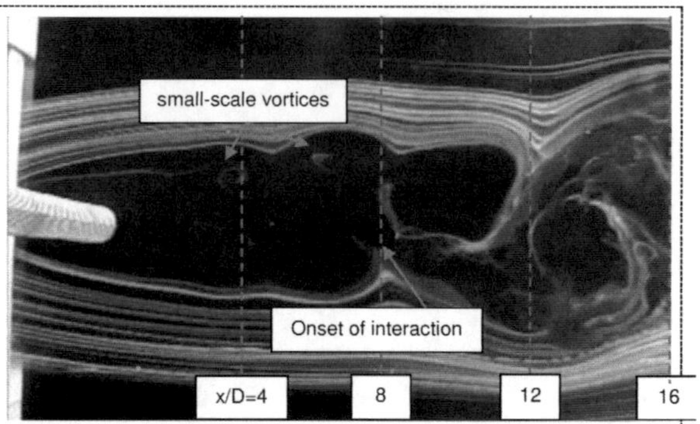

Figure 4. Flow visualization in the wake of a screen-shrouded cylinder (screen 2).

Figure 5 and 6 (based on screen 3 and 4, respectively) clearly shows that the undulation occurred latter in the wake and the formation length was delayed. Small-scale vortices are formed at about greater than $8D$. they do not roll up quickly to form the large-scale structures in comparison to the previous two cases. As evolving downstream, the small-scale structures become larger in size. When their sizes are large enough, they engulf the one next to them, resulting in amalgamation (Figure 5). The onset of the interaction between the two shear layers occurred at about $x/D = 14$ and 18 (not shown) for screen 3 and 4 respectively, where the large-scale structures start to form. Obviously, the shrouded cylinders wake show similar behaviour of growth of small-scale vortices and the formation of the large-scale ones further downstream, with the difference being in the formation length.

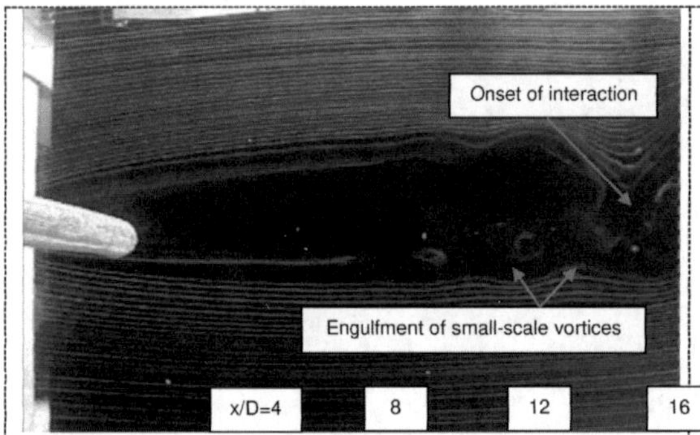

Figure 5. Flow visualization in the wake of a screen-shrouded cylinder (screen 3).

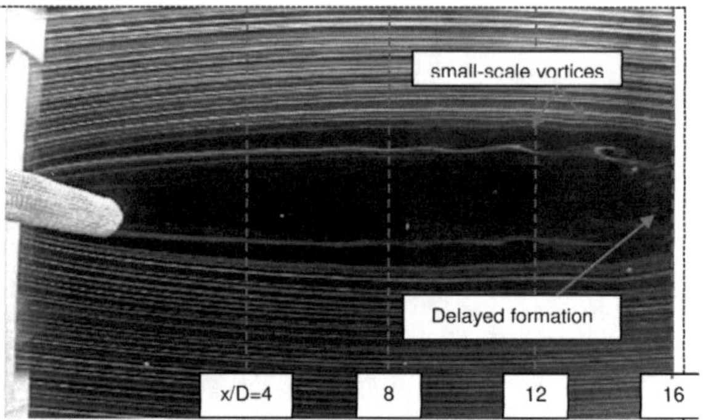

Figure 6. Flow visualization in the wake of a screen-shrouded cylinder (screen 4).

Table 1 summarizes the observed trend in the formation length as a function of the screen parameter and this trend is also plotted in Figure 7. Figure 7 shows that the non-dimensional parameter seems to have significant effect on the vortex formation length. As $\beta d_w/D$ increases, the vortex formation length significantly increases, particularly for $\beta d_w/D > 0.6$, where the formation length increases exponentially. For $\beta d_w/D$ less than 0.6, the formation length seems to lie in the same range. It needs to be noted that, this result may be applicable for the studied Reynolds number range and for the cylinder gap ratio of 2.0. The flow would behave differently, for other Reynolds number and gap ratio.

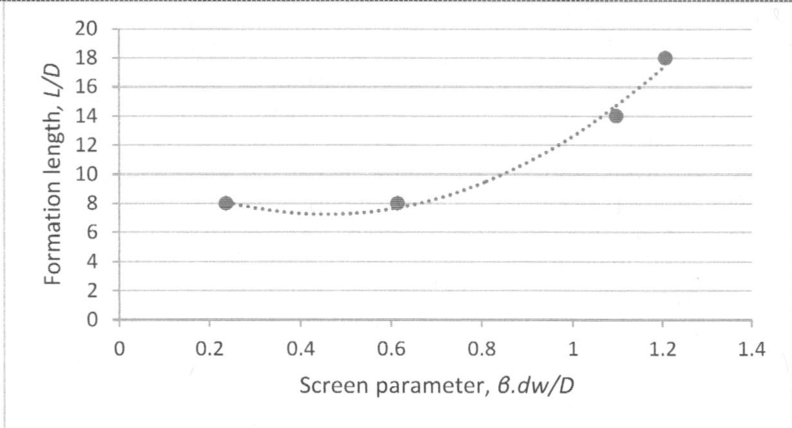

Figure 7. Relationship between screen parameter and formation length.

4. Conclusion

Flow visualization using smoke wire have been conducted to understand the flow behaviour behind screen-shrouded cylinders. The effect of screen parameter on the vortex development in the cylinder wakes was studied. From the visualization results, it can be concluded that with higher screen parameter value, the formation length became larger. For the non-dimensional parameter of greater than 0.6, the effect was more significant, where the formation length increased exponentially. Two flow regimes have been identified i.e. the vortex merging process and the delay of the large-scale organized structures in the screen-shrouded cylinder wake. The formation length is expected to depend on factors such as the screen parameter presented here, the gap ratio as well as Reynolds number.

Acknowledgments

AMA wishes to acknowledge financial support from Universiti Teknologi MARA and the Ministry of Higher Education Malaysia (600-RMI/FRGS 5/3 (0100/2016)).

References

[1] Zdravkovich MM 1981 Review and classification of various aerodynamic and hydrodynamic means for suppressing vortex shedding *J. Wind Eng. Ind. Aerodyn.* **7(2)** 145-189

[2] Price P 1956 Suppression of the fluid-induced vibration of circular cylinders *J. Eng. Mech.* **82(3)** 1-22

[3] Zdravkovich MM, Volk JR 1972 Effect of shroud geometry on the pressure distributed around a circular cylinder *J. Sound Vibration* **20(4)** 451-455

[4] Azmi AM, Zhou T, Cheng L, Wang H, Chua LP 2012 On the Effectiveness and Mechanism of Vortex-induced Vibration Suppression using a Screen Cylinder *The Twenty-second International Offshore and Polar Engineering Conference* International Society of Offshore and Polar Engineers

[5] Azmi AM, Zhou T, Zhou Y, Chen J, Cheng L 2015 The effect of a screen shroud on vortex-induced vibration of a circular cylinder and its wake characteristics *The Twenty-fifth International Ocean and Polar Engineering Conference* International Society of Offshore and Polar Engineers

[6] Takeuchi H, Tasaka T, Murai Y, Takeda Y, Tezuka H, Mori M 2007 Particle image velocimetry for air flows behind permeable cylinders. *Proc. of the 5th ASME–JSME Joint Fluids Engineering Conference* San Diego USA vol 37035 pp. 1-8

[7] Azmi AM, Zhou T, Zhou Y, and Cheng L 2016 Statistical analyses of a screen cylinder wake *Fluid Dyn Res*, **49(1)** 015506

[8] Gansel LC, McClimans TA, Myrhaug D 2012 Flow around the free bottom of fish cages in a uniform flow with and without fouling *J. Offshore Mech. Arct.* **134(1)** 011501
[9] Levy B, Friedrich H, Cater JE, Clarke RJ, Denier JP 2014 The impact of twine/mesh ratio on the flow dynamics through a porous cylinder *Exp. Fluids.* **55(10)** 1829
[10] Azmi AM, Zhou T, Cheng L 2014 On the Effect of Various Shrouds on the Wake of a Circular Cylinder. *ASME 2014 4th Joint US-European Fluids Engineering Division Summer Meeting collocated with the ASME 2014 12th International Conference on Nanochannels, Microchannels, and Minichannels 2014* pp. V01BT14A016-V01BT14A016
[11] Unal MF, Rockwell D 1988 On vortex formation from a cylinder. Part 1. The initial instability. *J. Fluid. Mech.* **190** 491-512

IOP Conf. Series: Materials Science and Engineering **280** (2017) 012029 doi:10.1088/1757-899X/280/1/012029

Engine-start Control Strategy of P2 Parallel Hybrid Electric Vehicle

Xu Xiangyang[1], Zhao Siqi[1] and Dong Peng[1]

[1] School of Transportation Science and Engineering, Beihang University, 100191, Beijing, China

zhaosiqi1212@126.com

Abstract. A smooth and fast engine-start process is important to parallel hybrid electric vehicles with an electric motor mounted in front of the transmission. However, there are some challenges during the engine-start control. Firstly, the electric motor must simultaneously provide a stable driving torque to ensure the drivability and a compensative torque to drag the engine before ignition. Secondly, engine-start time is a trade-off control objective because both fast start and smooth start have to be considered. To solve these problems, this paper first analyzed the resistance of the engine start process, and established a physic model in MATLAB/Simulink. Then a model-based coordinated control strategy among engine, motor and clutch was developed. Two basic control strategy during fast start and smooth start process were studied. Simulation results showed that the control objectives were realized by applying given control strategies, which can meet different requirement from the driver.

1. Introduction

As the energy security and environmental pollution issues are getting increasingly severe in the world, it's imperative to develop hybrid electric vehicles (HEVs), which can reduce the greenhouse gases emission and enhance the vehicle's system efficiency [1]. HEVs must shift working modes frequently to accommodate the complex road condition and enhance the driveline's efficiency. So the engine-start process is of vital significance, as it affects the vehicle's drivability and driving comfort. Studies have been made to use a single electric motor (EM) to start the engine, while some challenges exist in this process. The single EM must meet the torque demand for driving wheels and simultaneously provide an extra torque to start the engine by an engine-disconnect clutch (EDC) within an acceptable level of noise, vibration and harshness (NVH).

Several contributions have been made in this areas. Zhonghua Lu took the jerk as the single objective, and developed a coordinated control strategy about automatic clutch, ISG and engine, which is proved to be effective to reduce the jerk of vehicle [2]. Anthony Smith devoted to applying a closed-loop slip control and an open-loop torque control to EDC, but ignored the internal combustion engine (ICE) speed under different conditions [3]. Dongsuk Kum used Dynamic Programming to solve the nonlinear constrained optimal control problem, which meant to minimize engine-start time while accurately supplying the driver torque demand simultaneously [4]. Another research conducted by Koos van Berkel presented a new controller design with a single calibration parameter which was used to tradeoff between fast and smooth clutch engagement [5].

This paper built a physic model in MATLAB/Simulink, then developed a coordinated control method among ICE, EM and EDC, which included the torque compensation of EM, the torque

IOP Conf. Series: Materials Science and Engineering **280** (2017) 012029 doi:10.1088/1757-899X/280/1/012029

adjustment of ICE, speed modification of ignition, and the pressure to torque control of EDC. This model-based controller dedicated to optimizing the tradeoff between comfort and drivability according to various engine-start condition. It's proved that the controller successfully reduced vibration within an applicable start time by means of adjusting the main control objective under a certain situation. Which is of great importance to improve both stability and drivability of engine start process.

2. P2 configuration construction

Figure 1 shows the powertrain architecture of HEV P2 configuration, including an internal combustion engine (ICE), engine-disconnect clutch (EDC), a dual mass flywheel (DMF), an electric motor (EM), an automatic transmission (AT) without torque converter, and a transaxle and wheels.

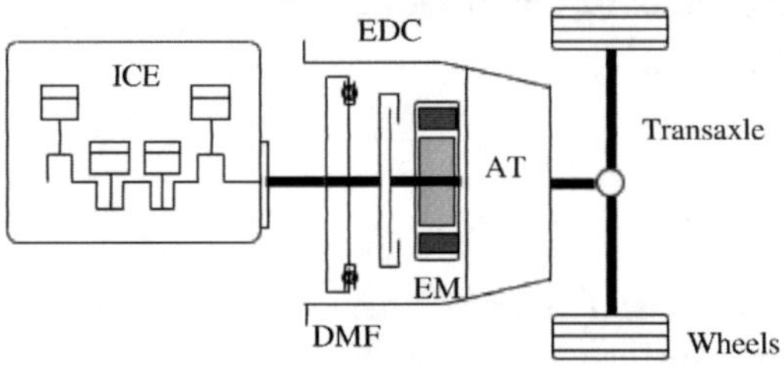

Figure 1. HEV P2 Configuration

ICE is mechanically connected to EDC hub through DMF, while EM is connected to EDC friction disc. Both ICE and EM can output torque to the front shaft (drive shaft) of the vehicle through AT. When the EDC is separated with ICE, the vehicle works under the pure electric mode, with EM as the only power source. When there is a need to switch mode, EDC engages and transfers torque form EM to ICE, dragging the engine to ignition speed. During this engine-start process, it is hard to control the engine output torque, especially when igniting, so a dual mass flywheel is employed to decrease vibration in the front of EDC. After start-up, the vehicle may enter the engine-drive mode or the parallel drive mode, as well as the battery-charge mode.

3. Engine-start dynamic analysis

During the engine start process, the resistant torque is consist of three parts: rotary inertia torque (T_r), gas compression torque (T_g), and friction torque (T_f).

3.1. Rotary inertia torque

The rotary inertia resistant torque results from the reciprocating movement of the piston and linkage, and it equals to [6]

$$T_r = \mathrm{r}F_r(\sin\theta + \frac{\sin\theta\cos\theta}{\sqrt{(\frac{L}{R})^2 - sin^2\theta}}) \tag{1}$$

Where, r is the radius of crank (m), θ is the crank rotary angle (°), F_r is the inertia force, and l is the length of the linkage (m).

3.2. Gas compression torque

The compressed gas in the cylinder mainly acts on the top of the piston and pushes the piston to do the reciprocating movement. During the compression stroke and exhaust stroke, the compressed gas does negative work, serving as a resistant torque of engine-start; during the intake stroke and power stroke, the compressed gas does positive work, driving the piston to move. The gas compression resistant torque (T_g) is calculated by

$$T_g = (P_{gas} - P_0)S_p rK \tag{2}$$

In equation (2), P_{gas} is the pressure of compressed gas in the cylinder (Pa), P_0 is the atmosphere pressure (Pa), S_p is effective area of the piston top (m^2), and K is a conversion factor, which equals to

$$K = \sin\theta + \frac{r}{l}\sin\theta\cos\theta \Big/ \sqrt{1 - (\frac{r}{l}\sin\theta)^2} \tag{3}$$

3.3. Friction torque

The friction torque results from the friction between the piston ring and cylinder, the piston skirt and the cylinder, also the other frictions among the attachments.

3.3.1. Piston ring. During start process, the friction resistant torque can be calculated by

$$T_{frh} = a_1\left[\mu v\left(P_e + P_{gas}\right)w\right]^{0.5} d(n_0 + 0.4n_c)r|K| \tag{4}$$

Where, μ is the kinetic viscosity coefficient of lubricating oil [*kg/(m·s)*], v is the speed of piston (*m/s*), P_e is the pre-tightening force (*Pa*), w is the width of the piston ring (*m*), d is the diameter of the cylinder (*m*), n_0 is the number of oil rings, n_c is the number of gas ring, and a_1 is a shape coefficient of piston rings.

3.3.2. Piston skirt. The friction between cylinder liner and piston skirt is much smaller than that resulted from the piston ring, and is calculated by

$$T_{fs} = a_2(\mu\frac{v}{h})dMr|K| \tag{5}$$

In which, M is the length of the piston skirt (*m*).

3.3.3. Attachment friction. The attachment friction of the engine operation includes the resistance from water pump, oil pump, fuel pump, etc. The resistant torque is proportioned to the speed of the crankshaft, and can be expressed by

$$T_a = a_3\mu w \tag{6}$$

Where w is the speed of the crankshaft (*rad/s*).

4. Engine-start control strategy

As shown in Fig.2, ICE is controlled by engine control unit (ECU), and EM is controlled by motor control unit (MCU). The actuation of AT and EDC are both controlled by transmission control unit (TCU). While, all the ECU, MCU and TCU coherently follow the order of hybrid control unit (HCU), which is determined by an explicit control law according to the operating condition.

Fast responsiveness and driving comfort are the two major objectives of the engine start control. A large engine-start torque is needed to ensure fast responsiveness, however, to meet the comfort demand, it's necessary to avoid the sharp increase of the input torque on AT shaft. So a coordinate control strategy among the engine, motor and clutch is applied to a certain extent.

Two basic control methods were developed in this paper, both of them are divided into five phases as shown in Figure 2.

4.1. Fast start control strategy

Fast start control strategy is suitable to urgent torque demand conditions such as overtaking and the control method is described as follow.

Phase 1: HCU is awaiting the starting ICE command.

Phase 2: while starting ICE=1, EDC starts to fill oil, which process last for t1 seconds (here t1=0.2) to ensure the clearance between EDC can be eliminated, and the oil pressure finally comes to kisspoint pressure.

Phase 3: EDC begins to slip and transmit torque from EM to ICE, dragging the ICE to accelerate. During this phase, to ensure a constant torque outputting to axel, it's necessary to increase the output torque of EM to compensate EDC's friction torque.

Phase 4: when ICE speed is dragged to n1 (here n1=800r/min), HCU sends a command to ignite ICE, then ICE starts to work. In this way the ICE continues to accelerate via its own output torque as well as EDC' drag torque. Besides compensate torque acting on EM is also a must.

Phase 5: once the speed ratio (*nu*) of ICE and EM is more than *a* (here *a*=0.8), the control pressure of EDC increases dramatically to lock the clutch.

Figure 3 shows the control pressure of EDC during fast start process.

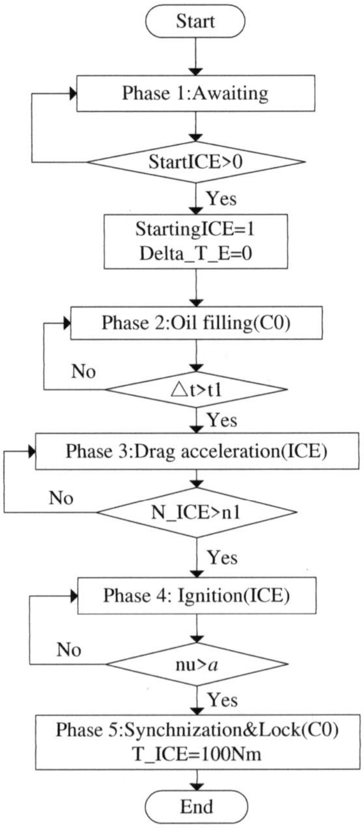

Figure 2. Engine-start control flow

4.2. Smooth start control strategy

When the vehicle is climbing on a slope or the SOC value is low, it's applicable to start ICE. Since the torque demand is not emergent, so driving comfort is the priority to ensure, rather than start time.

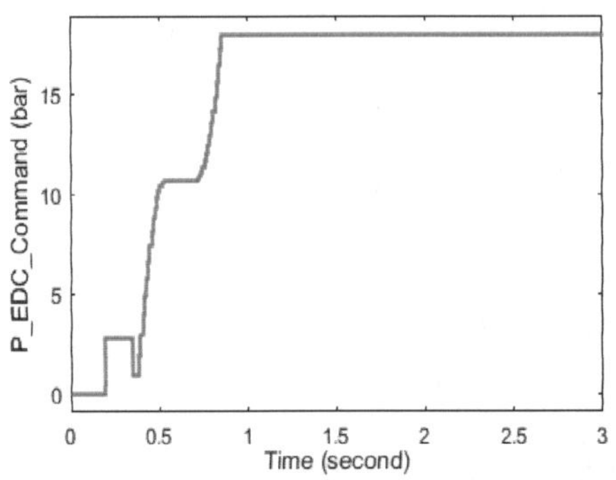

Figure 3. Fast start control pressure of EDC

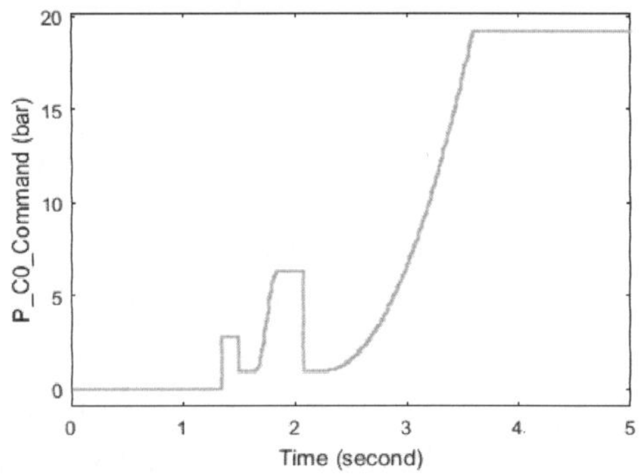

Figure 4. Smooth start control pressure of EDC

The control process is similar to fast control. First, t1seconds (here t1=0.3) is required to finish oil filling phase, then the clutch starts to slip and drag the ICE to accelerate. When the speed of ICE is more than n1 (here n1=850rpm), ICE starts igniting and the control pressure of EDC drops to kisspoint pressure as soon as possible to open the clutch. This process is very different from fast start control, because it's critical to ensure the driving comfort by cutting off the vibration from ICE. Since EDC is open and not available to transmit torque, ICE then accelerates itself. When the speed ratio (*nu*) of ICE and EM is more than *a* (here a=0.85), EDC begins to lock.

Compared with fast start control, Figure 4 shows the control pressure of EDC during smooth start process.

5. Simulation results

The simulation results of fast start control and smooth start control are shown respectively as follow.

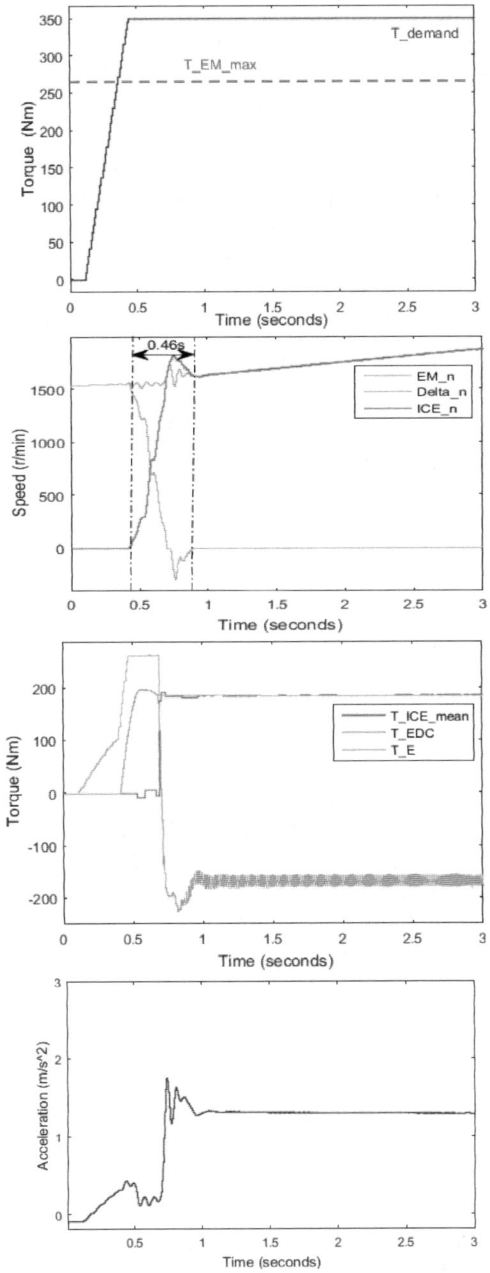

Figure 5. Fast start control results

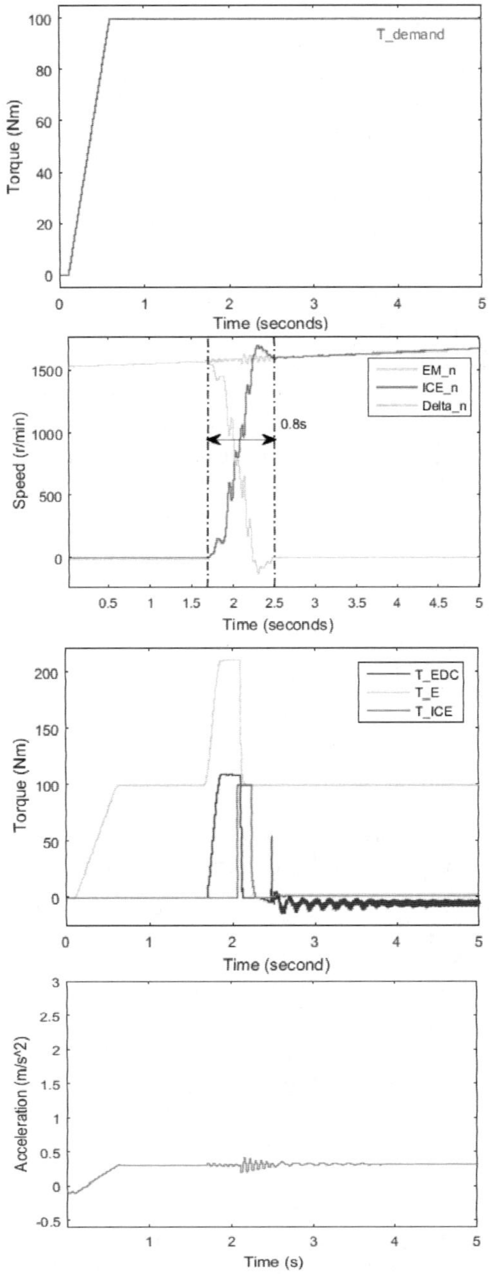

Figure 6. Smooth start control results

5.1. Fast start control results

Figure 5 shows the simulation results of fast start control. The vehicle is assumed to be costing at 60km/h in the 5th gear, ICE is turned off. Suddenly the driver pushes the accelerator to the maximum pedal position in 0.3 seconds. This is interpreted as an input torque request of 350 Nm, while the EM can generate only 260 Nm and only for a short time. So the ICE has to be started. To start the ICE as soon as possible, the control pressure of EDC maintains at a high level and the EDC keeps slipping to accelerate the ICE. ICE ignites at around 0.7s when the speed is over 800r/min. The whole start process last for only 0.46s, which can meet the sudden torque requirement form driver. On the other hand, this sharp increase of torque cause some vibration to the drivetrain, which is still acceptable.

5.2. Smooth start control results

Figure 6 shows the simulation results of smooth control results. During this situation, the vehicle is driven purely by the electric motor at 60 km/h in 5th gear with an input torque at 100Nm. However when there is a decreasing SOC (state of charge) of the battery, HCU will start the ICE without changing the input torque. Because this situation should not cause the attention from driver. To start the ICE smoothly, EDC opens when the ICE start igniting as mentioned in the smooth start control strategy. The whole process last for 0.8s, and the vibration is limited within 0.4m/s^2. When the start process finishes, the ICE maintains at the idle condition, and the car is still driven by the EM.

Compare the results above with each other, it's proved that the control strategy can make different compromise between start speed and stability. During fast start process, only 0.46s is needed, which direnctly responds to the driver's demand, twice as fast as smooth start. While the jerk on the other side is relatively obvious compared with smooth start. During smooth start, the driver doesn't not send out any demand and he doesn't have preparation for the coming jerk during engine start process, so any disturbance will cause discomfort. Therefore, in the smooth control strategy, stability is set as the first priority. From the results above, it's found that the vibration during smooth start process is ignorable compared with fast start process, but at the expense of extending the start time. In a word, through the comparison, the effectiveness of control strategy is clearly proved.

6. Conclusions

This paper analysed the resistant torque during the engine start process, and establishes the physical model in MATLAB/Simulink. Considering the problem is a multi-objective optimization, different control laws were developed to suit various conditions. Both start time and driving comfort are considered during the engine start process and a compromised control strategy is established.

The simulation results shows that the control strategy is effective to balance the start time and the driving comfort under some typical condition.

As future work, a more detailed control strategy will be developed, other influence elements such as the SOC value and the start temperature will be included during the control, then, more driving conditions will be studied, and the fuel economy will be considered as well.

References

[1] Ehsani M, Gao Y and Emadi A 2010 *Mod. Elec. Hyb. Ele. & Fue. Cel. Veh. Fund. Theo. & Des.* **23** 949
[2] Lu Z, Cheng X and Feng W 2010 *Coordinated control study in engine starting process of hybrid vehicle* (Harbin: IEEE) p 2111
[3] Smith A, Bucknor N, Yang H and He Y 2011 *Controls Development for Clutch-Assisted Engine Starts in a Parallel Hybrid Electric Vehicle* (Detroit: SAE 2011 World Congress & Exhibition)
[4] Kum D, Peng H and Bucknor N K 2013 *J. Dyn. Sys. Meas. & Con.* **135** 450
[5] Berkel K V, Hofman T, Serrarens A and Steinbuch M 2014 *Con. Engi. Prac.* **22** 57
[6] Liu H Q, Chalhoub N G and Henein N 2001 *J. Engi. F. G. Tur. & Pow.* **123** 121

A NOVEL APPROACH FOR ANKLE FOOT ORTHOSIS DEVELOPED BY THREE DIMENSIONAL TECHNOLOGIES

R M Belokar [1], H K Banga[2] and R Kumar[3]

[1, 2]Production & Industrial Engineering Department,
PEC University of Technology, Chandigarh, 160-012, India
[3]Mechanical Engineering Department, UIET, Panjab University, Chandigarh, 160-025, India

rmbelokar@pec.ac.in

Abstract. This study presents a novel approach for testing mechanical properties of medical orthosis developed by three dimensional (3D) technologies. A hand-held type 3D laser scanner is used for generating 3D mesh geometry directly from patient's limb. Subsequently 3D printable orthotic design is produced from crude input model by means of Computer Aided Design (CAD) software. Fused Deposition Modelling (FDM) method in Additive Manufacturing (AM) technologies is used to fabricate the 3D printable Ankle Foot Orthosis (AFO) prototype in order to test the mechanical properties on printout. According to test results, printed Acrylonitrile Butadiene Styrene (ABS) AFO prototype has sufficient elasticity modulus and durability for patient-specific medical device manufactured by the 3D technologies.

1. Introduction
Orthosis are assistive medical devices to support patients' biomechanical needs and contribute their quality-of-life as a result of aging and accidents. Each orthosis is used to control extremity to recover functionality and named according to location of injured body part. The most common orthosis used in orthopedics can be classified as illustrated in Figure 1.

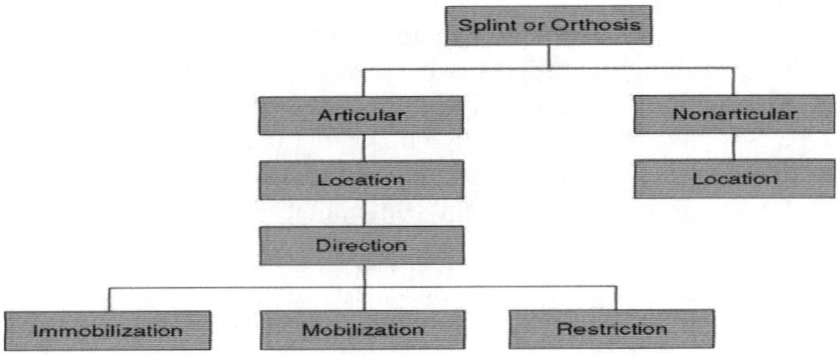

Figure 1. Orthosis classification in orthopedics

AFOs are used to control motion of the ankle and prevent deformities. AFOs can generally be used for following purposes: brace weak limbs, grasp ankle & foot in the right position and correct the foot drop. Since the users have different sizes of ankles and different types of deformities, an AFO must be produced to match in size to the patient. Mass production AFOs may not suitable for each patient and may deteriorate the illness. Considering these factors, 3D printing and 3D scanning technologies are of great importance for producing individualized biomedical equipment [1].

1.1. Scanning Process

Nowadays there are numerous 3D scanning technologies such as laser triangulation, structured light, photogrammetry, and contact and laser pulse. These technologies digitize the real objects to form the raw geometry. 3D scanning is a process to obtain the geometry of an object by using 3D scanner. The resolution of 3D scanning is as important as the 3D scanner type and hardware.

3D scanning technologies applied outside of the real objects are generally used in industry, medicine, military and education. These technologies have promising solutions for medical needs such as anatomical models, biocompatible implants, orthotic & prosthetic models, plastic surgery, biomedical engineering and tissue engineering [2]. 3D scanning can be classified as explained in Table 1.

Table 1. 3D scanning usage classification [3]

Field	Section	Solution
Industry	Reverse Engineering	Remanufacturing old parts
	Rapid Prototyping	Prototype manufacturing
	Aerospace	Part replacement
Medicine	Orthopedics & Prosthetics	Patient-specific medical devices
	Anatomical Bio Models	Anatomical to assist-surgical models
	Plastic Surgery	Prediction-planning plastic surgery
	Implants Dental	implants
Military	Mission Planning	Terrain-building reconstruction
	Crime Scene	Criminal investigation
	Prototype Manufacturing	Gun-vehicle parts
Education	Training	Educational models
	Research	Experimental studies
Art	Architecture	Environmental-heritage
	Fashion	Clothes-shoes

New 3D scanners can create mesh geometry of objects directly with real-time scanning. However most of the 3D scanners used in medicine still require manual surface reconstruction to finalize the missing meshes. Some details on the surface can be distortable and pattern failure can occur in raw model after reconstruction process. In order to prevent this distortion, 3D scanning process should be carried out in real-time with relevant types of 3D scanners. Thus designers can achieve more precise results.

1.2. Printing Process

3D printing is generally known as Rapid Prototyping (RP) or Additive Layer Manufacturing (ALM). RP and ALM are the fabrication approaches that build two dimensional (2D) layers to form almost any shape of three dimensional (3D) solid objects. A 3D printer can be accepted as a type of industrial robot that prints almost anything from a toy or a gun to manufacturing prototypes with outstanding resolution and surface quality. Prototype manufacturing especially in conventional methods is time consuming, expensive and requires specific skills which few engineers have. However, 3D printers reduce time and cost of designing new prototypes [4]. In general, 3D printing procedure of any object including AFOs is illustrated on Figure 2. A solid object can be fabricated by utilizing a 3D printer in four particular stages to be specific; demonstrating stage, cutting stage, printing stage and completing stage.

IOP Conf. Series: Materials Science and Engineering **280** (2017) 012030 doi:10.1088/1757-899X/280/1/012030

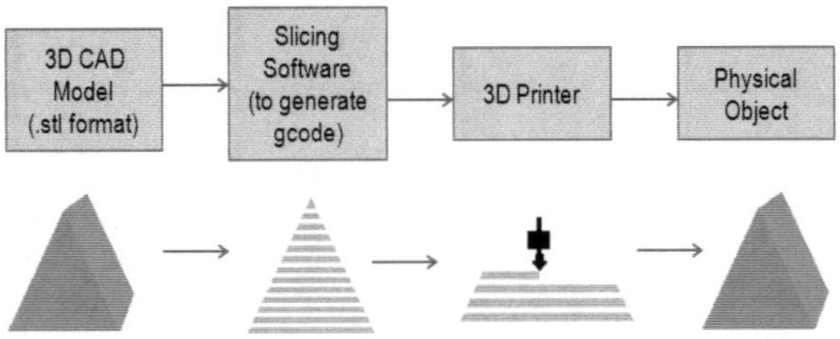

Figure 2. Basics of 3D printing process

3D printers are quickly creating advancements and progressively being utilized as a part of a few fields, for example, car, assembling, workmanship and building. A considerable amount of studies have been completed on the execution of car, assembling, workmanship and building fields [5]. Notwithstanding, utilization of 3D imprinting in medicinal division is toward the starting stage and has an energizing vision that will bring numerous open doors for humankind [6, 7]. As of not long ago, extremely constrained executions utilizing 3D printers have been led on biomedical division. This usage can be given as titanium pelvis embedded, bring down jaw transplant, plastic tracheal prop and prosthetics utilized for debilitated creatures [8, 9]. Figure 3 shows real-time scanning in progress of foot deformities in patients with the help of human scanner.

2. Methodology

In this study, a hand-held type 3D laser scanner with structured light scanning technology is used to obtain mesh geometry of the patient's limb. This scanning technique is an automotive process of attaining digital input data of patient's anatomy. The 3D scanner is rotated manually around patient's limb to create the template model in just one minute. This raw template is used directly for the modelling process of personalized orthotic design. This approach can be accepted as a time-saving approach when digitizing.

3. Results

Poly (Acrylonitrile-butadiene styrene, ABS) is a popular thermoplastic used widely in industry and science. It has some important specialties such as high impact strength, easy processability, smooth surface characteristics, dimensional stability and low price. ABS is also important for patient-specific medical orthosis due to its mechanical properties mentioned above and easy printability in 3D printers. Mechanical properties of an AFO prototype produced from ABS by means of a 3D printer are tested first time in the literature. Dog bone samples have been printed by a 3D printer for tensile test to analyse material strength of AFO prototype. Figure 4 illustrates dog bone samples that have been produced with the same printing parameters of AFO due to having meaningful results. The test results present patient-specific printing parameters that are very useful for computing material cost and measuring material strength. When external force is exerted on a material, they change shape and size. ABS is stiff, rigid and tough material. It provides the necessary support under a certain load because of having high impact resistance. A tensile test is performed in order to identify the resistance of the 3D printed AFO prototype. At his tensile strength, the AFO prototype has maximum %6.8 strain. At his tensile strength, the AFO prototype has maximum %6.8 strain. If the applied tensile stress goes beyond this peak value, the material has permanent deformation on its polymer structure and finally material is ruptured. The same AFO prototype produced in this paper is experimentally applied to an end-user with 65 kg weight male in order to observe durability.

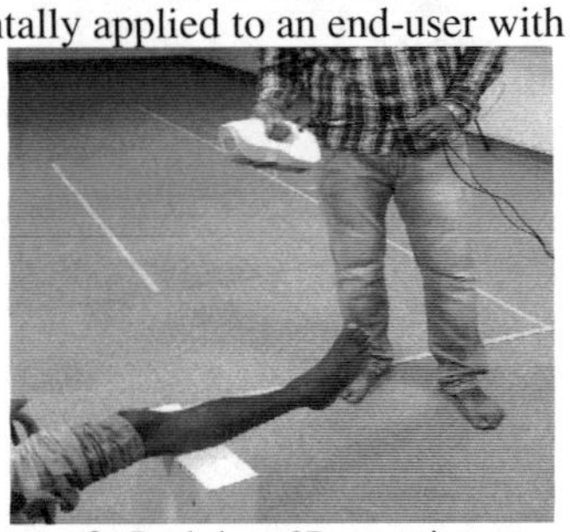

Figure 3. Real-time 3D scanning process

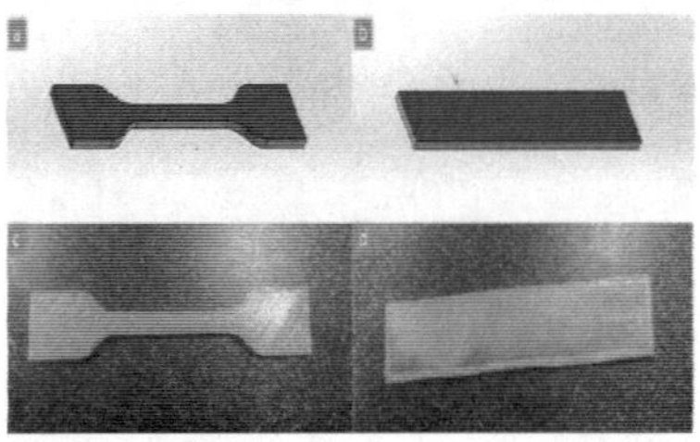

Figure 4. a) Solid model of tensile-strength test part, b) Solid model of charpy impact test part, c) 3D printed tensile-strength test model, d) 3D printed charpy impact test model

IOP Conf. Series: Materials Science and Engineering **280** (2017) 012030 doi:10.1088/1757-899X/280/1/012030

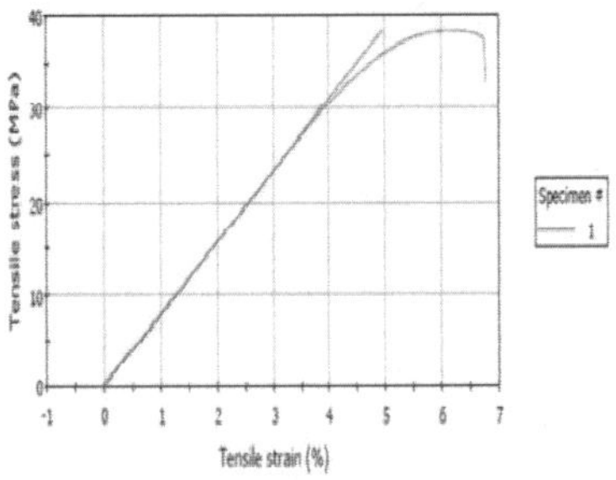

Figure 5. Tensile test results

Table 2. Tensile and Charpy impact test results

Tensile test	Tensile strength	38.4 MPa
	Max strain %	6.8
	Elasticity modulus	767.6 MPa
Charpy test	Sample	13x80x2 mm
	Energy	14.96 kJ/m^2

The tensile and charpy impact test data of AFO prototype is given in Table 2. The sample ABS material (Figure 5) having dimensions of 13mm (width) x 80mm (length) x 2mm (height) is ruptured when 14.96kJ/m^2 impact is exerted on its surface. The AFO made from ABS has sufficient elasticity modulus for patient-specific medical device manufactured by means of three-dimensional technologies. In addition to tensile and charpy impact test, a static load test is applied on the surface of medical orthosis by means of Universal Test Machine to observe compression durability. In order to perform this test, AFO prototype is first cut by an electrical saw and then mounted on the test machine. Figure 6 illustrates the prepared orthosis for compression test. Two different tests are performed in order to have more reliable results First test: A load is applied to the inner surface of AFO prototype up to 5 kN with 0.1 kN/s increment. Second test: A load is applied to the same surface of AFO prototype up to 15 kN with 0.1 kN/s increment. Figure 6 shows that 3D printed ankle foot orthosis for foot deformities patients. Successful results are obtained from these tests such that AFO prototype withstand loads without deformation. Figure 7 illustrates the compression test versus load and time.

4. Discussion

Recovery process can take several weeks to several years depending on the damaged tissues or disease. As 3D technologies are time and cost saving when manufacturing, it is important to analyze mechanical properties of printed medical orthosis. These test approach can show the mechanical properties of adhesion. The adhesion is depending on the slicing parameters, printing parameters and the 3D printer. Thus, these tests are necessary to obtain results of printed medical orthosis. It is important to know the quality of manufactured medical orthosis by means of three-dimensional technologies. Solid Works simulation tests also can be applicable at this stage.

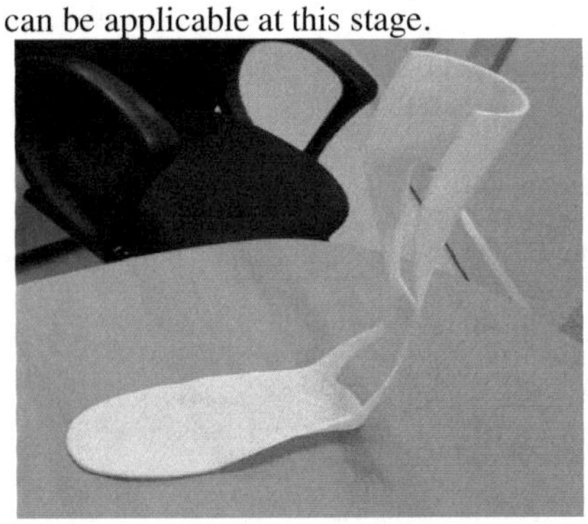

Figure 6. AFO prototype for compression test

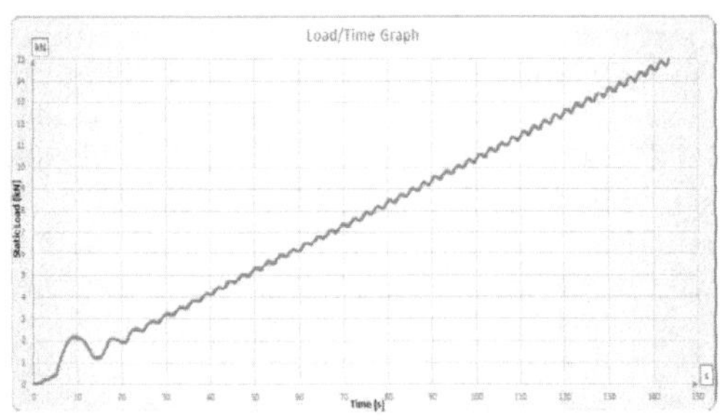

Figure 7. Compression test versus load and time.

5. Conclusion

Printing parameters such as layer height and infill rate plays crucial role to determine the mechanical properties of medical orthosis. Four different tests namely tensile test, charpy impact test, static load test and hydraulic press test are performed to analyze the AFO prototype made from ABS. The tests and its

IOP Conf. Series: Materials Science and Engineering **280** (2017) 012030 doi:10.1088/1757-899X/280/1/012030

results can be summarized as follows: A tensile test is carried out for identifying the resistance of AFO prototype. According to this test, maximum %6.8 strain is observed when 38.4 MPa tensile strength is exerted on the AFO prototype. Charpy impact test is performed for observing durability of AFO prototype. According to this test, when 14.96kJ/m2 impact is exerted on the surface of sample ABS material with dimensions of 13mm x 80mm x 2mm, a certain rupture is occurred on its surface. Two static load tests are performed to measure durability of orthosis and obtain more reliable results. In order to perform this test, AFO prototype is first cut by an electrical saw and then mounted on the test machine. A load is applied to the inner surface of AFO prototype up to 5 kN with 0.1 kN/s increment as a first test. Then a load is applied to the same surface of AFO prototype up to 15 kN with 0.1 kN/s increment. Successful results are obtained from these tests without deformation. Hydraulic press test is applied on the orthosis to observe maximum durability. Maximum 10 tons load is applied to the AFO prototype which passed again successfully without any deformation.

Acknowledgment

This work supported by Prof. (Dr.) Parveen Kalra Coordinator Centre of Excellence (Industrial & Product Design) in the Department of Production & Industrial Engineering of PEC University of Technology Chandigarh ,160012 India.

References

[1] Mavroidis, C. Ranky, R. G. Sivak, M. L. Patritti, B. L. DiPisa, J. Caddle, A. and Drillio, R. "Patient specific ankle- foot orthosis using rapid prototyping", Journal *of neuro engineering and rehabilitation*, **8**(1), 1-9. 2011.

[2] Murphy, S. V. and Atala, A. "3D bioprinting of tissues and organs", *Nature biotechnology*, **32**(8), 773-785. 2014.

[3] Banga H. K, Belokar R.M. , Madan R. , Dhole S. Three dimensional Gait assessments during walking of healthy people and drop foot patients, *Defence Life Science Journal* **2**, No.1, 14-20, 2017

[4] Berman, B. "3-D printing: The new industrial revolution", *Business horizons*, **55**(2), 155-162. Low-Cost 3D Printing, 2012.

[5] Gibson, I. Kvan, T. and Wai Ming, L. "Rapid prototyping for architectural models", *Rapid prototyping journal*, **8**(2), 91-95. 2002.

[6] Giannatsis, J. and Dedoussis, V. "Additive fabrication technologies applied to medicine and health care: a review", *The International Journal of Advanced Manufacturing Technology*, **40**(1-2), 116-127. 2009.

[7] Melchels, F. P. Feijen, J. and Grijpma, D. W. "A review on stereo lithography and its applications in biomedical engineering", *Biomaterials*, **31**(24), 6121-6130. 2010.

[8] Beltran-Fernandez, J. A. Garibaldi, P. M. Saucedo, F. L. Escalante, E. R. Hernandez-Gomez, L. H. Calderon, G. U. and Camacho, N. "Design and manufacturing of prosthesis of a jaw for a young patient with articular ankyloses", in Applications of Computational Tools in Biosciences and Medical Engineering , *Springer International Publishing*, 73-87, 2015.

[9] Chang, J. W. Park, S. A. Park, J. K. Choi, J. W. Kim, Y. S. Shin, Y. S. and Kim, C. H. "Tissue Engineered Tracheal Reconstruction Using Three Dimensionally Printed Artificial Tracheal Graft: Preliminary Report", *Artificial organs,* **38**(6), E95-E105. 2014.

IOP Conf. Series: Materials Science and Engineering **280** (2017) 012031 doi:10.1088/1757-899X/280/1/012031

Effect of Mechanical Heterogeneity on the Crack Driving Force of a Reactor Pressure Vessel Outlet Nozzle DMW Joint

Zhao Lingyan[1], Cui Yinghao[2] and Xue He[2]

[1]School of Science, Xi'an University of Science and Technology, Xi'an 710054, China
[2]School of Mechanical Engineering, Xi'an University of Science and Technology, Xi'an 710054, China

gloomy2@foxmail.com

Abstract. The welding mechanical heterogeneity, load complexity, material and geometrical structure makes it very difficult to assess the structural integrity of dissimilar metal weld (DMW) joints. Based on a numerical simulated approach of the continuous change of material mechanical property in the buttering layer, a reactor pressure vessel (RPV) outlet nozzle DMW joint with service loads is studied, effect of mechanical heterogeneity on the stress-strain field and stress triaxiality at the semi-elliptical surface crack front are discussed. The analyses show that once the crack extends into the high hardness zone of Alloy 182 buttering, the strain decreases sharply, the strain gradient increases and the crack propagation slows down. The influence of strength mismatch on the stress triaxiality at the shallow crack front is greater than that at the deep crack front. The interaction between strength mismatch and crack depth directly affects the crack growth direction.

1. Introduction

DMW joints are widely used in primary water systems of pressurized water reactors (PWR) [1]. Generally, nickel-based alloy is pre-deposited on the ferritic RPV nozzle face. Then welding is carried out to form a DMW joint with nickel-based alloy. Stress corrosion cracking (SCC) of DMW joints has been paid more attention in the nuclear power industry [2-4]. As a weld filler metal, the high-temperature yield strength of Alloy 182 makes it more susceptible to SCC [5].

The complexity of load, material and geometrical structure also sets obstacles for the structural integrity assessment of DMW joints. And the welding mechanical heterogeneity makes it very difficult to determine the macroscopic fracture parameters at the crack tip [6]. How to simplify the geometry and material properties of DMW joints to increase the assessment reliability is still being studied.

A full-scale 3D finite element model of a RPV outlet nozzle DMW joint under service load, is established based on a continuously changing material model of the buttering layer. Fully considering the mechanical heterogeneity, the stress-strain field and stress triaxiality at the crack front with different crack depth are discussed in detail. The influence of mechanical heterogeneity on the SCC behavior of DMW joints is analyzed, which is of certain engineering value.

2. Finite element modeling

2.1. Specimen model

IOP Conf. Series: Materials Science and Engineering **280** (2017) 012031 doi:10.1088/1757-899X/280/1/012031

A typical nozzle safe-end DMW joint to connect the low alloy steel (LAS) nozzles to austenitic stainless steel (SS) pipes is shown in Fig.1, which consists of A508, Alloy 182 buttering, Alloy 182 weld and 304SS. For an axial semi-elliptic crack, c is one half of the crack length and a is the crack depth , the crack angle θ varies from 0° to 180°. The crack parameter c/a is set to be 1.5, and the crack depth a is taken as 5,7.5 and 10 mm, respectively. Preliminary computation results by the elastic-plastic finite element method (EPFEM) show that the Alloy 182 weld and buttering are the high stress zone of the whole component exposed to all the service loads, so it is assumed that a axial semi-elliptical surface crack is located in the Alloy 182 weld.

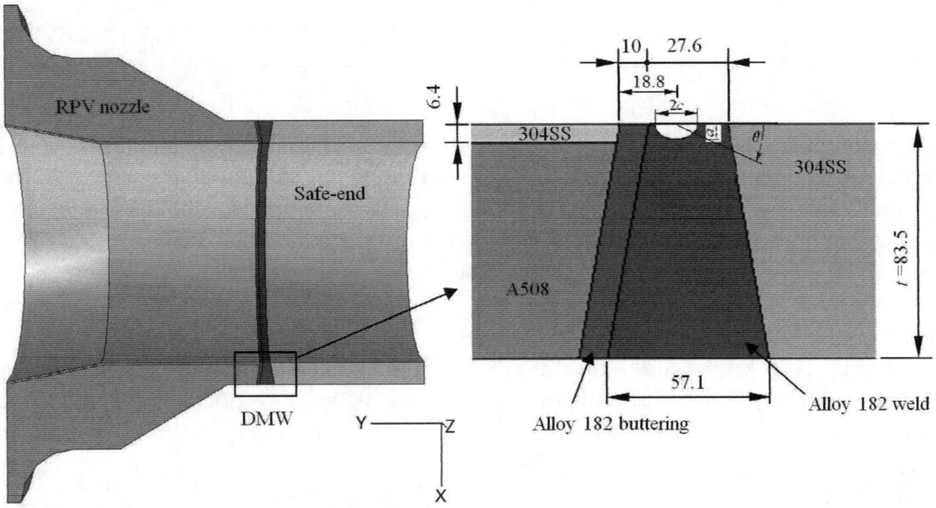

Figure 1. Geometry and material configuration of a DMW joint

2.2. Material model

A typical DMW joint consists of four materials. Their non linear relationship between stress and strain beyond yield is described by Ramberg-Osgood equation in this simulation [7]:

$$\frac{\varepsilon}{\varepsilon_0}=\frac{\sigma}{\sigma_0}+\alpha\left(\frac{\sigma}{\sigma_0}\right)^n \tag{1}$$

Where ε is the strain, including elastic and plastic strain. σ is the total stress; ε_0 is the yield strain of the material, σ_0 is the yield stress of the material, and n is the strain hardening exponent of the material, α is the offset coefficient of the material.

The mechanical properties of the four materials are given in Tab. 1.

Table 1. Material mechanical performance parameters of a DMW joint at 343℃

Material	E /MPa	ν	σ_0 /MPa	n	α
A508	193 000	0.288	440	5.333	1.0
Alloy 182 buttering	193 000	0.288	385~550	4.779~6.617	1.0
Alloy 182 weld	193 000	0.288	385	4.779	1.0
304SS	193 000	0.288	254	4.402	1.0

Following the determination of yield strength, the strain-hardening exponent(n) are estimated according to Eq. (2) [8]:

$$n=\frac{1}{\kappa\ln(1390/\sigma_0)} \tag{2}$$

IOP Conf. Series: Materials Science and Engineering **280** (2017) 012031 doi:10.1088/1757-899X/280/1/012031

Where κ=0.163. By using the predefined field module of ABAQUS [9], the material properties of Alloy 182 buttering, for example, yield stress and strain-hardening exponent, become continuously changing parameters [10, 11].

Besides the component gravity force, normal service loads of the DMW joint for RPV nozzles are listed in Tab. 2 [11].

Table 2. Service loads of the DMW joint

Service temperature /℃	Internal pressure /MPa	Axial stress /MPa	Bending moment /kN·m
343	15.59	44	2492

2.3. FE Model

To reduce the computation time, half of the actual structure is selected to establish the 3D finite element model. A symmetry constraint is set on the symmetry plane of the pipe. To reduce the influence of the edge effect on the analysis results, the left and right side of the model are properly extended and fixed constraints are set at the left end according to the Saint-Venant's principle.

Fig.2 gives the mesh of a DMW joint specimen (global model and sub-mode), where X-Y datum is the crack surface and Z-axis is the crack growth direction. 80543 and 41260 8-node linear brick elements are adopted in the global model and sub-model, respectively. A more refined mesh is adopted near the crack front, in order to obtain a more detailed and accurate data. There are 32020 elements in the vicinity of 2 mm around the crack front in the sub-model, more than the half the element total number. The minimum size of the element is about 0.02 mm in the sub-model.

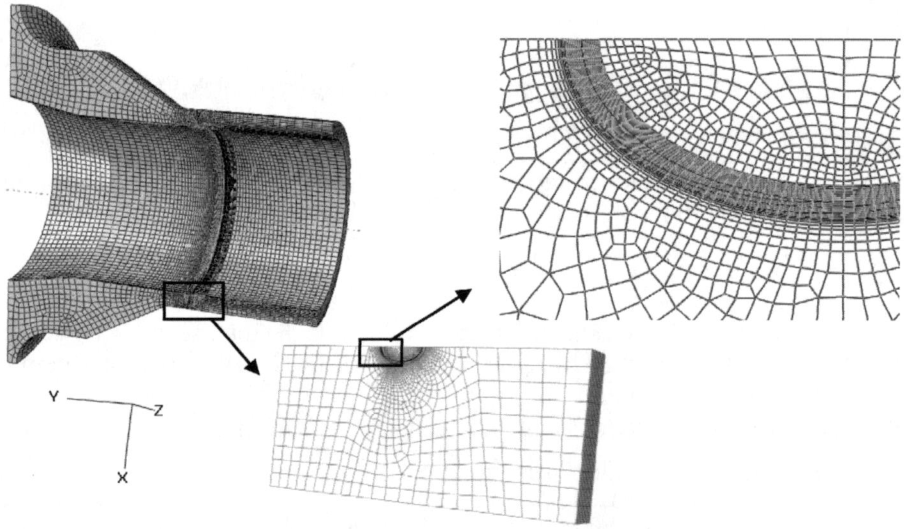

Figure 2. Mesh of the DMW joint

3. Results and Discussion

3.1. Stress and strain around crack front

The tensile stress distribution at the crack front is shown in Fig.3. The stress varies a lot with different crack depths. Due to the yield stress of stainless steel is the smallest among the four material, the high stress zone is close to the stainless steel pipe. Stress of the shallow crack front is the smallest, while high stress area of the deep crack increases greatly, which indicates that the crack has experienced rapid expansion and high stress area is gradually away from the stainless steel pipe.

The tensile plastic strain distribution at the crack front is shown in Fig.4. Similar to the distribution of the open stress, due to low plasticity constraint of the low strength material the high strain zone is close to the stainless steel pipe.

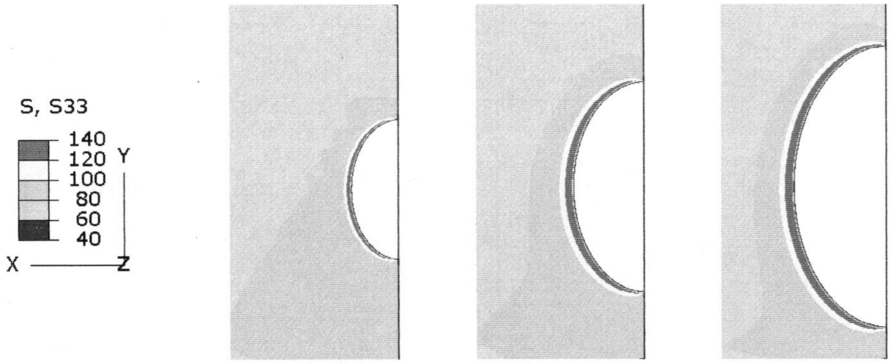

(a) *a*=5 mm (b) *a*=7.5 mm (c) *a*=10 mm

Figure 3. Tensile stress around crack front (MPa)

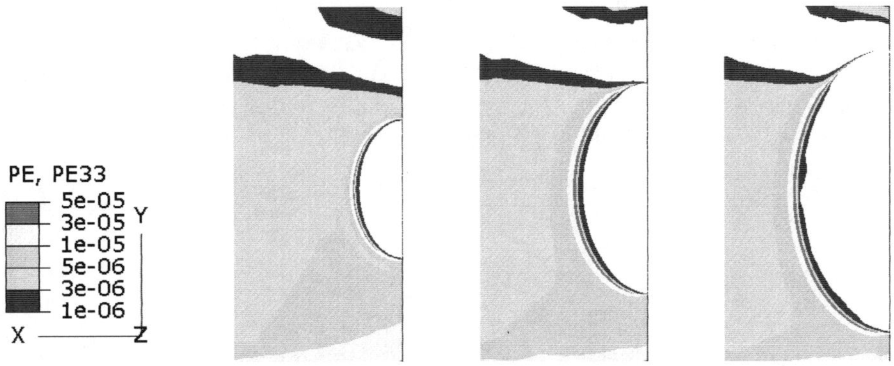

(a) *a*=5 mm (b) *a*=7.5 mm (c) *a*=10 mm

Figure 4. Tensile plastic strain around crack front

Strain at shallow crack front is the smallest and the low strain region appears on the side of the 182 alloy isolation layer. High strain region of the shallow crack front increases much, indicating that the crack has experienced rapid expansion and the high strain zone is close to the stainless steel pipe. Because of stainless steel yield strength is the smallest of all the materials, the largest strain also appears near the side of the stainless steel pipe when the crack depth is10 mm.

3.2. Stress and strain at a certain distance ahead of crack front

Strain and strain rate quickly decrease with the increasing distance ahead of crack tip. A critical issue is to estimate the characteristic distance (*r*) ahead of crack tip when SCC growth rate is predicted by FRI model [12]. Considering the the stress-strain singularity and comparing the crack tip plastic zones where the equivalent plastic strain is 0.2% with different crack depths, the reasonable distance from crack front is designated as 60 μm in this paper.

The tensile stress at the characteristic distance (r=60 μm) ahead of crack front is shown in Fig.5. Crack propagates gradually and enters the high hardness region of the Alloy 182 buttering with the increase of crack depth. Squeezed by the high strength material. the original flat strain curve appear a peak value, and the stress value is much greater when the crack is deeper. However, the tensile stress ahead of crack front only increases a bit in the Alloy 182 buttering.

The tensile plastic strain at the characteristic distance (r=60 μm) ahead of crack front is shown in Fig.6. The minimum strain occurs when the crack depth is 10 mm. The strain curve ahead of the shallow crack front extending in the weld metal is flat. The stress increased when the semi-elliptical surface crack is deeper, but the strain reduced drastically and the strain gradient is large, which indicates that the crack propogation in Alloy 182 buttering will slow down.

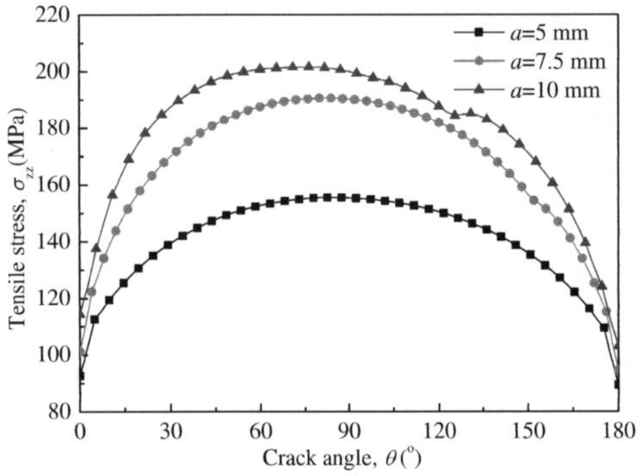

Figure 5. Tensile stress ahead of crack front (MPa)

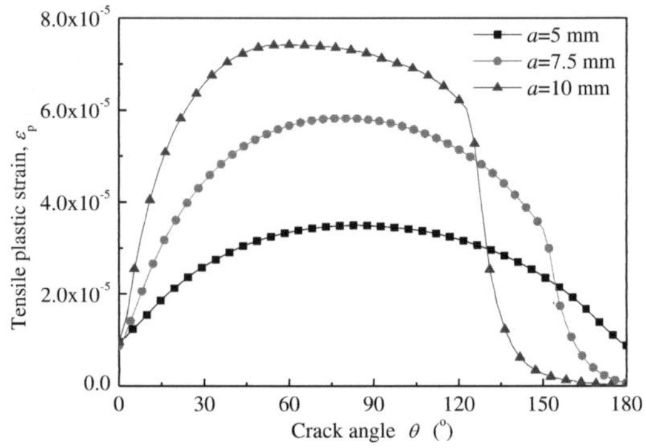

Figure 6. Tensile plastic strain ahead of crack front

3.3. Stress triaxiality ahead of crack front

Research shows that the distribution of the stress triaxiality in the crack front zone may have a great influence on the fracture behavior [13]. The stress triaxiality is described by σ_h/σ_e,, where σ_h denotes the hydrostastic stress and σ_e denotes the equivalent Von-Mises stress.

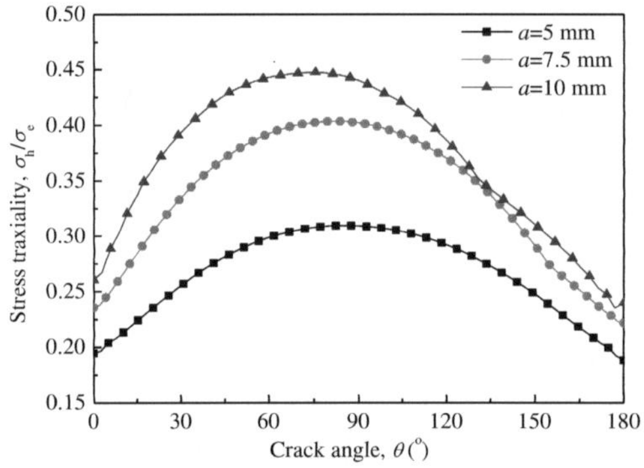

Figure 7. Stress triaxiality ahead of crack front

Stress triaxiality at the characteristic distance (r=60 μm) ahead of crack front is shown in Fig.7. At the same stress loading level, The maximum stress triaxiality occurs when the crack depth is 10 mm.. The high stress triaxiality at the interface region leads to a decrease of the local region toughness. A high stress triaxiality occurs in the low-strength material region due to its low plastic constraint. The high plastic strain zone is consistent with the high stress triaxiality. The local high plasticity strain is coincidence with local high stress triaxiality, which leads to the change of crack propagation direction.

4. Conclusion

Numerical simulation technology is commonly used to analyze piping welded joints that contain irregular crack under service loads. Based on a numerical simulated approach of the continuous change of material mechanical property in Alloy 182 buttering, the stress-strain field and stress triaxiality along the inner surface axial crack front of a RPV outlet nozzle DMW joint are simulated by using EPFEM, conclusions are as follows:

(1) The stainless steel pipe side is the potential direction of crack growth. Once the crack extends into the high hardness zone of Alloy 182 buttering, the strain decreases sharply, the strain gradient increases and the crack propagation slows down.

(2) Due to its low plastic constraint, high stress triaxiality, which occurs in the low-strength material region, leads to a decrease of the local region toughness at the interface region.

(3) The influence of strength mismatch on the stress triaxiality at the shallow crack front is greater than that at the deep crack front. The interaction between strength mismatch and crack depth directly affects the crack growth direction.

Acknowledgments

This work is financially supported by the National Natural Science Foundation of China (Grant nos. are 11502195 and 51475362).

References

[1] Bamford W and Hall J. A review of Alloy 600 cracking in operating nuclear plants including Alloy 82 and 182 weld behavior[C]. *ASME 12th International Conference on Nuclear Engineering*, 2004: 131-139.

[2] Li G F and Congleton J. Stress corrosion cracking of a low alloy steel to stainless steel transition weld in PWR primary waters at 292℃[J]. *Corros. Sci.*, 2000, 42(6): 1005-1021.

[3] Li G F, Li G J, Fang K and Peng J. Stress corrosion cracking behavior of dissimilar metal weld A508/52M/316L in high temperature water environment[J]. *Acta. Metall. Sin.*, 2011, 47(7): 797-803.

[4] Bamford W, Newton B and Seeger D. Recent experience with weld overlay repair of indications in Alloy 182 butt welds in two operating PWRS[C]. *ASME 2006 Pressure Vessels and Piping/ICPVT-11 Conference*. 2006: 427-434.

[5] Andresen P, Young L, Horn R and Emigh P. Stress corrosion crack growth rate behavior of Ni Alloys 182 and 600 in high temperature water[J]. *Corros.*, 2002.

[6] He X, Ogawa K and Shoji T. Effect of welded mechanical heterogeneity on local stress and strain ahead of stationary and growing crack tips[J]. *Nucl. Eng. Des.*, 2009, 239(4): 628-640.

[7] Ramberg W and Osgood W R. Description of stress-strain curves by three parameters[J]. *Technical Report Archive & Image Library*, 1943(902).

[8] Ueda Y, Shi Y, Sun S and H Murakawa. Effect of crack depth and strength mis-matching on the relation between J-integral and CTOD for welded tensile specimens[J]. *Trans. JWRI*, 1997, 26(1): 133-140.

[9] ABAQUS. ABAQUS/Standard User's Manual Version 6.14[Z]. ABAQYS Inc, 2015.

[10] Peng Q, Xue H, Hou J, Sakaguchi K, Takeda Y, Kuniya J and Shoji T. Role of water chemistry and microstructure in stress corrosion cracking in the fusion boundary region of an Alloy 182-A533B low alloy steel dissimilar weld joint in high temperature water[J]. *Corros. Sci.*, 2011, 53(12): 4309-4317.

[11] Zhao L, Xue H, Jiao K and Tang W. Analysis of Interface Fracture on Piping Welded Joint with Continuous Material Properties[C]. *Fourth International Conference on Digital Manufacturing and Automation. IEEE*, 2013: 207-210.

[12] Peng, Q, Kwon, J, Shoji T. Development of a fundamental crack tip strain rate equation and its application to quantitative prediction of stress corrosion cracking of stainless steels in high temperature oxygenated water[J]. *J. Nucl. Mater.*, 2004, 324(1): 52-61.

[13] Kim Y J, Kim J S, Cho S M and Kim Y J. 3-D constraint effects on J, testing and crack tip constraint in M(T), SE(B), SE(T) and C(T) specimens: numerical study[J]. *Eng. Fract. Mech.*, 2004, 71(9): 1203-1218.

IOP Conf. Series: Materials Science and Engineering **280** (2017) 012032 doi:10.1088/1757-899X/280/1/012032

Automatic Adjustments of a Trans-oesophageal Ultrasound Robot for Monitoring Intra-operative Catheters

Shuangyi Wang[1], James Housden[1], Davinder Singh[2] and Kawal Rhode[1]

[1] Division of Imaging Science and Biomedical Engineering, King's College London, 4th Floor North Wing, St Thomas' Hospital, UK
[2] Xtronics Ltd., Gravesend, Kent, UK

Shuangyi.wang@kcl.ac.uk

Abstract. 3D trans-oesophageal echocardiography (TOE) has become a powerful tool for monitoring intra-operative catheters used during cardiac procedures in recent years. However, the control of the TOE probe remains as a manual task and therefore the operator has to hold the probe for a long period of time and sometimes in a radiation environment. To solve this problem, an add-on robotic system has been developed for holding and manipulating a commercial TOE probe. This paper focuses on the application of making automatic adjustments to the probe pose in order to accurately monitor the moving catheters. The positioning strategy is divided into an initialization step based on a pre-planning method and a localized adjustments step based on the robotic differential kinematics and related image servoing techniques. Both steps are described in the paper along with simulation experiments performed to validate the concept. The results indicate an error less than 0.5 mm for the initialization step and an error less than 2 mm for the localized adjustments step. Compared to the much bigger live 3D image volume, it is concluded that the methods are promising. Future work will focus on evaluating the method in the real TOE scanning scenario.

1. Introduction

Catheter-based procedures for minimally invasive surgeries have been widely used in the last two decades as a replacement for open-surgery to treat various types of heart disease. In these procedures, catheters are inserted into a large vein through a small incision, made usually in the groin area, and then are advanced into the heart. With the development in recent years of real-time three-dimensional (3D) ultrasound, guiding of catheter-based procedures can be performed using a 3D TOE probe [1] and this approach has been gradually investigated by more organizations ever since the first experience reported by Perk *et al* [2]. 3D TOE has its unique advantage in visualizing catheters and 3D heart structures due to its volumetric imaging nature [3]. Studies have also indicated that the entire scenario including the catheter, devices, and heart structures (e.g. septum, valve, and appendage) in most catheter-based procedures can be imaged in a single 3D view [4]. Because of this, 3D TOE has great potential to be used more widely in the future for guiding catheter-based procedures and researchers have already started pursuing standardized 3D protocols for specific procedures [5]. Despite the advancement of the imaging techniques, 3D TOE is still a manually controlled approach where a skilled operator is required to hold and manipulate the probe. This is tedious and harmful for the duration of the longer interventional procedures, especially when TOE is used in conjunction with X-ray. Heavy protection clothes and long periods of standing are required which can lead to several occupational diseases [6]. Considering these challenges, we designed an add-on robotic system [7] which allows remote control of a commercial TOE probe (x7-2t, Philips, The Netherlands). The robot has four degrees-of-freedom (DOFs), which allow the translation and rotation of the probe shaft, as well as the bi-directional bending of the probe head.

In this paper, we propose a new concept using the designed TOE robot to automatically monitor intra-operative catheters and place the catheter in the centre of the Field of view (FOV). To our knowledge, the device we designed is the first published system for robotically manipulating a TOE probe. The automatic positioning of the probe for catheter monitoring is a unique topic which has not been reported before, although detailed methods employed in this paper are related to research works for other types of endoscopes in different applications, such as described in [8, 9]. The automatic adjustments of the probe are done by a pre-planning method for initialization and a numerical servo loop based on differential kinematics for localized adjustments. This paper focuses on reporting the detailed methods relating to the robotic automation and presents simulation studies for the validation. Details of the related existing image processing methods will not be discussed in detail.

2. Overview of the robotic TOE system

The developed add-on TOE robot (figure 1a) holds the probe handle and manipulates four mechanical DOFs that are available in manual handling of the probe. The robot comprises three structures: the handle control structure to rotate the two knobs with belt mechanisms built into an actuating chamber; a probe rotation mechanism with multiple gears actuated by a pair of motors for rotating the shaft about the long axis of the probe; and a linear belt mechanism mounted on a rail for providing linear translation of the probe and all other structures along the long axis of the probe. With these mechanisms, the probe tip where the ultrasound transducer is located can be translated, rotated, and bent bi-directionally according to a set of four robotic parameters. As human assistance is usually required for the translation axis to insert the probe and guide it down to the oesophagus, the translational axis was designed to allow both robotic and manual controls. More details of the design, along with the method of operation and safety concerns, can be found in our previous work [7]. The whole system is designed to work as an individual piece next to the surgical bed (figure 1b).

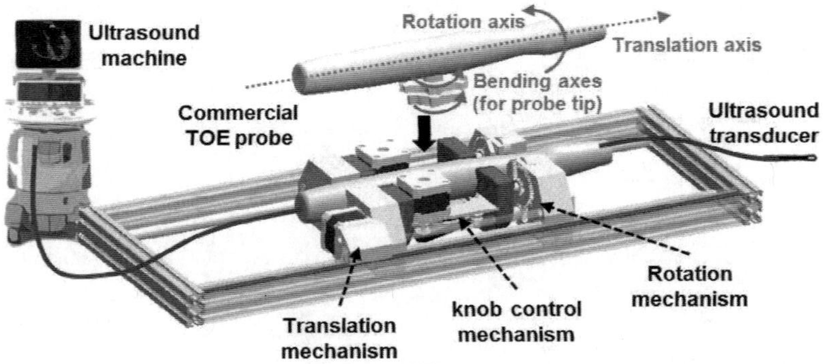

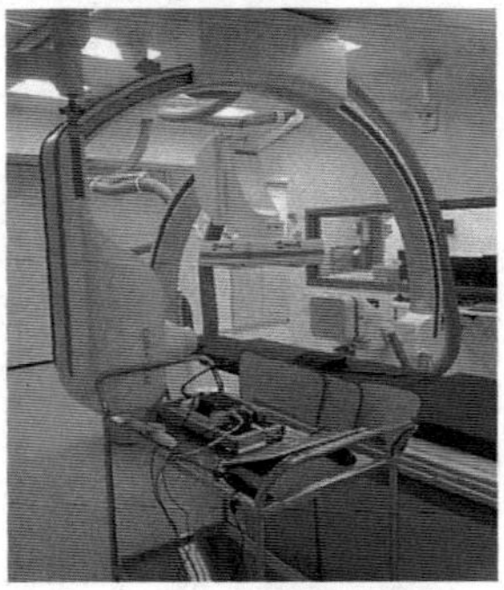

Figure 1. (a) Overview of the proposed add-on robotic 3D TOE system. (b) Possible clinical setup for the TOE robot in the surgical room with the robot sitting on a trolley.

3. Pre-planning method for initialization

The pre-planning method is intended to provide the user guidance for the initial probe positioning during a procedure. In a view-planning platform designed for pre-planning the 3D TOE view (figure 2a), an automatically segmented 3D heart model from a pre-scanned MR image, a manually segmented oesophagus centre line, and the model of the TOE probe head can be loaded and viewed intuitively. The kinematics of the probe are modelled and the corresponding virtual 2D centre slice and the cone-sized 3D volume FOV are displayed based on the given probe parameters [10]. In the platform, a number of markers representing the location of the target catheter that is to be monitored are defined from the pre-scanned MR image and used to determine the necessary monitoring range of a TOE view. The desired robotic parameters for the robotic TOE system to image the catheter are determined by the middle of the complete movement range of the catheter. This is defined as the target and its position in the ultrasound image coordinates is used for pre-planning.

An objective function based on the X- and Y-axial distances from the origin of the ultrasound image coordinates to the target is defined for the kinematics to optimize the robotic parameters using the gradient-descent search strategy based on the forward kinematics [7]. The Z-distance between the probe's transducer face and the target, measured along the Z-axis of the ultrasound image coordinates, is constrained by the heart-to-oesophagus distance and therefore the Z-distance optimization is not planned to be optimized. As reported previously, the forward kinematic model F(**p**), where **p** is the 4-DOF robotic parameter set, gives the transformation between the patient coordinates and the predicted probe tip coordinates when a heart model and the corresponding oesophagus model are given:

$$^{\text{Patient}}\mathbf{T}_{\text{ProbeTip}}(\mathbf{p}) = \mathrm{F}(\mathbf{p}) \tag{1}$$

The patient coordinates are defined based on the MR image coordinates. The position vector of the target defined in the patient coordinates is known as $^{\text{Patient}}\mathbf{T}_{\text{Target}}$. Therefore, the position of the target in the ultrasound image coordinates based on the predicted probe tip pose is calculated using a known calibration between the probe and ultrasound image [11]:

$$^{\text{US}}\mathbf{T}_{\text{Target}}(\mathbf{p}) = {}^{\text{US}}\mathbf{T}_{\text{ProbeTip}}\ {}^{\text{ProbeTip}}\mathbf{T}_{\text{Patient}}(\mathbf{p})\ {}^{\text{Patient}}\mathbf{T}_{\text{Target}} \tag{2}$$

The objective function for the search strategy is then defined based on the target position in the ultrasound image coordinates. Since both X- and Y- components of this position vector are expected to be zero to locate the target at the centre, the objective function is defined as the root sum square of the X- and Y-components extracted from $^{\text{US}}\mathbf{T}_{\text{Target}}(\mathbf{p})$. The search strategy for the probe robotic parameters in a defined range uses gradient-descent iteration and results in the best-fit probe robotic parameters $\mathbf{p}^*$ to represent the desired probe pose $^{\text{Patient}}\mathbf{T}_{\text{ProbeTip}}{}^*$. For the initialization, 4-DOF movements are utilized giving the best-fit probe robotic parameters suggested to the user. This process is performed by the operator using the robot and the initial view will then be pointing to the approximate centre of the catheter's working range.

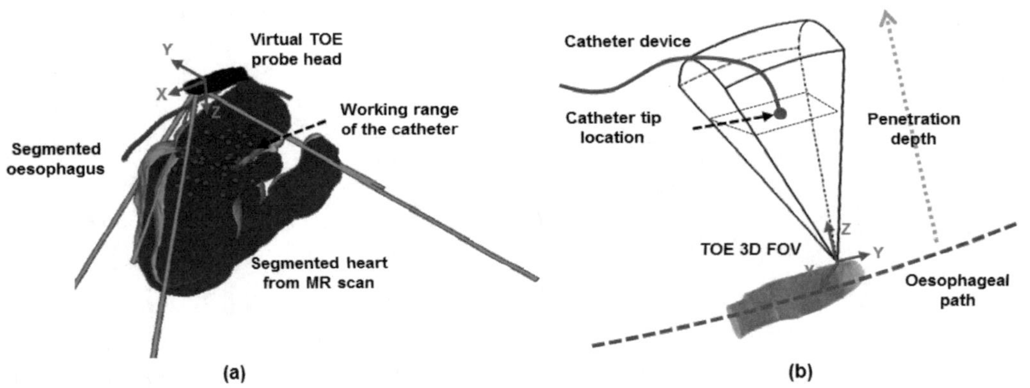

Figure 2. (a) Illustration of the strategy for probe initialization based on the pre-planning platform to locate the probe in the approximately correct region. (b) Illustration of the strategy for localized probe adjustment in order to image the device accurately in the centre of the FOV.

4. Differential kinematics for localized adjustments

For localized adjustments of the TOE probe (figure 2b) to monitor the catheter after the initialization, differential kinematics and image servoing methods are utilized to adjust the 2-DOF bi-directional bending. This starts from investigating the forward kinematics of the bi-directional bending, which has been previously described in [7]. The inputs are the rotation angle of the first bending knob, φ_x, which controls the bending tip pitch in the posterior-anterior plane; and the rotation angle of the second bending knob, φ_y, which controls the bending tip yaw in the left-right plane. The geometric illustration of TOE probe tip bending is shown in figure 3a. The translation vector and rotation matrix between the probe tip coordinates and the probe base coordinates are expressed as:

$$^{\mathbf{ProbeBase}}\mathbf{t}_{\mathbf{ProbeTip}} = \begin{bmatrix} \frac{L_f}{\beta}(1-C_\beta)C_\alpha + L_u S_\beta C_\alpha \\ \frac{L_f}{\beta}(1-C_\beta)S_\alpha + L_u S_\beta S_\alpha \\ \frac{L_f}{\beta}S_\beta + L_u C_\beta \end{bmatrix} \quad ^{\mathbf{ProbeBase}}\mathbf{R}_{\mathbf{ProbeTip}} = \begin{bmatrix} S^2_\alpha + C_\beta C^2_\alpha & -S_\alpha C_\alpha(1-C_\beta) & C_\alpha S_\beta \\ -S_\alpha C_\alpha(1-C_\beta) & C^2_\alpha + C_\beta S^2_\alpha & S_\alpha S_\beta \\ -C_\alpha S_\beta & -S_\alpha S_\beta & C_\beta \end{bmatrix} \tag{3}$$

where α is the angle between the bending plane and the X-Z plane, β is the bending angle in the bending plane, L_f is the length of the bending section of the probe tip, and L_u is the length of the rigid section of the probe tip. In the matrix, Sx and Cx, respectively, denote sin(x) and cos(x).

For the differential kinematics, the robotic joint space parameters are defined as $\mathbf{R} = [\alpha, \beta]^T$. The velocity skew of the probe tip $^{\mathbf{ProbeTip}}\mathbf{V} = [\mathbf{v}, \boldsymbol{\omega}]^T$, expressed in the probe tip coordinates, is determined by the robotic Jacobian $\mathbf{J_R} = [\mathbf{J_v}, \mathbf{J_\omega}]^T$, where $^{\mathbf{ProbeTip}}\mathbf{V} = \mathbf{J_R}\dot{\mathbf{R}}$. The resulting Jacobian is expressed as:

$$\mathbf{J_v} = \begin{bmatrix} -\frac{L_f}{\beta}S_\alpha(1-C_\beta) - L_u S_\alpha S_\beta & \frac{L_f}{\beta^2}C_\alpha(1-C_\beta) + L_u C_\alpha \\ \frac{L_f}{\beta}C_\alpha(1-C_\beta) + L_u C_\alpha S_\beta & \frac{L_f}{\beta^2}S_\alpha(1-C_\beta) + L_u S_\alpha \\ 0 & -\frac{L_f}{\beta^2}S_\beta + \frac{L_f}{\beta} \end{bmatrix} \quad \mathbf{J_\omega} = \begin{bmatrix} -C_\alpha S_\beta & -S_\alpha \\ -S_\alpha S_\beta & C_\alpha \\ -1 + C_\beta & 0 \end{bmatrix} \tag{4}$$

The image servoing control for adjusting the probe bi-directional bending pointing to the catheter can be further determined by the image Jacobian, which relates differential changes in the image features to differential changes in the configuration of the transducer, i.e. the velocity skew $^{\mathbf{ProbeTip}}\mathbf{V}$. Let $^{\mathbf{US}}\mathbf{p} = [^{\mathrm{US}}x,\ ^{\mathrm{US}}y,\ ^{\mathrm{US}}z]^T$ be the tracked catheter tip position in the 3D ultrasound image space. The coordinates of the catheter tip in the probe tip coordinates $^{\mathbf{ProbeTip}}\mathbf{p}$ are decided by the fixed calibration $^{\mathbf{ProbeTip}}\mathbf{p} = {}^{\mathbf{ProbeTip}}\mathbf{T}_{\mathbf{US}}{}^{\mathbf{US}}\mathbf{p}$. In this study, the catheter tip position is assumed to be tracked either by image-based processing methods for $^{\mathbf{US}}\mathbf{p}$ or sensor based tracking methods for both the TOE probe and the catheter, giving $^{\mathbf{ProbeTip}}\mathbf{p}$ as the result. The following relation describes the coordinates of the catheter tip in the probe base coordinate frame $^{\mathbf{ProbeBase}}\mathbf{p}$:

$$^{\mathbf{ProbeBase}}\mathbf{p} = {}^{\mathbf{ProbeBase}}\mathbf{t}_{\mathbf{ProbeTip}} + {}^{\mathbf{ProbeBase}}\mathbf{R}_{\mathbf{ProbeTip}}\ {}^{\mathbf{ProbeTip}}\mathbf{p} \tag{5}$$

As the probe shaft is not expected to move during the localized adjustment, the catheter tip is considered not moving in the probe base coordinates for a given catheter tip position when bi-directional bending is applied (oesophagus movements are ignored). The rest of the components in (5) will all change when the bi-directional bending is applied. Taking the time derivative of (5), the following relationship is obtained according to the visual servoing method summarized in [12]:

$$^{\mathbf{ProbeTip}}\dot{\mathbf{p}} = \begin{bmatrix} -\mathbf{I} & S(^{\mathbf{ProbeTip}}\mathbf{p}) \end{bmatrix} {}^{\mathbf{ProbeTip}}\mathbf{V} = \begin{bmatrix} -\mathbf{I} & S(^{\mathbf{ProbeTip}}\mathbf{p}) \end{bmatrix} \mathbf{J_R}\dot{\mathbf{R}} \tag{6}$$

where $S(^{\mathbf{ProbeTip}}\mathbf{p})$ is the skew-symmetry matrix of $^{\mathbf{ProbeTip}}\mathbf{p}$. The final Jacobian $\mathbf{J}$ relating the changes of robotic joint parameters to the changes of the catheter tip position in the probe tip coordinates is therefore obtained. With the differential kinematics, we have considered a simple controller as shown in figure 3b, where $^{\mathbf{ProbeTip}}\mathbf{p}*$ is the ideal location of the catheter in the probe tip coordinates. This can be determined in the same way as described in the previous section so that the X- and Y-components for the $^{\mathbf{US}}\mathbf{p}*$ are zero, while the Z-component remains at the same value for its current position in the ultrasound image coordinates. The corresponding desired position $^{\mathbf{ProbeTip}}\mathbf{p}*$ can then be calculated based on the fixed calibration $^{\mathbf{ProbeTip}}\mathbf{T}_{\mathbf{US}}$ and used for the bi-directional position controller. A simple relationship then relates the resulting $\mathbf{R} = [\alpha, \beta]^T$ to the desired joint parameters.

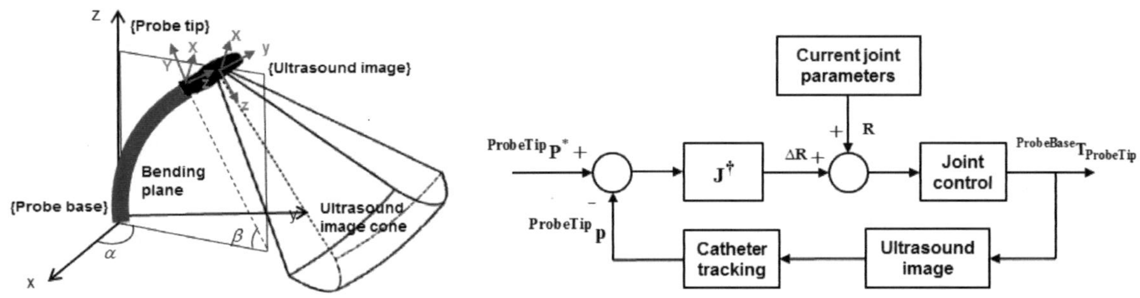

Figure 3. (a) Geometric illustration of the TOE probe tip bending with the coordinates of the probe base, probe tip, and ultrasound image shown. The probe tip coordinate frame is defined on the rigid section of the probe head (based on a Nano-CT scan) and the ultrasound image coordinate frame was defined by a fixed calibration. (b) Jacobian position control loop for localized adjustments of the probe to monitor the catheter. In the diagram, $\mathbf{J}^{\dagger}$ is the pseudo-inverse of the final Jacobian $\mathbf{J}$.

5. Simulation experiments

5.1. Positioning accuracy of the initialization method

To verify the correct working of the pre-planning strategy for the initialization step, a simulation experiment has been performed using the view-planning platform. 100 random points are defined within a heart segment obtained clinically to simulate the planned targets defined by the user. The automatic probe positioning method described in Section 3 is applied to calculate the desired probe tip pose to image the defined target, which is the centre of the catheter's moving range, at the centre of the ultrasound FOV. Based on the resulting target's position in the ultrasound image coordinates ${}^{US}\mathbf{T}_{Target}$, the error function is defined as the root sum square of the X- and Y-components extracted from the matrix. This error function quantifies the offset error from the centre of the ultrasound FOV. The evaluation indicates a mean error less than 0.5 mm for the pre-planning method.

5.2. Positioning accuracy of the Jacobian control loop

To verify the correct working of the Jacobian control loop for the localized adjustments, a simulation experiment was performed by simulating the catheter's position in the ultrasound image coordinates. Since most structures in a TOE scan are located at a depth of 5–6 cm, the Z-component ${}^{US}z$ is simulated with a random depth in this range. For the X- and Y-components indicating the catheter's movement range in the plane perpendicular to the ultrasound image penetration direction, uniformly distributed simulated positions in the range -5–5 cm were assigned. This is to cover the potential movement range based on the size of a normal heart. In the experiment, 100 target catheter positions were generated for the analysis. The corresponding resulting catheter tip positions, after applying the Jacobean control loop, were calculated and the X- and Y-components were compared with the desired values, in which case both distances are expected to be zero.

The median absolute errors of the X- and Y-components are shown in table 1 and the histogram of the absolute errors is shown in figure 4a. Since the FOV of 3D TOE extends with increasing ultrasound penetration depths, we further analyzed the centralization of the catheter tip positioning in the X–Y planes at different penetration depths taking the size of the X–Y planes into consideration. This was done by calculating the ratio of the X- and Y-components to the half-length of the X–Y plane measured along the X- and Y-axis of the live 3D TOE image coordinates. These two ratios are defined as the normalized X- and Y-distance errors. The median normalized errors were calculated and are also shown in table 1 and the histogram of the normalized errors is shown in figure 4b. For the Z-component, i.e. depth component in the ultrasound image space, the resulting mean value for the 100 simulated positions is 6.40 cm, which is within the efficient range of the TOE scan.

Table 1. Result summary for the simulation experiment using the control loop

	Absolute error (Median ± IQR)	Normalized Error (Median ± IQR)
X-axis	1.48 ± 2.80 mm	4.07 ± 8.51 %
Y-axis	1.02 ± 3.18 mm	3.21 ± 7.15 %

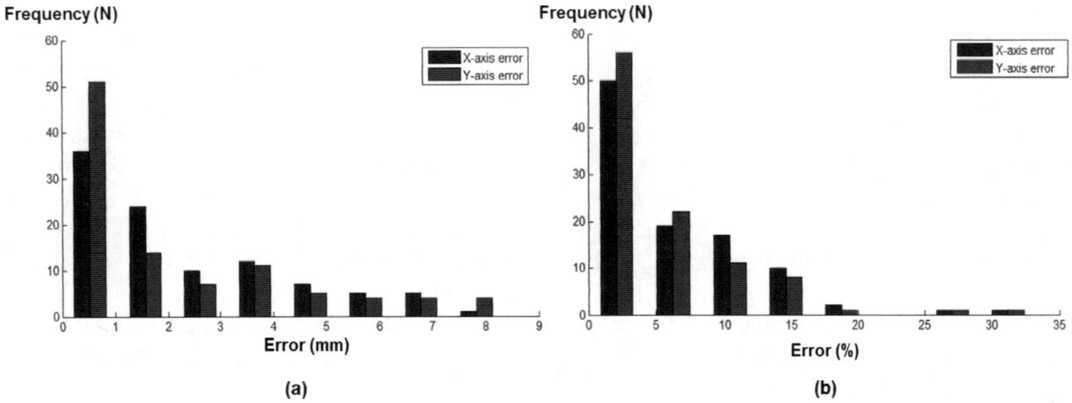

Figure 4. Histograms of the (a) absolute positioning errors and (b) the normalized positioning errors.

6. Discussion and conclusion

This paper has described a method for controlling a recently developed TOE robotic system to adjust the pose of the probe and therefore accurately monitor intra-operative catheters used in cardiac surgeries. The method has been divided into two stages with an approximate pose of the probe determined using a pre-planning method and an accurate pose of the probe adjusted based on a Jacobian position control loop. Accordingly, all of the four DOFs are employed for the once-only initialization when human assistance should be included for the probe insertion, and only two bending DOFs are utilized for the localized adjustments when the robot could be controlled automatically. In the paper, both the initialization and the localized adjustment methods have been described in detail. The results from the simulation of the initialization process indicate the correct working of the method with a submillimetre probe positioning error identified. The results from the simulation of the localized adjustment process indicate that all absolute errors are less than 2 mm as summarized in table 1 and the majority of errors are less than 1 mm as shown in figure 4a. The normalized positioning errors, as shown in table 1 and figure 4b, indicate that all catheter tip positions are located in the centre area of the FOV. Based on both results from the initialization and localized adjustments simulation, it is concluded that the proposed methods are correct and accurate for this application using the TOE probe to monitor the catheters considering the size of the FOV.

Though the simulation experiments described in this paper have verified the accurate working of the proposed methods, a number of differences between the simulation environments and the real scanning scenarios could potentially influence the performance of the methods. For the initialization, though the pre-planning method can accurately find an ideal probe pose, guiding the probe either manually or robotically to the exact location without real-time tracking of the TOE probe would be difficult. The issues for probe tracking have been discussed in detail in our previous work [10, 13] where an image-based approach and a sensor-based tracking approach were investigated. For the localized adjustments, the performances of the bi-directional bending are in reality likely to be different to the prediction of the kinematic model due to the complexities of the real environment. The tracking of the catheter, either based on an image processing technique or an additional sensor, is assumed to be known in the simulation experiments. Additionally, the catheter tip is assumed to be not moving in the probe base coordinates for a given catheter tip position when bi-directional bending is applied. It is assumed that the oesophagus's movements are negligible, while in reality this may not be true and a time-dependent oesophagus movement might need to be included into the model. With all these differences, the final performances of the methods in real scanning scenarios remain unknown. Thus, our future works will focus on developing a simple ultrasound imaging phantom to simulate the trans-oesophageal approach, allowing both the TOE probe and a catheter to work inside the phantom in order to further test the proposed methods.

References

[1] Vegas A and Meineri M 2010 Core review: three-dimensional transesophageal echocardiography is a major advance for intraoperative clinical management of patients undergoing cardiac surgery: a core review *Anesthesia and analgesia* **110** 1548-73
[2] Perk G, Lang R M, Garcia-Fernandez M A, Lodato J, Sugeng L, Lopez J, Knight B P, Messika-Zeitoun D, Shah S, Slater J, Brochet E, Varkey M, Hijazi Z, Marino N, Ruiz C and Kronzon I 2009 Use of Real Time Three-Dimensional Transesophageal Echocardiography in Intracardiac Catheter Based Interventions *Journal of the American Society of Echocardiography* **22** 865-82
[3] Faletra F F, Ho S Y and Auricchio A 2010 Anatomy of right atrial structures by real-time 3D transesophageal echocardiography *JACC. Cardiovascular imaging* **3** 966-75
[4] Altiok E, Becker M, Hamada S, Grabskaya E, Reith S, Marx N and Hoffmann R 2010 Real-time 3D TEE allows optimized guidance of percutaneous edge-to-edge repair of the mitral valve *JACC. Cardiovascular imaging* **3** 1196-8
[5] Faletra F F, Pedrazzini G, Pasotti E, Muzzarelli S, Dequarti M C, Murzilli R, Schlossbauer S A, Slater I P and Moccetti T 2014 3D TEE During Catheter-Based Interventions *JACC: Cardiovascular Imaging* **7** 292-308
[6] Goldstein J A, Balter S, Cowley M, Hodgson J and Klein L W 2004 Occupational hazards of interventional cardiologists: prevalence of orthopedic health problems in contemporary practice *Catheterization and cardiovascular interventions : official journal of the Society for Cardiac Angiography & Interventions* **63** 407-11
[7] Wang S, Housden J, Singh D, Althoefer K and Rhode K 2016 Design, testing and modelling of a novel robotic system for trans-oesophageal ultrasound *The international journal of medical robotics + computer assisted surgery : MRCAS* **12** 342-54
[8] Ott L, Nageotte F, Zanne P and Mathelin M d 2011 Robotic Assistance to Flexible Endoscopy by Physiological-Motion Tracking *IEEE Transactions on Robotics* **27** 346-59
[9] Abolmaesumi P, Salcudean S E, Wen-Hong Z, Sirouspour M R and DiMaio S P 2002 Image-guided control of a robot for medical ultrasound *IEEE Transactions on Robotics and Automation* **18** 11-23
[10] Wang S, Singh D, Johnson D, Althoefer K, Rhode K and Housden R J 2016 Robotic Ultrasound: View Planning, Tracking, and Automatic Acquisition of Transesophageal Echocardiography *IEEE Robotics & Automation Magazine* **23** 118-27
[11] Gao G, Penney G, Ma Y, Gogin N, Cathier P, Arujuna A, Morton G, Caulfield D, Gill J, Aldo Rinaldi C, Hancock J, Redwood S, Thomas M, Razavi R, Gijsbers G and Rhode K 2012 Registration of 3D trans-esophageal echocardiography to X-ray fluoroscopy using image-based probe tracking *Medical image analysis* **16** 38-49
[12] Hutchinson S, Hager G D and Corke P I 1996 A tutorial on visual servo control *IEEE transactions on robotics and automation* **12** 651-70
[13] Wang S, Singh D, Lau D, Reddy K, Althoefer K, Rhode K and Housden R J 2016 Probe Tracking and Its Application in Automatic Acquisition Using a Trans-Esophageal Ultrasound Robot. In: *International Workshop on Computer-Assisted and Robotic Endoscopy*: Springer) pp 14-23

IOP Conf. Series: Materials Science and Engineering **280** (2017) 012033 doi:10.1088/1757-899X/280/1/012033

From GCode to STL: Reconstruct Models from 3D Printing as a Service

Felix W Baumann[1,2,3], Martin Schuermann[1], Ulrich Odefey[1] and Markus Pfeil[1]

[1] TWT GmbH Science & Innovation, D-70565 Stuttgart, Germany
[2] University of Stuttgart, Institute of Computer-aided Product Development Systems, D-70569 Stuttgart, Germany

[3] Author to whom any correspondence should be addressed: felix.baumann@twt-gmbh.de

Abstract. The authors present a method to reverse engineer 3D printer specific machine instructions (GCode) to a point cloud representation and then a STL (Stereolithography) file format. GCode is a machine code that is used for 3D printing among other applications, such as CNC routers. Such code files contain instructions for the 3D printer to move and control its actuator, in case of Fused Deposition Modeling (FDM), the printhead that extrudes semi-molten plastics. The reverse engineering method presented here is based on the digital simulation of the extrusion process of FDM type 3D printing. The reconstructed models and pointclouds do not accommodate for hollow structures, such as holes or cavities. The implementation is performed in Python and relies on open source software and libraries, such as Matplotlib and OpenCV. The reconstruction is performed on the model's extrusion boundary and considers mechanical imprecision. The complete reconstruction mechanism is available as a RESTful (Representational State Transfer) Web service.

1. Introduction

With 3D printing it is possible to create physical objects from a variety of materials using only a digital model and a specialised tool, the 3D printer. The manufacturing of the models works by extrusion of material, curing of material, fusion of material or other generative material manipulation techniques. It is carried out by a special machine that executes a machine-dependent instruction set. This instruction set, also known as machine code, is commonly used in a standardized format called GCode [1]. The machine code is generated by a software called slicer. Numerous slicers exist, as open-source software, such as *Slic3r*, freeware, such as *Cura*, or commercially available software, such as *Simplify3D*. For an overview of the availability and influence of this software see [2].

The resulting GCode is partially machine or vendor specific, as vendors implement parts of the instruction set or extensions thereof. It is furthermore dependent upon the slicing software. The proposed implementation is developed for machine code generated with *Slic3r*, version 1.2.9.

For Fused Deposition Modeling (FDM) type 3D printing, the machine code contains information as to where the printhead moves, what the movement speed is, and how much material is to be extruded along the movement path. This code is generated from an intermediary file format, commonly STL (Stereolithography, file format) or AMF (Additive Manufacturing File Format). For this paper, STL and AMF are considered as equal for the purpose presented and only STL is discussed where necessary. The STL file format contains the surface representation of the original model as oriented triangles. STL is the resulting file format for the reconstruction method of this paper. The details of the file transformations are described in Sec. 2. In 3D printing the resulting physical object is an approximation to the original digital 3D model. The latter model is often created using CAD software. The object is only an approximation as information is lost during the file conversion steps. Information is also lost due to 3D printing strategies. Furthermore, information is lost through strategies to compensate for technology specific alterations such as shrinkage. These strategies are employed by the slicing software to various degrees. The physical creation is, furthermore, limited by mechanical constraints. In cases where the creation of an object must be certified to adhere to the original template, i.e., the digital model, it is required to make statements on the degree of conformity or identity. Alterations to the model are not possible when only the GCode is available. Reverse engineering does not completely recover the original model, as information is lost due to approximations.

Currently, a limited number of implementations for model reconstruction from GCode exists. One such solution is commercially available at *Makeprintable*, another one is presented by [3]. To the knowledge of the authors, no open source service is available.

In this paper, a method to reconstruct the 3D model based on the available GCode is presented. This method is implemented and the implementation simulates the extrusion process of FDM type 3D printing graphically. From the slice-wise graphical extrusion, the layer geometry is reconstructed and then triangulated to create the respective STL model. The resulting model is suitable to be compared to the original model, and to calculate difference or identity metrics. The implementation of the method for simulation and the model reconstruction is performed in *Python* and relies on open source software as described in Sec. 4. The reconstruction mechanism is available as a RESTful Web service. The focus is on GCode files created for FDM 3D printing with a Cartesian geometry using *Slic3r*. Model reconstruction is important for two reasons. Firstly, the question whether the fabricated model is still corresponding to the original model or if it was altered during the transfer from the user to the printer can answered. Secondly, model reconstruction is important for reverse-engineering of existing GCode files for which no corresponding digital model is available.

The remainder of this paper is structured as follows: In Sec. 2, an explanation and introduction to 3D printing is provided. In Sec. 3 relevant and related work to this approach is examined and discussed. The main work, i.e., the presentation of the reconstruction approach, is outlined in Sec. 4. The implementation overview for the Web service is provided in Sec. 5. The proposed approach is finally discussed in Sec. 6, whereas a conclusion and an outlook are provided in Sec. 7.

This paper is an extension to [4] focusing on the reconstruction aspect implemented as a Web service.

2. Additive Manufacturing

Additive Manufacturing (AM) and 3D printing are used synonymously in the literature. Further synonyms include Rapid Manufacturing, Generative Manufacturing, Additive Layer Manufacturing, Layered Manufacturing or Rapid Prototyping, with an additional meaning of the last term for a broader method of prototyping. 3D printing has the tendency to be used for consumer-focused additive fabrication, whereas AM commonly denotes the industrial application of these technologies. Both terms describe a number of different technologies that are used to create physical objects directly made from different materials using only a 3D printer as a manufacturing tool. Furthermore, robots or CNC machines [5] equipped with special actuators can be used for 3D printing. The materials available for processing [6] include plastics, metals, ceramics, bio-inspired or bio-compatible materials, and glass [7]. The various technologies and techniques differ fundamentally from each other so that files that must be prepared in the 3D printing process differ as well. The 3D printing process, see also [8-10], can be described in seven steps with the first step being the model generation in a CAD (Computer Aided Design) environment or another software. The second step is the conversion stage from the accurate digital 3D CAD model to an AM intermediary file format such as STL or AMF. These files are digital representations with reduced accuracy and information, suitable for 3D model interchange and interoperability. The third step is the placement and orientation of the 3D model within the virtual build environment of the 3D printer. The fourth step is the creation of the specific machine code from the digital model. This machine code adheres mostly to GCode [1] standard. The 5th step is the actual fabrication of the object in the 3D printer which can be error prone, especially in consumer-grade 3D printers, due to unaligned axes or build plate, imprecise parts or lack of adhesion of the object. Following step five, the object must be post-processed, with varying intensity and goals. The seventh step is the testing of the object according to a test protocol. In literature, an eighth step, which is the actual usage, is sometimes regarded as being part of the 3D printing process.

3. Related Work

Tsoutsos et al. [3] presented a very detailed description of the distributed process chain for AM to help secure 3D printing. The authors proposed a reverse engineering mechanism to reconstruct digital 3D models from acquired GCode toolpaths. In their work, the authors also analyse the resulting models using finite element analysis (FEA). Their simulation creates basic geometric shapes according to the extrusion which are then fused and processed to result in an approximation of the original model. The difference between their approach and the one proposed here is that they use solid geometry generation directly on the GCode simulation, whereas the approach in this paper creates an intermediary point cloud first. Gao et al. [11] proposed a reverse engineering method for blade reconstruction in the aerospace industry. In their work, they reconstructed a worn-out turbine blade of an engine. The reconstruction utilised a digitiser and a CAD model reconstructed with *CATIA V5*. The authors implemented their approach with an integrated additive and machining based manufacturing. The work by Thrimurthulu et al. [12] discussed the simulation of FDM type 3D printing to optimise surface roughness and part orientation. Relevant to the work presented herein is the discussion on process parameters and influences on the material deposition. Okarma and Fastowicz [13] presented an approach to assess the quality of FDM printed objects based on feature similarities from image analysis. The authors did not reconstruct the shape or model of the original but compared the resulting surfaces graphically. Zhang and Chout [14] developed an FEA-based simulation method for FDM 3D printing. The simulation was used for a specific specimen to assess the influence of process parameters on the resulting object. Shrivastava and Modi [15] presented an approach to reconstruct the digital model from a point cloud acquired by a coordinate measuring machine. The authors used the *CATIA V5* software to construct the CAD representation of the physical model. This method, like the approach presented here, is only suitable for the reconstruction of the outer shape, with limitations on holes and other concave structures. The difference is that Shrivastava and Modi relied heavily on the reconstruction via the CAD software and did not discuss reconstruction principles.

4. Model Reconstruction

Figure 1 shows the proposed concept of the model reconstruction. GCode files are used as input and no additional comments in the file are considered. The input file is preprocessed to remove comments and unexpected instructions, and to rewrite certain instructions. A dialect that is compatible with the *Sailfish* firmware is implemented in the simulation core.

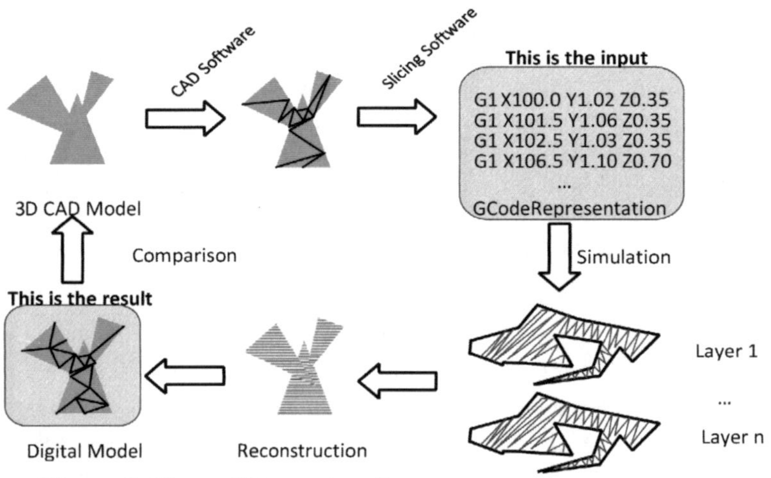

Figure 1. Flow of Reconstruction

4.1. GCode and Simulation Basics

The simulation core mainly processes **G1** (linear movement) instructions that contain information about the **X**, **Y**, and **Z** coordinates to which the printhead and the build plate are moved. Furthermore, these instructions contain information about the movement speed (**F**) towards these coordinates which are given as target movement speeds in mm/min. Information on the extrusion material, material extruded in mm along the path, is given in the **G1** commands per separate extruder, denoted by **A** and **B**, or cumulatively, denoted as **E**. In case of using the E parameter, the GCode allows for the switching of extruders which are also known as tools. The following two lines demonstrate the structure of the **G1** and **G0** commands.

G1 F1800 X89.999 Y89.150 E49.60291 G0 X89.999 Y89.716 F9000

G0 is used for rapid linear movement and **G1** for normal linear movement. Most implementations ignore this difference and treat them the same. Our system rewrites **G0** instructions to **G1** instructions. As shown in the example, the position of the parameters for each instruction is flexible. Instructions with partial parameter sets, such as *G1 X89.111* preserve parameters set previously, such as the **Y** and **Z** coordinates missing in this example. Every instruction parameter is active until set otherwise.

In GCode, all given coordinates are target coordinates, meaning that they must be interpreted within a context, e.g., the last target, to infer information about path lengths and extrusion volumes. The extrusion volume is also dependent on the filament diameter. Common diameters for FDM 3D printers are either 3.0 mm or 1.75 mm. A variation of 10 % of the diameter of the filament can result in approximately 20 % variation of the resulting material extrusion, for an exemplary extrusion of 10 mm material over 10 mm.

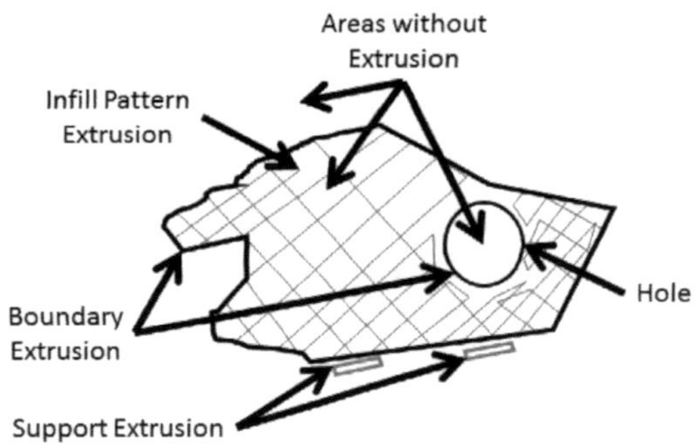

Figure 2. Extrusion Information per Layer

IOP Conf. Series: Materials Science and Engineering **280** (2017) 012033 doi:10.1088/1757-899X/280/1/012033

The simulation is based on physical processes to describe the movement of and within the 3D printer. All axes are moved with pre-defined acceleration profiles and can be exercised at a specific maximum velocity. The extrusion simulation is performed on a virtual canvas provided by *Matplotlib*. This extrusion is then visually thresholded with *OpenCV* to fill the resulting holes. These internal holes are either from gaps between extruded material, or from internal structures of the model, or from support structures within the model introduced by the slicing software. Boundary detection is then performed using *OpenCV* functionality. The boundaries are collected for each layer and exported to a file containing X, Y and Z coordinates in a Cartesian coordinate system. To reconstruct surfaces that are flat in one respective layer correctly, random sampling of points is performed. A percentage (5 %) of the area of the total image area is populated with random points. These points are further for structures and areas between individual layers to avoid random points inside the reconstructed model. *MeshLab* is used to process the resulting XYZ files via batch processing. The following operations are used to produce a mesh representation that is then exported as an STL file:

1. Poisson-disk sampling to reduce the number of points to operate on (Param.: 150 000 samples, Base mesh subsampling enabled, other parameters are at their default value),
2. surface reconstruction by ball pivoting (Param.: Ball radius 15 %, other parameters at their default value),
3. hole closing (Param.: Max size to be closed 40, other parameters are at their default value), and
4. removal of zero area, unconnected, null, and duplicate vertices and faces.

In Figure 2, the extrusion structure is depicted. The model is bounded by an outer shell that is indicated by a thick black line. This line also indicates where in reality material is extruded. The blue criss-cross pattern inside the object is for stability, as the object is otherwise printed hollow. The material extrusion of the outer shell and the inner infill-pattern is hard to distinguish as the GCode does not differentiate between them, and the amount of material dispensed is identical per the same length. In this figure, the object contains a hole in this layer which can be going through the complete object or be just part of this specific layer. The differentiation between designed holes, as in this case, and holes or other structures created by the slicing software for stability purposes is indistinguishable, as described earlier. The current implementation does, therefore, not support internal structures, holes or other concave structures. The figure also shows support structure extrusion indicated by red coloured lines, which are also indistinguishable in the GCode due to the reasons discussed above.

4.2. Reconstruction Algorithm

The simulation adds slight random errors in the positioning of the coordinates to simulate machine imprecision due to underlying hardware.

The following algorithmic description details the implementation of the reconstruction method.

1. Simulate printhead movement and material extrusion layer-wise, with random errors introduced to simulate machine imprecision.
2. From each simulated layer, extract geometrical information by:
   (a) Threshold the graphical representation to identify extruded material against the background;
   (b) Perform hole closing to ignore infill strategy;
   (c) Find contours of the resulting graphical representation;
   (d) Identify suitable contours according to their hierarchy and size;
3. Export of X, Y, Z information to a separate file
4. Perform mesh reconstruction in *MeshLab* via batch processing
5. Compare distances from created mesh to the STL representation to create a metric for identity, see [13, 16, 17] for information on similarity metrics.

In Figure 3, the intermediary reconstruction phases are depicted on a model acquired from *thingiverse*.

The leftmost image (a) depicts the simulated extrusion path, based on the GCode. Subfigure (b) is the thresholded representation of the extrusion path with holes filled. Based on Figure 3(b), the boundaries are extracted using the *OpenCV* findContours function, see Figure 3(c). Part (d) of the

Figure shows the extrusion process by placing coloured disks at the individual target points and smaller coloured disks on the respective movement in between the GCode instructions.

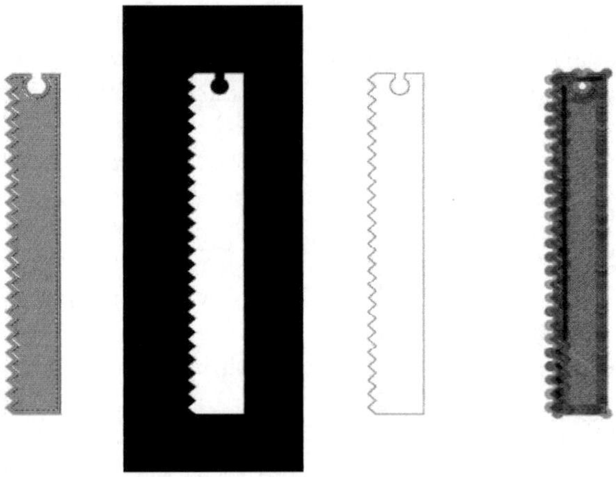

Figure 3. Reconstruction Stages for Model "weigh spoolholder" by Nefi Velarde, https://www.thingiverse.com/thing:1000042; (a) Simulated Extrusion (b) Thresholded Representation (c) Boundary Detection (d) Simulated Movement

5. Webservice Implementation

The implementation of the reconstruction method is performed in *Python* utilising the *web.py* framework. With RESTful Web services, every object is considered a digital resource that is accessible through a URL (Uniform Resource Locator). Based on the implemented and exposed methods a number of operations is facilitated. The resources in this case are the original GCode file, the resulting STL model and digital renderings of the model. User management and security are not considered at this point. User management and security are highly dependent on the infrastructure and environment where such an implementation is deployed and requires adaption. The operations or methods described in the following are invoked by a user, either programmatically or using a Web browser. For example, the following URL is exposed by the service:

https://www.example.com/ModelRecSrv/GCode/10442/convertToSTL

In this example, *https://www.example.com* denotes the transport protocol and base host name for the service. Furthermore, *ModelRecSrv* denotes the name of the service. The following part, *GCode*, denotes the realm of the objects to operate on, here it is GCode. Furthering this example, calling *ModelRecSrv/Rendering/10442* results in the identical response, a digital rendering of the model that is stored under the identifier 10442 in the PNG format, as a call to *ModelRecSrv/GCode/10442/renderModel*. The identifier part is unique per uploaded GCode. In future, more security-aware implementations are required. In the example, *convertToSTL* is the name of the operation that is called. **POST** and **GET** are different ways for a user to call the Web service. The service implements the following **POST** operations:

– **uploadGCode**; A method that expects the GCode file as payload using multipart/form-data encoding and stores the file on the server. A unique identifier is responded to the uploader.

– **convertGCodeToSTL**; A method that expects a GCode file as payload using multipart/form-data encoding and returns the reconstructed STL model to the uploader.

The service also implements the following **GET** operations:

– **convertToSTL**; A method that is expected to be called on a file object identified by the unique identifier to initiate the conversion to a STL file. If the file has previously been converted, the reconstructed model is sent back to the caller from local storage or cache. If the file is not converted yet, the conversion is performed and the result is send back to the caller when available.

– **GCodeStatistics**; A method that is expected to be called on a file object identified by the unique identifier, for detailed information on the original GCode file. This includes information on the number of instructions, file size, and extrusion length per extruder

– **renderModel**; A method that is also expected to be called on a file object identified by the unique identifier that returns the rendering of the resulting STL model as a PNG (Portable Network

IOP Conf. Series: Materials Science and Engineering **280** (2017) 012033 doi:10.1088/1757-899X/280/1/012033

Graphics) file. This method takes parameters to adjust the canvas size of the rendering and the rotation of model along its three axes. Furthermore, methods to delete models, files, and renderings from the server are implemented. The uploaded and rendered files will remain on the server indefinitely when not manually removed. The local caching and storage is implemented using a *SQLite* database.

6. Discussion

To discuss this approach, a number of problems encountered in the creation of this method are presented:

– Support material: Support material is indistinguishable in GCode, especially in case of one extruder. With two extruders, it could be derived from the amount of material extruded and then ignored. With a single extruder, heuristics are possible that would work on structural information, e.g., thin, column-like structures that are connected to the base plate and the main object with little surface area.

– Additional material: Some slicing programs instruct to extrude material in a specific pattern to clean the nozzle before the actual 3D print. With multi nozzle printing, it is also common to have "wiping" structures for nozzle cleaning. These structures are indistinguishable in the GCode.

– Simulation of mechanical imprecision: The actual fabrication of objects is limited by the mechanical precision of the AM machine and material behaviour, such as semi-molten flow.

– Adaptive and uneven layer height: Currently, the algorithm expects uniform layer heights for the mesh reconstruction.

– Internal structures: These structures are currently unsupported as the reconstruction is based on boundary reconstruction. Internal structures, such as holes, are hard to distinguish from artificial internal structures constructed by the slicing software, e.g., internal support structures.

– Objects printed larger than supported by the build envelope: The algorithm works by simulating the extrusion on a virtual build plate that is mapped to a digital canvas of predefined dimensions. Reconstructing such large models fails in the current stage. Actual 3D printing of such models would also fail.

– Flat surfaces: The current implementation fails in some cases to correctly close the models with flat surfaces. In future iterations, detection of flat bottom and top structures is improved by the introduction of randomly sampled points on these surfaces.

The rationale to create an intermediary pointcloud instead of layer-wise meshing is as follows:

– With layer-wise meshing, (internal) supporting structures will be meshed too. This will lead to a flawed reconstruction as most objects, in FDM 3D printing, are printed partially hollow, with specific support or infill structure provided by the slicing software, e.g., honey-comb structure or criss-cross pattern, see Figure 1.

– The meshing of the object's surface structure is made complicate with layer-wise meshing as per-layer meshing is discontinous.

– With layer-wise meshing, most of the generated mesh structure must be removed in the aggregation step as it would describe internal surface structures that are not surfaces in reality, in the model or in reconstruction resulting in illegal STL structures.

7. Conclusion & Outlook

In this paper, the authors demonstrated a method to reconstruct the digital model from a GCode file which is implemented as a RESTful Web service. This reconstruction is based on the virtual simulation of the extrusion process. The reconstruction is intended for FDM type 3D printing and GCode generated for a Cartesian geometry. The underlying simulator is capable of interpreting *Sailfish* targeted machine code. Mechanical accuracy of the fabrication is simulated in the reconstruction via random inaccuracies. The reconstruction uses existing open source software and libraries, such as *OpenCV* and *MeshLab* to reconstruct the model based on the simulated boundaries. The method is unable to reconstruct internal structures. Improvements of the method and its implementation are discussed in the previous Section. The future implementation will support the

IOP Conf. Series: Materials Science and Engineering **280** (2017) 012033 doi:10.1088/1757-899X/280/1/012033

detection of internal structures, detection of support material and supplemental material extrusions. This work is intended to be part of a complete model reconstruction framework that will help to create a chain of trust in AM. This will allow to answer the question whether a specific machine code represents an original 3D model or not. This reconstruction is not only reasonable for reverse engineering purposes but also within the context of the proposed chain of trust. Model reconstruction is important to certify that a machine code instruction file corresponds to its respective digital model.

References

[1] ISO 6983-1:2009 2009 Automation systems and integration - Numerical control of machines - Program format and definitions of address words

[2] Baumann F W, Bugdayci H, Grunert J, Keller F and Roller D 2016 Influence of slicing tools on quality of 3D printed parts *Computer-Aided Design and Applications* **13** 1 pp 14–31

[3] Tsoutsos N G, Gamil H and Maniatakos M 2017 Secure 3D printing: reconstructing and validating solid geometries using toolpath reverse engineering *Proc. of the 3rd ACM Workshop on Cyber-Physical System Security (CPSS '17)* pp 15–20

[4] Baumann F W, Schuermann M, Odefey U and Pfeil M 2017 Model reconstruction from machine code instructions *Proc. of the 6th International Conference on Mechanical Engineering & Mechanics: Industrial Engineering, Mechatronics and Industry Future (ICMEM)* In Preparation.

[5] Chakraborty D, Reddy B A and Choudhury A R 2008 Extruder path generation for curved layer fused deposition modeling *Computer-Aided Design* **40** 2 pp 235–43

[6] Wong K V and Hernandez A 2012 A review of additive manufacturing *ISRN Mechanical Engineering* **2012**

[7] Klein J, Stern M, Franchin G, Kayser M, Inamura C, Dave S, Weaver J C, Houk P, Colombo P, YangM Oxman N 2015 Additive manufacturing of optically transparent glass *3D Printing and Additive Manufacturing* **2** 3 pp 92–105

[8] Baumann F W and Roller D 2016 3D printing process pipeline on the internet *Proc. of the 8th Central European Workshop on Services and their Composition (ZEUS 2016)*, pp 29–36

[9] Gibson I, Rosen D and Stucker B 2015 Additive Manufacturing Technologies - 3D Printing, Rapid Prototyping, and Direct Digital Manufacturing. Springer New York, 2 edition

[10] Zeltmann S E, Gupta N, Tsoutsos N G, Maniatakos M, Rajendran J and Karri R 2016 Manufacturing and security challenges in 3D printing *JOM* **68** 7 pp 1872–81

[11] Gao J, Chen X, Yilmaz O and Gindy N 2008 An integrated adaptive repair solution for complex aerospace components through geometry reconstruction *The International Journal of Advanced Manufacturing Technology* **36** 11 pp 1170–79

[12] Thrimurthulu K, Pandey P M and Reddy N V 2004 Optimum part deposition orientation in fused deposition modeling *International Journal of Machine Tools and Manufacture* **44** 6 pp 585–94

[13] Okarma K and Fastowicz J 2017 Quality assessment of 3D prints based on feature similarity metrics *Proc. of the International Conference on Image Processing and Communications* pp 104–11

[14] Zhang Y and Chou K 2008 A parametric study of part distortions in fused deposition modelling using three-dimensional finite element analysis *Proc. of the Institution of Mechanical Engineers, Part B: Journal of Engineering Manufacture* **222** 8 pp 959–68

[15] Shrivastava R and Modi Y K 2015 Reverse engineering approach for rapid manufacturing of freeform components *Proc. of the 2nd International Conference on Science, Technology and Management* pp 2086–95

[16] Adan A and Adan M 2004 A flexible similarity measure for 3d shapes recognition *IEEE Transactions on Pattern Analysis and Machine Intelligence* **26** 11 pp 1507–20

[17] Shum H-Y, Hebert M and Ikeuchi K 1996 On 3d shape similarity *Proc. of the IEEE Conference on Computer Vision and Pattern Recognition (CVPR '96)* pp 526–31

IOP Conf. Series: Materials Science and Engineering **280** (2017) 012034 doi:10.1088/1757-899X/280/1/012034

Experimental Study of Autoignition Characteristics of Ethanol-Methylcellulose Gel Droplet

Donggi Lee[1], Jonghan Won[1] and Seung Wook Baek[1*]

[1]Department of Aerospace Engineering, Korea Advanced Institute of Science and Technology (KAIST), 291 Daehak-ro, Yuseong-gu, Daejeon 34141, Republic of Korea

[*]swbaek@kaist.ac.kr

Abstract. In present study, the autoignition of ethanol based gel propellant, which used methylcellulose as a gellant, has been investigated experimentally. The content of gellant is 9 wt%. The initial droplet diameter range of ethanol gel droplet was 2.5±0.3 mm, and suspended on 0.335 mm quartz fiber. The experiment was conducted at 500, 600, and 700°C under atmospheric pressure, considering that the autoignition temperature of ethanol is 365°C. High speed camera was used for analysis with 100 images per second. General combustion behavior of gel droplet were observed such as swelling, micro explosion, and vapor jetting. It was confirmed that ignition did not occur at 500°C. At 600°C and 700°C, it had short lifetime, and the size of the droplet decreased linearly.

1. Introduction

Gel propellants can be obtained by agitating liquid fuels with suitable gellants, which have the merit of being able to achieve different properties depending on the combination. It has the advantages of liquid propellant, which is capable of re-ignition and thrust control when the propulsion system is operating. As it is in gel form, it shows high stability and easy handling in case of storage, which is the advantages of solid propellant [1]. Because of these advantages, right after the suggestion of concept of the gel propellant, it has been actively studied from the 1940s to 1970s. After some visible results had been shown in the mid-1980s, such as the success of the flight test, related research is underway again in the 2000s.

So far, studies on gel propellants have been divided into three categories: rheological properties, atomization with spray behavior, and combustion characteristics. In order to use the gel propellant practically, it is generally necessary to have shear thinning properties. To confirm this, rheological characterization should be performed. Spray and atomization studies are then carried out at the injector exit to see if it can be utilized at levels similar to conventional liquid fuels. In the combustion characteristics study, the performance was estimated by checking the ignition delay, combustion heat and burning rate etc. Most of researchers observed combustion characteristics after igniting the gel droplet from the external ignition source.

Evaporation and ignition of fuel droplets occurs when exposed to a high temperature environment inside of the combustion chamber. Autoignition take place if the amount of droplet vapor, the oxidant concentration and the temperature around the droplet are sufficiently high. Since the gel droplet is ignited because of the high temperature in the real application, in this study, the experiment had been performed to grasp autoignition characteristics.

2. Experimental setup

2.1. Ethanol gel

To synthesize the gel used in the experiment, ethanol was selected as parent fuel which has very low toxicity with easy handling, and methylcellulose was used as gellant. When the content of gellant is 9~11 wt%, a gel having excellent rheological properties is synthesized. In this experiment, 9 wt% gel was used. The ethanol gel had shear thinning property, and its calorific value was 95% of liquid ethanol.

2.2. Constant volume combustion chamber

A constant volume combustion chamber was used to create autoignition and is shown in the fig. 1 below [2]. As shown in the fig. 1, the electric furnace is moved up and down to expose the gel droplets to the high temperature environment. It maintains constant temperature by PID control.

The initial droplet diameter range was 2.5±0.3 mm, and suspended on 0.335 mm quartz fiber. Experiments were conducted under atmospheric pressure at 500, 600, and 700°C, considering that the autoignition temperature of ethanol is 365°C.

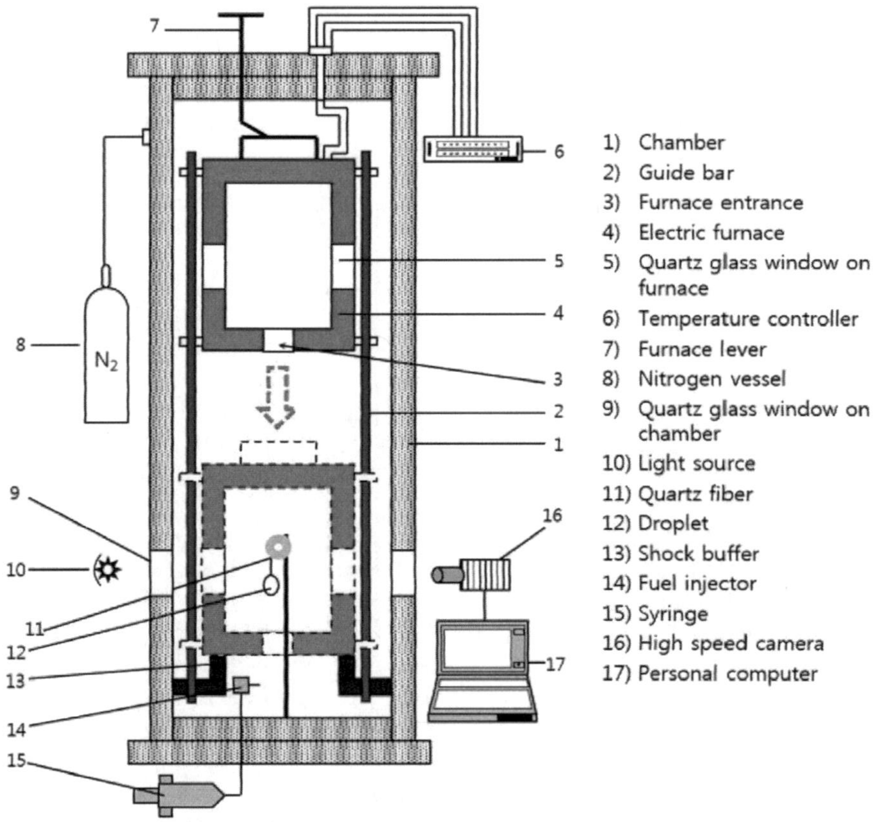

Figure 1. Schematic diagram of experimental setup [2]

2.3. High speed camera

To observe the combustion process of the ethanol gel, high speed camera (x-stream 3, Integrated Design Tools) was used. Although it can obtain up to 50,000 images per second, 100 images per second with high resolution were acquired for more accurate analysis. Analysis of recorded images for the calculation of droplet diameter was done by own developed program based on Visual Basic. This program was validated from previous study [3].

3. Results and discussion

Overall experimental setup is shown in the fig. 2. The ethanol gel droplet is suspended on quartz fiber in the middle of fig. 2. The electric furnace is moved to its bottom position and created high temperature environment. Bright part at the backside is light source, which helps the high speed camera to get clear image.

The fig. 3 below depicts the autoignition and combustion process of a 9 wt% ethanol gel droplet with a high speed camera. No flame was observed at 500°C, and ignition occurred at 600 and 700°C case. Since there was no combustion at 500°C, residues were found but volume of the droplet was reduced overall, and partial swelling was observed. This is because the parent fuel on the surface of the droplet evaporates first and the rest of gellant forms a thin layer [4]. On the other hand, at 600°C, the volume change was severe immediately after the droplet was exposed to the high temperature. As the flame was generated, the volume rapidly decreased and all of it burned without any residue. The volume change of the droplet was not large at 700°C, but the ignition and combustion process proceeded faster than 600°C and no residue was observed. The particularly remarkable thing is that there is a possibility of presence of a flame which cause linear and rapid decrease of the droplet volume, before the orange flame appears which is presented in the fig. 3. It can be considered as implicit evidence of two stage combustion of ethanol gel droplet. The fig. 4 shows vapor jetting during an active combustion process. At the left image, gellant layer got ruptured where an arrow points. Next moment, hot ethanol vapor inside of the droplet came out swiftly and ignited from the downstream when it met oxygen. At the right image, it formed a flame instantaneously. Micro explosion and collapse of flame envelope appeared also.

Figure 2 Ethanol gel droplet inside of the combustion chamber with electric furnace

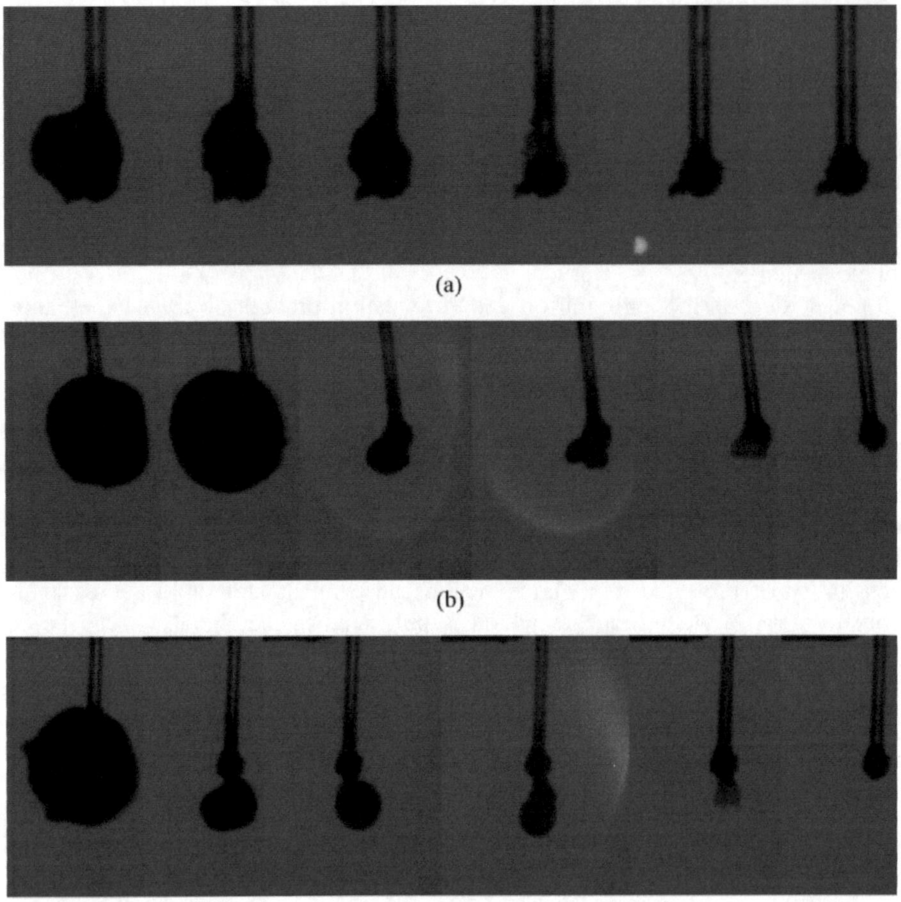

(a)

(b)

(c)

Figure 3 Autoignition behaviour of ethanol gel droplet under high temperatures. (a) at 500℃ (b) at 600℃ (c) at 700℃

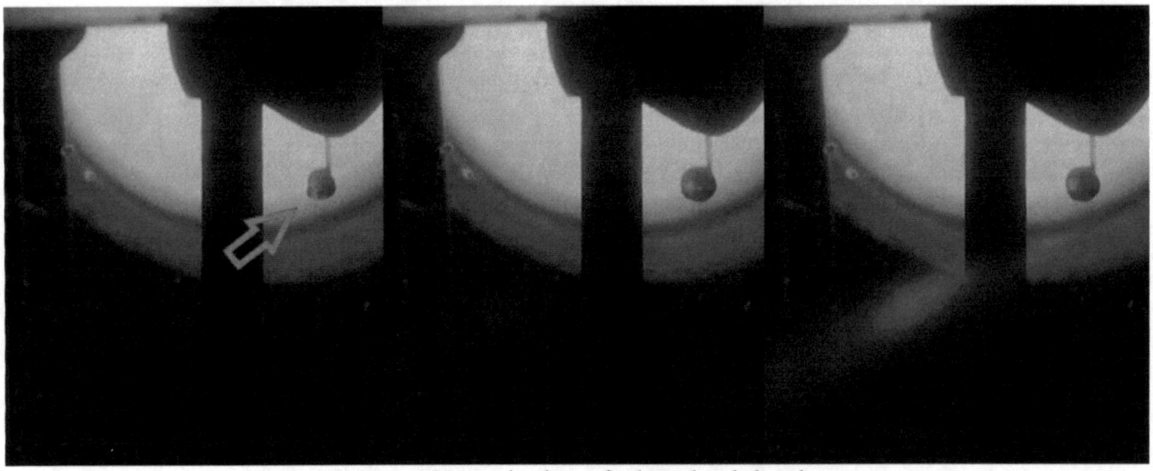

Figure 4 Vapor jetting of ethanol gel droplet

The fig. 5 represents the droplet size change by temperature with time. The vertical axis is a nondimensionalized value obtained by dividing the square of the diameter of the droplet with respect to time by the square of the initial diameter of the droplet. The abscissa means the lifetime of droplet divided by the initial diameter squared of the droplet. The droplet size change is shown where the leftmost first abscissa is 700°C case, the second abscissa is 600°C case, and the last abscissa is 500°C case. Compared with the 500°C droplet, which has not burned, 600 and 700°C droplets have a very short lifetime. In addition, at 600°C and 700°C cases, the size of the droplet decreases linearly. In this section, the d2-law using for the analysis of general droplet seems to be applicable [5].

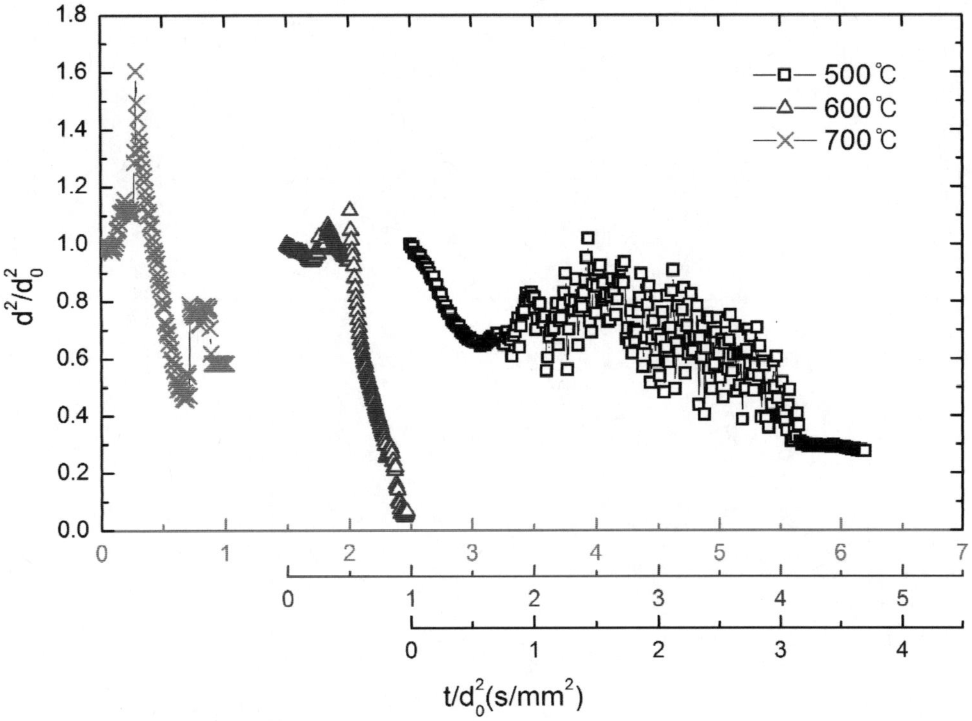

Figure 5 Ethanol gel droplet size change by temperature with time

4. Conclusion

In this study, the autoignition of ethanol based gel propellant, which used methylcellulose as a gellant, has been investigated experimentally. General combustion behavior of gel droplet were observed such as swelling, micro explosion, and vapor jetting. It was confirmed that ignition did not occur at 500°C. However, additional research is required to accurately observe and identify the cause of the double burning phenomenon.

Acknowledgments

This work was supported by the Korean Research Fellowship Program funded by the Ministry of Science, ICT and Future Planning through the National Research Foundation of Korea (No. 2015H1D3A1061637).

Reference

[1] Natan B and Rahimi S 2002 The status of gel propellants in year 2000 *Combustion of Energetic Materials* (Boca Raton: Begel House) pp 172-194

[2] Kang H, Won J, Baek S W and Kwon S 2017 Autoignition and combustion characteristics of sodium borohydride-based non-toxic hypergolic fuel droplet at elevated temperatures *Combust. Flame* **181** p 149-56

[3] Javed I, Baek S W and Waheed K 2013 Evaporation Characteristics of Heptane Droplets with the Addition of Aluminum Nanoparticles at Elevated Temperatures *Combust. Flame* **160** pp 170-83

[4] Nachmoni G and Natan B 2007 Combustion characteristics of gel fuels *Combust. Sci. Technol.* **156** pp 139-157

[5] Mishra D P and Patyal A 2012 Effects of initial droplet diameter and pressure on burning of ATF gel propellant droplets *Fuel* **95** pp 226-233

IOP Conf. Series: Materials Science and Engineering **280** (2017) 012035 doi:10.1088/1757-899X/280/1/012035

An Investigation to Optimize the Layout of Protective Blast Barriers Using Finite Element Modelling

V Lawrence[1], C Ngamkhanong[1,2] and S Kaewunruen[1,2]

[1]Department of Civil Engineering, School of Engineering, The University of Birmingham, Birmingham, B15 2TT, United Kingdom
[2]Birmingham Centre for Railway Research and Education, The University of Birmingham, Birmingham, B15 2TT, United Kingdom

CXN649@student.bham.ac.uk

Abstract. The past has shown that the risk of terrorism is on the rise as can be seen in many events. Terrorist attacks such as the 2004 Madrid bombings, the 2005 London bombings, the 2013 Boston bombing and the 2017 Manchester bombings have shown the impact acts of terrorism has on the public. Nowadays, terrorist attack is likely to increase in the wider area in the future together with a higher density in train passengers. Previous researches into protective measures have been focused on the structure. However, the impacts of terrorist attack on human have not been fully investigated. The needs for protective measures for the public have never been greater. This research aims to investigate the optimum layout for protective blast barriers situated on a train station platform using finite element analysis. The 3-dimensional structure is modelled and analysed using LS-DYNA.The focus is placed on an island platform at Birmingham New Street Station, which is one deemed to be at high risk of terrorist threats. Two shapes of barriers were tested, straight and angled. A total of six models were created and tested against two scenarios. Scenario one is a bomb placed on the ground, scenario two is a bomb being carried. The results focus on the impact the pressure created from the blast has on a person's lungs and head. Both can cause the most fatalities due to bombings. The results demonstrated that the shape of the barriers had no effect on the pressure. However, it can be concluded that an increase in the number of barriers, reduced the pressure below the critical amount for lung damage. Increasing the number of intervening objects between the bomb and target has a positive effect on the reduction of blast pressure. The insight into this study will help railway and structural engineers to establish strategic preventing methods to minimise catastrophic damage to and potential losses of the public.

1. Introduction

In the past fifty years the risk of terrorist bombings in public places has increased. The Madrid train bombings in 2004, London bombings in 2005 and Manchester bombings in 2017 demonstrated the impact terrorism can have on the public. Many governments are introducing various protective measures to important structures within their country. Much research has already been performed on the protection of structures. However, it is clear from recent terrorist attacks that the public is the target. Though research is being done into protection for the public, there is very little implementation in the real world. Railway stations are a highly congested and crowded place many hours of the day. Making railway stations a target for terrorism. The finite element modelling was used to determine the

IOP Conf. Series: Materials Science and Engineering **280** (2017) 012035 doi:10.1088/1757-899X/280/1/012035

effects of an explosion within a railway station [1-3]. Considering the results only small amounts of research have been done on protective blast barriers for the public. The concrete blast barrier was tested considering barrier shapes and layouts [4-5]. The blast barrier was investigated by changing the material of the barrier to glass fibre reinforced polymer [6], but the results showed that concrete has a higher resistance. With such little research done on this subject, it is unclear on the effectiveness and benefits of protective blast barriers.

This paper will investigate the optimal layout of protective blast barriers, to reduce the blast wave pressure of an explosion at railway stations. A finite element model of a Birmingham New Street Station platform will be created. Each blast barrier arrangement will be subjected to an explosion from the surface and free air. Both scenarios are similar to terrorism tactics. The blast wave pressure will be recorded to determine the effects of each arrangement.

2. Methodology

A blast wave is in the "form of a shock wave composed of a high-intensity shock front which expands outward from the surface of the explosion into the surrounding air" [7]. The pressure exerted from a standard blast wave is shown below (Figure 1). The sudden increase in pressure from the detonation of the explosion to the peak pressure is defined as the overpressure. In Figure 1, the pressure plotted is the incident pressure. The incident pressure is the pressure on the structure from the incident wave.

This report will not focus on incident pressure due to the reflected pressure being much greater.

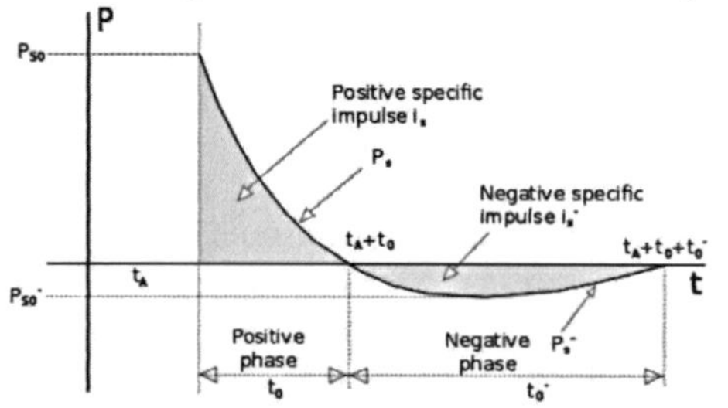

Figure 1. Blast wave pressure plot [8]

LS-DYNA software was used to simulate the air blast created from TNT. LS-DYNA has an empirical function LOAD_BLAST_ENHANCED which computes to a high degree of calibration, the pressure exerted on a Lagrangian structure from an air blast. LS-DYNA calculates pressure by measuring the distance from segment to charge, and the angle of incidence from the segments normal. To identify the blast load criticality, the document Unified Facilities Criteria (UFC), Structures to Resist the Effects of Accidental Explosions [7] has been reviewed.

$$Z = \frac{R}{\sqrt[3]{W}} \quad (1)$$

where R is the distance between the point of the detonation and the structure, W is the weight of the charge.

The scaled distance is computed. LOAD_BLAST_ENHANCED uses both of these variables at every cycle to calculate the pressure. The pressure can also be calculated by hand using this formula to validate the pressure computed by LS-DYNA. LOAD_BLAST_ENHANCED enables the blast to be located at any point throughout the model. The function can allows for four different types of blast shape. This research uses the hemispherical blast with reflected waves, as well as a free air blast with reflective waves. The model was created through LS-DYNA Prepost and is based on Birmingham New Street Station. The floor slab and roof slab were made completely rigid to simulate a blast underground.

The platform is an island platform with a width of 16 metres. Columns are situated 12 metres apart. The columns are 3 metres long and 1 metre wide. The epicentre of the air blast was situated in the middle of the platform. Two scenarios were tested, both using 15kg of TNT.

- Case 1: The bomb is placed in a bag and left on the floor of the platform (0m above G.L)
- Case 2: The bomb is carried by a person in a rucksack. (1.6m above G.L)

The barrier design and orientations from the previous study were used [4], as shown in Figure 2.

a) b)

Figure 2. Dimensions of barrier designs a) corner b) straight

The internal holes in the barriers were designed to increase the porosity of the barrier such that they did not fail. The holes aided the diffusion of the total blast pressure. The internal holes in the barriers were designed to increase the porosity of the barrier such that they did not fail. The holes aided the diffusion of the total blast pressure. Figure 3 represent the barrier arrangements. The sensors were placed such that to model pedestrians waiting for a train. To calculate a valid air blast pressure at a targets lungs and head, the sensor were positioned 1.4 metres above G.L. Three finite element models were created, each model contained two different arrangements as can be seen in Figure 3.

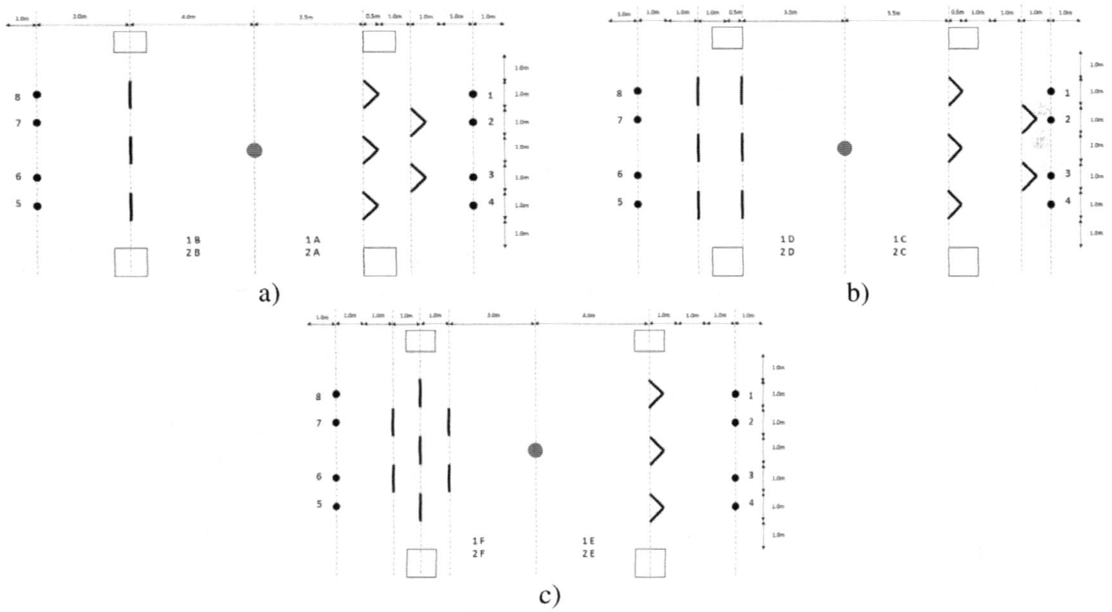

Figure 3. Barrier arrangement
a) A – Five Corner Barriers in close arrangement and B – Three Straight Barriers
b) C – Five Corner Barriers with larger separation and D – Six Straight Barriers
c) E – Three Corner Barriers and F– Seven Straight Barriers in tight arrangement

IOP Conf. Series: Materials Science and Engineering **280** (2017) 012035 doi:10.1088/1757-899X/280/1/012035

The platform and columns are to be modelled as reinforced concrete. The function CONSTRAINED_LAGRANGE_IN_SOLID is a validated solution to model the rebar within the concrete mesh. The concrete material uses the function MAT_CSCM_CONCRETE. Through which C32/40 concrete properties are imputed. The rebar and steel barriers are both defined to be MAT_PLASTIC_KINEMATIC. Transverse rebar is 16 millimetres diameter, longitudinal rebar is 10 millimetres diameter. The steel properties of the barrier are shown in Table 1.

Table 1. Steel Properties

Mass Density	7850 kg/m^3
Young's Modulus	2.1 x 10^5 MPa
Poisson's Ratio	0.3
Yield Stress	400 MPa

3. Results and discussions

3.1. Case 1: Near Surface Blast

Table 2 shows the pressure on the sensors due to near surface blast. When comparing the control sensor values with arrangements 1B and 1E the two central sensors have an increased pressure. The increased pressure is caused by the blast pressure reflecting and intensifying as it is transmitted through the barriers. It is noted that the flow of a shock wave expands directly after the barrier gap, resulting in the velocity of the wave intensifying [9]. The two external sensors of both arrangements were predicted to act the same due to the findings of recent study [4]. Concluding that the shape of the barriers placed in identical orientation has minimal effect on mitigating the blast wave. It is also concluded that a blast wave travels over a barrier, reforming a reduced pressure wave on the other side.

Table 2. Near Surface blast pressure (KPa)

Sensor Number	1A	1B	1C	1D	1E	1F
1 & 8	254	289	238	248	289	211
2 & 7	279	339	276	315	339	237
3 & 6	279	339	276	315	339	237
4 & 5	266	305	250	261	305	221

Note: Sensors 1-4 for 1A, 1C, 1E. Sensors 5-8 for 1B, 1D, 1F

Arrangement 1A and 1C were created to investigate this conclusion. Arrangement 1C is the same orientation as 1A but the barriers have a greater distance between them. From the results 1C reduced the overall pressure of the blast wave. The two external sensors of 1C both have a decreased pressure of 16 KPa than 1A. As found by [9], the reformed blast wave travels a longer distance until reaching the second row of barriers. Within this distance the wave is dissipating, thus reducing in force. However, the two central sensors show a very minimal decrease in pressure. The cause of both outcomes are as a result of the shape of the transmitted shock wave. As demonstrated in [9] the peak of a transmitted shock wave is created at the centre point between two barriers. The remaining transmitted wave travels in a 'close to' hemispherical shape (Figure 4), depending on the angular degree of barrier shape.

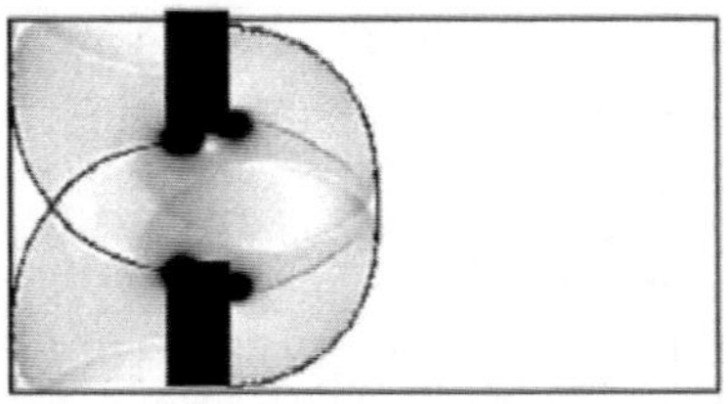

Figure 4. Mapping of a shock wave travelling through a barrier [10]

1C allows for the transmitted blast wave to dissipate and reduce more, before impacting the external sensors. The peak transmitted blast wave travels identically in both 1A and 1C resulting in minimal reduction at the central sensors. Arrangement 1F insured the two most vulnerable sensors (central) were protected from the peak transmitted blast wave. 1F increased the number and density of the barriers. Comparing the results with the other arrangements, 1F has significantly decreased the pressure over all sensors.

3.2 Case 2: Free Air Blast

Table 3 shows the pressure on the sensors due to free air blast. The free air blast results all have decreased pressure when compared with the identical arrangement subjected to a surface blast. As explained in [11], compared with a surface blast, free air blast waves amplify less from the ground due to the distance they must first travel.

Table 3. Free Air blast pressure (KPa)

Sensor Number	1A	1B	1C	1D	1E	1F
1 & 8	218	271	227	233	271	174
2 & 7	278	309	255	299	309	193
3 & 6	278	309	255	299	309	193
4 & 5	229	284	233	245	284	182

Note: Sensors 1-4 for 1A, 1C, 1E. Sensors 5-8 for 1B, 1D, 1F

Comparing 1A, 1C, 2A and 2C, a larger gap between the barrier rows has increased the overall pressure on the sensors subject to a free air blast. The opposite occurred in a surface explosion. The free air blast waves travel at a steeper angle over the barrier due to the height of the epicentre. Resulting in the reflection point behind the barrier being nearer. 2C's larger gap allows for the wave to be reflected and reformed into a Mach front before the second row of barriers. However, the two central sensor of 2C are recorded to have a decreased pressure than 2A. The blast wave projection has changed [9,12]. With a larger gap between rows the central sensors are protected due to being located directly behind the barriers. The external sensors however, are positioned far enough for the reflected wave to impact them. With a reduced gap between rows the central sensor are subjected to the full Mach front. The external sensors are protected due to the dissipation angle of the blast wave. In case of B and E, the layout arrangements are identical, with only the shape of the barrier changing. The results showed very minimal difference between the corner and straight barrier shapes.

4. Conclusion

This research can conclude that the layout of blast barriers does have an impact on the reduction of blast pressure. Increasing the amount of barriers between the explosion and target, increases the chance of survival. The results show that layouts 1F and 2F both had very successful effects compared with other layouts. The positive effects of increasing the distance between barrier rows is subject to the height of the bomb. Moreover, it can be concluded that the shape of the barrier provides minimal effect on the mitigation of blast pressure. Due to similar effects seen in both experiments. The omnidirectional nature of a blast wave causes the pressure to travel over the barrier as well as through them. In addition, the gap between barriers row also plays a role in pressure reduction. Further research is need on whether full optimization of the entire space available for blast protection is a viable option. Increasing the height, thickness and amount of barriers could all aid in the reduction of blast pressure. This research only accounted for the blast pressure cause from an explosion. The heat and fragmentation of explosions must also be tested. The psychology of reducing the space of an already highly congested area is important. Though people may want to be protected from bombings, limiting space could be putting people in more danger. The insight into these blast behaviours will help railway and structural engineers to establish strategic retrofitting methods to minimise catastrophic damage to and potential losses of train passengers, the public and railway assets.

Acknowledgement
The second author gratefully appreciates the School of Engineering and Birmingham Centre for Railway Research and Education of his PhD scholarship. The last author gratefully acknowledges the Japan Society for the Promotion of Science (JSPS) for his JSPS Invitation Research Fellowship (Long-term), Grant No L15701, at Track Dynamics Laboratory, Railway Technical Research Institute and Concrete Laboratory, the University of Tokyo, Tokyo, Japan. The authors are sincerely grateful to European Commission for the financial sponsorship of the H2020-RISE Project No. 691135 "RISEN: Rail Infrastructure Systems Engineering Network," which enables a global research network that tackles the grand challenge in railway infrastructure resilience and advanced sensing in extreme environments (www.risen2rail.eu).

References

[1] Larcher M, Casadei F, Giannopoulos G, Solomos G, Planchet J and Rochefrette A 2010 *Proceedings of the Institution of Mechanical Engineers, Part F: Journal of Rail and Rapid Transit* **225** p 373-382
[2] Solomos, G., Casadei, F., Giannopoulos, G. and Larcher, M. (2011) *Internal Security,* **3** 49.
[3] Larcher M, Forsberg R, Björnstig U, Holgersson A and Solomos G 2016 *Journal of Transportation Safety & Security* **8** 83
[4] Hajek R and Foglar M *Structures Under Shock and Impact XIII* **141** 265-275
[5] Hajek R, Foglar M and Fladr 2016 *Construction and Building Materials* **120** 54-64
[6] Asprone D, Prota A, Manfredi G and Nanni A 2015 *International Journal of Polymer Science* 11
[7] Unified Facilities Criteria (UFC) 2008 *Structures to Resist the Effects of Accidental Explosions, U. S. Army Corps of Engineers,* Naval Facilities Engineering Command, Air Force Civil Engineer Support Agency, UFC 3-340-02, 5 December 2008.
[8] Bermejo M, Santos A and Goicolea J 2014 *The 9th International Conference on Structural Dynamics.*
[9] Berger S, Ben-Dor G. and Sadot O 2015 *Journal of Fluids Engineering* **137** 41203-1-41203-11
[10] Berger S, Ben-Dor G. and Sadot O 2014 *European Journal of Mechanics / B Fluids* **50** 60-70
[11] Koccaz Z, Sutcu F and Torunbalci N 2008 *The 14th World Conference on Earthquake Engineering.*
[12] Remennikov A M and Rose T A *International Journal of Impact Engineering* **34** 1907-1923

IOP Conf. Series: Materials Science and Engineering **280** (2017) 012036 doi:10.1088/1757-899X/280/1/012036

Nonlinear Dynamic of Curved Railway Tracks in Three-Dimensional Space

X Liu[1], C Ngamkhanong[1,2] and S Kaewunruen[1,2]

[1]Department of Civil Engineering, School of Engineering, The University of Birmingham, Birmingham, B15 2TT, United Kingdom
[2]Birmingham Centre for Railway Research and Education, The University of Birmingham, Birmingham, B15 2TT, United Kingdom

E-mail: CXN649@student.bham.ac.uk

Abstract. On curved tracks, high-pitch noise pollution can often be a considerable concern of rail asset owners, commuters, and people living or working along the rail corridor. Inevitably, wheel/rail interface can cause a traveling source of sound and vibration, which spread over a long distance of rail network. The sound and vibration can be in various forms and spectra. The undesirable sound and vibration on curves is often called 'noise,' includes flanging and squealing noises. This paper focuses on the squeal noise phenomena on curved tracks located in urban environments. It highlights the effect of curve radii on lateral track dynamics. It is important to note that rail freight curve noises, especially for curve squeals, can be observed almost everywhere and every type of track structures. The most pressing noise appears at sharper curved tracks where excessive lateral wheel/rail dynamics resonate with falling friction states, generating a tonal noise problem, so-call 'squeal'. Many researchers have carried out measurements and simulations to understand the actual root causes of the squeal noise. Most researchers believe that wheel resonance over falling friction is the main cause, whilst a few others think that dynamic mode coupling of wheel and rail may also cause the squeal. Therefore, this paper is devoted to systems thinking the approach and dynamic assessment in resolving railway curve noise problems. The simulations of railway tracks with different curve radii will be carried out to develop state-of-the-art understanding into lateral track dynamics, including rail dynamics, cant dynamics, gauge dynamics and overall track responses.

1. Introduction

Curve squeal is a strongly tonal noise emitted from wheel/rail contact caused by the passage of the train in tight curve tracks with a low speed [1]. The occurrence of squeal induces significant environmental impacts immensely annoying people living nearby due to its high frequencies characteristics. Based on previous studies, unsteady lateral creepage of the wheel is thought to be the prime reason of squeal, while other mechanisms such as longitudinal creepage and flange contact, do not necessarily eliminate squeal noise thereby are determined to be of secondary importance [2-4]. A number of researches have been carried out to demonstrate the mechanism of squeal noise by means of site investigation and numerical modeling. Negative damping mechanism has been primarily proposed by Rudd [5]. The friction coefficient, which is relevant to the sliding velocity, falls after creepage exceeds the saturation point. This falling characteristic of creep force at high creepage is believed to be the main reason of squeal. However, the mode coupling instability as a new mechanism between normal and tangential dynamic is getting growing support while the negative friction mechanism is no longer emphasized in research field [6]. It is still questionable at which mechanism results in wheel/rail squeal.

The variation of squeal noise has been observed on-site in conjunction with many critical parameters including curve radii, rail dynamics, cant dynamics and overall track response during operation. Previous work showed that squeal only occurs when the curve radius is smaller than 100b, where b is the wheelbase of the bogie [5]. The results of on-site measurements also presented that there is no substantial reduction in wheel squeal associated with limiting operation speed. The investigation implemented by J.Jiang [7] conducted the track dynamic in terms of fluctuation of curve noise after upgrading from timber sleepers to concrete sleepers. But the question remains as for how other parameters change the lateral track dynamic. In fact, data collected in the field suggest a diverse range of curving behavior, which is largely relevant to curve radii [8,9]. However, it is still uncertain to what extent the track lateral response is affected by rail radius. The occurrence of cant is believed to have positive impacts on reducing curve squeal due to its counteraction with creep force, an analytic approach has to be implemented for the determination of the effective range of cant. In reality, stable and unstable regimes of track deflection are predicted in operation condition due to high lateral load. The track dynamic response characteristics need to be taken into account to justify the track dynamic under different lateral load.

The critical review reveals that lateral track dynamic characteristics have not been fully understood, especially when it comes to various curve radii, cants and lateral loads. The aim of this study is to investigate the lateral track behavior into the effects of their properties. The present analyses are based on numerical modeling, using Timoshenko beams in a general-purpose finite element package STRAND7 to build a three-dimensional track model. Both static and nonlinear transient analysis are considered. Dynamic displacement and acceleration have been evaluated at rail over sleeper, rail at midspan and sleeper at midspan under different curve radii and load cases. Dynamic characteristics are then highlighted in this paper. The findings will provide some key parametric insights into fundamental dynamics of track in the lateral direction and establish the development of the dynamic design of curve track.

2. Track model

The track model consists of a two-dimensional Timoshenko beam, which has been validated to be one of the most suitable options for concrete sleepers modeling [10,11]. In this investigation, the finite element models of railway tracks have been previously developed and calibrated against the numerical and experimental modal parameters [12,13]. A general-purpose finite element package STRAND7 is used for finite element simulation. It is noted that beam elements, which take into account shear and flexural deformations, are considered to simulate sleeper and rails in terms of lateral response. Each sleeper consists of 60 beam elements and each rail consists of 200 beam elements. The overall model was represented by 1348 beam elements. The 60kg rail cross section was introduced in track model. The trapezoidal cross section was assigned to the sleeper elements. The rail pads at rail seats were simulated using a series of spring-dashpot elements. The distance offset between rail and sleepers was set to 100mm which does not have impacts on numerical results [13]. In this study, stiffness and damping values of high-density polyethylene pads were assigned to these spring-dashpot elements both in vertical and lateral direction. Ballast was placed under sleeper, supporting the rails and sleepers, which provides compression only. Tensionless beam support was used to achieve the support condition with an allowance of hover over the support while the tensile supporting stiffness is omitted. It was validated in previous studies, effects of length and boundary of track in this investigation (18 bays or 10.8m) on the computation and frequencies of interest are negligible [10-13].

In total, seven types of curve radii have been evaluated for dynamic investigations using curves ranging from 100m to 600m with tangent one. Also, four types of cants were determined with the range from 0mm to 300mm. Engineering properties have been validated by a parametric analysis which is implemented by consideration of different rail, sleeper and ballast properties. In order to adopt these, the partial support condition is believed to conform with real condition of standard gauge tracks vastly. The nodes at both ends of track were fixed in every direction to represent the track boundary. Four separated forces with various magnitudes have been used to simulate the load condition of a passenger train bogie (2 per each rail, 2 m apart). This load magnitude has been used for benchmarking purpose [14,15]. The nondimensional analyses have been carried out to investigate the dynamic amplification over curve radii and frequency domain.

3. Results and discussion

The Linear Static Solver of STRAND7 has been used to evaluate the static response of railway tracks. The Nonlinear Transient Solver is then used to get dynamic behaviors of railway tracks after being verified by the quasi-static condition. Two moving load envelope with a speed of 10m/s have been established to evaluate the dynamic response of the rail based on an impulse excitation of a period of 0.0001s starting at time t =0.005s. The calculation time step is set to be 5×10^{-5} so that it can include high-frequency modes for responses of curving. With this FE model, the eigenfrequencies and corresponding eigenmodes are calculated up to 10kHz, which is believed to cover modes of squeal noises vastly. The dynamic analyses are taken into account for various curve radii, cants and lateral loads. To appropriately take into consideration of dynamic effects of moving load, calculation time has been set to be 5s thus the entire process of dynamic responses can be adaptively reflected in simulation associated with the length of track and vehicle velocity. Typical deformation and stress distribution of a railway track under the lateral moving load are shown in figure 1. The rails present positive bending due to the lateral load applied perpendicular to load moving direction.

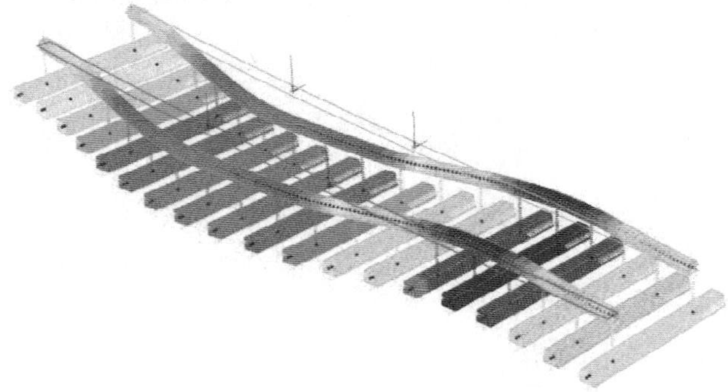

Figure 1. The example of the track's deformation under lateral moving load

3.1. Maximum displacement response

Figure 2 illustrates the dynamic lateral displacements of rail over sleeper at different curve radii. The L/V's are lateral to vertical load ratios where the vertical load was selected as 100KN as a benchmark for passenger train bogie. Varying load ratios were comparatively investigated, and similar trends to the results for the rail dynamic were exhibited. It is clear that the tracks with larger curve radii deform laterally more severe than tight curve tracks due to lower lateral resistance. The result can be effectively explained by arch mechanics that arches with larger curvature possess of higher compressive strength. Take load ratio of 0.2 as an example, displacement increases 75.8% for curve radius varying from 100m to 200m, while the increase rate is only 3.2% for radius varying from 500m to 600m. This implies that lateral displacement responses are more sensitive corresponding to low radii, which gave evidence on the appearance of squeal during train negotiating tight curves. It can also be observed from the graph that the lateral track displacement of the tangent track is similar to the value of track with a radius of 600m. For large curve radius, the lateral displacement of the track no longer changes significantly with increasing radius therefore the increase of radius plays a little role on the dynamic amplitude of track. The results above indicate that the increased track radius has positive effect on reducing curve squeal and squeal noise would disappear when the curve radius comes to a particular value.

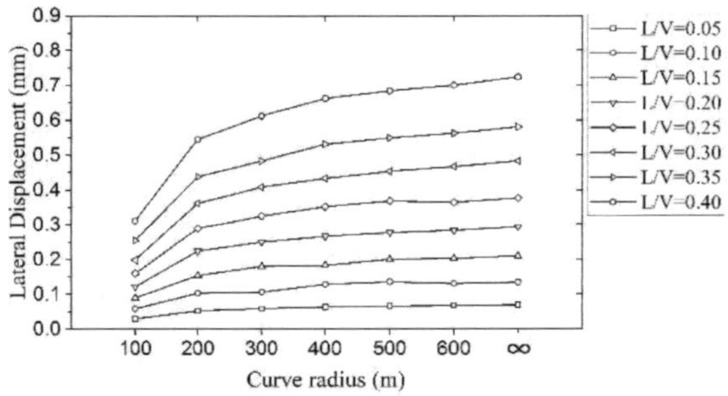

Figure 2. Dynamic lateral displacements of rail over sleeper at different curve radii, cant=100mm

The Rail cant is a crucial parameter in relation to rail noise vibration. To evaluate the effects of cants on dynamic responses, data from Figure 3 exhibits a comparison of dynamic lateral rail displacements based on the simulation performed with a vehicle velocity of V=10m/s for varying cants. The results apparently demonstrate the positive effect of including cant in curve track, which leads to a lower dynamic response when cant goes towards a higher value. The result can be inferred by the occurrence of centrifugal forces and compensative effect of cants. As expected, the railway track without a cant is more likely to cause severe damage to track components and even leads to derailment. Therefore, these negative effects, which could be avoided by adding suitable cants to the railway track, should anyhow not obstruct railway normal operation.

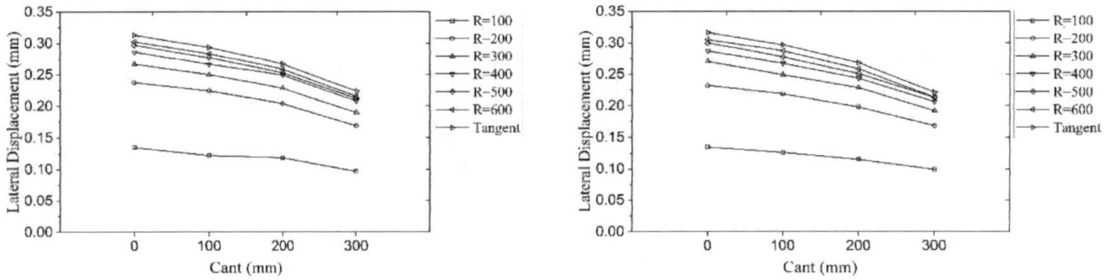

(a) Rail over sleeper (b) Rail at midspan

Figure 3. Maximum lateral displacements of track at a) Rail over sleeper b) Rail at midspan with varying cants

3.2. Lateral Acceleration

Figure 4 shows the corresponding Fourier spectrum for rail vibration at midspan with respect to curve radius of 200m, 500m and infinite. The most distinct resonance of 200m radius track occurs at 150Hz with the value of 131dB while the maximum response of 500m radius track happens at 251Hz with the value of 125.6dB. Therefore, the increase of track radius expands the resonance frequency and reduces the peak of vibration due to the time-variation of response spectrum. It can also be verified by the fact that the acceleration in the tangent track is much lower than that on curved tracks. The track responses are approximately equivalent in amplitude for various curve radius up to 1000Hz, although several discrete resonances and antiresonances appear in this region. On the contrary, the track acceleration in the frequency region of most interest for the squeal, 1000Hz-5000Hz, is much greater on 200m radius track than on 500m radius track and tangent track. This would lead to more unstable modes occur at squeal frequency region, which enhances the severity of squealing. When zoomed in, a comparison of acceleration levels between curve track and tangent track suggests that vibration in tangent track in high frequencies have little relation to curve squeal.

IOP Conf. Series: Materials Science and Engineering **280** (2017) 012036 doi:10.1088/1757-899X/280/1/012036

Figure 5 illustrates the effect of curve radius by showing the difference between the responses of rail with radius of 200m and 500m. Insertion gain as an indication of vibration mitigation has been used here to quantify the relative mitigation effects. In the overall frequency domain, the insertion gain distributes mainly in the range of -10dB-30dB. The region at frequencies below 1000Hz is much lower than that at frequencies between 1000Hz and 5000Hz where curve squeal is most likely to occur. Therefore, it is demonstrated that the responses of track in high frequencies are sensitive to track radius thus increasing of curve radius has significant positive effect on reducing squeal noise.

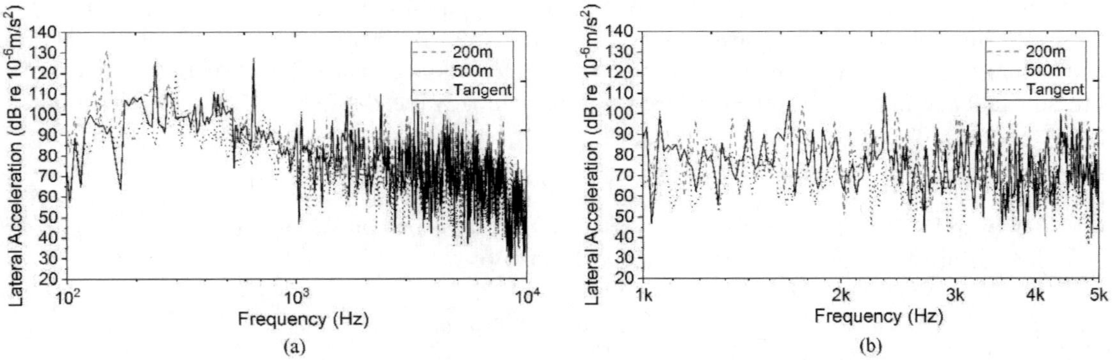

Figure 4. Spectra of the rail lateral acceleration at middle span of rail for varying types of track in the frequencies range of a) 100-10000Hz b) 1000-5000Hz

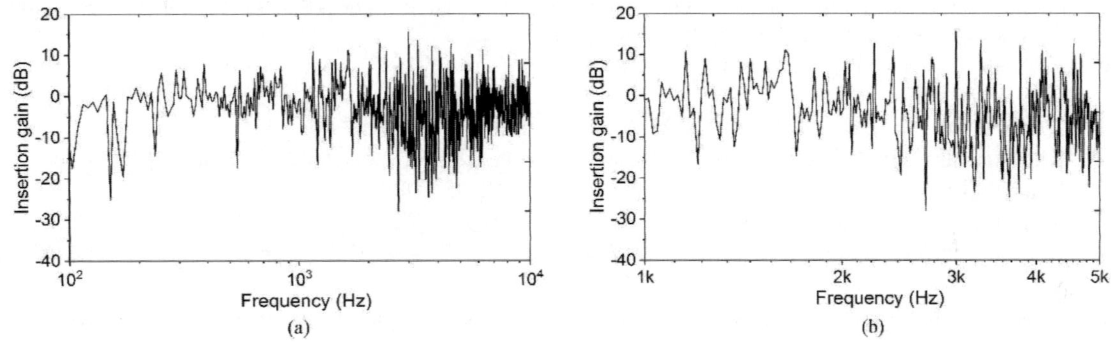

Figure 5. Insertion gain of track vibration due to change in track radius from 200m to 500m in the frequencies range of a) 100-10000Hz b) 1000-5000Hz

4. Conclusion

A numerical simulation has been implemented in this paper to identify the lateral dynamic characteristics of the curved track with many influencing parameters including track radii, cants and lateral to vertical load ratios. Parametric studies performed with curved track model have indicated many important results.

The relative rail displacement can be reduced by the increasing of curve radius. However, for the tight curve, relative displacement responses are more sensitive to the change of radius. This provides evidence on the occurrence of squeal in tight curve. Cant is believed to have a positive effect on dynamic responses. However, a high cant is demonstrated to incur severe dampening performance, therefore, a reasonable range of cants need to be established. Moreover, track acceleration is greater on track with lower radius in higher frequencies. The insertion gain in squeal region is two times of that in lower frequencies. When the results for different track types are compared, it is determined that vibration in the tangent track would not incur squeal noise.

Acknowledgement

The first author would like to thank Beijing Jiaotong University for his sponsorship and also a collaboration between Beijing Jiaotong University and University of Birmingham. The second author gratefully appreciates the School of Engineering and Birmingham Centre for Railway Research and Education of his PhD scholarship. The last author gratefully acknowledges the Japan Society for the Promotion of Science (JSPS) for his JSPS Invitation Research Fellowship (Long-term), Grant No L15701, at Track Dynamics Laboratory, Railway Technical Research Institute and Concrete Laboratory, the University of Tokyo, Tokyo, Japan. The authors are sincerely grateful to European Commission for the financial sponsorship of the H2020-RISE Project No. 691135 "RISEN: Rail Infrastructure Systems Engineering Network," which enables a global research network that tackles the grand challenge in railway infrastructure resilience and advanced sensing in extreme environments (www.risen2rail.eu).

References

[1] Thompson D 2009 *Railway noise and vibration: mechanisms, modeling and means of control* 1st ed (United Kingdom: Elsevier Science Technology)
[2] Vincent N, Koch J R, Chollet H and Guerder J Y 2006 *Journal of Sound and Vibration* **293** 691
[3] Koch J R, Vincent N, Chollet H and Chiello O 2006 *Journal of Sound and Vibration* **293** 701
[4] Pieringer A 2014 *Journal of Sound and Vibration* **333-18** 4295
[5] Rudd M J 1976 *Journal of Sound and Vibration* **46-3** 381
[6] Jiang J, Anderson DC and Dwight R 2015 *Proc.Int.Conf. Noise and Vibration for Rail Transportation Systems* (Berlin: Springer) p 587
[7] Jiang J, Ying I, Hanson D and Anderson D C 2015 *Proc.Int.Conf. Noise and Vibration for Rail Transportation Systems* (Berlin: Springer) p 441
[8] Kaewunruen S and Remennikov A M 2010 *Insight-Non-Destructive Testing and Condition Monitoring* **52-1** 20
[9] Kaewunruen S and Remennikov A M 2007 *Ndt&E International* **40-7** 510
[10] Kaewunruen S and Remennikov A M 2008 *Engineering Failure Analysis* **15-8** 1065
[11] Grassie S L 1995 *Journal of Sound and Vibration* **187-5** 799
[12] Kaewunruen S and Remennikov A M 2008 *Experimental Techniques* **32-4** 29
[13] Kaewunruen S, Remennikov A M 2009 *Proc.Int.Conf. Mechanical Vibration* (New York: Nova Science) p 621
[14] Mirza O, Kaewunruen S and Galia D. Seismic 2016 *Structural Engineering and Mechanics* **57-3** 569
[15] Kaewunruen S and Chamniprasart K 2016 *Proc.Int.Conf. on Computational Mechanics* (Gothenburg: Chalmers University Press)

IOP Conf. Series: Materials Science and Engineering **280** (2017) 012037 doi:10.1088/1757-899X/280/1/012037

A ECG Signal Gathering and Displaying System Based on AVR

LiNing[1], Zhang Ruilan[1,*],Liu Jian[2],Wang Xiaochen[1],Chen Shuying [1] and Lang Zhuolin[1]

[1]Basic Medical of Beihua University,Jilin,132013; 2. Jilin People's Congress, Jilin,132000

*corresponding author,Email: bhdxzrl@163.com

Abstract: This article introduces a kind of system which is based on the AVR to acquire the data of ECG. Such system using the A/D function of ATmega8 chip and the lattice graph LCD to design ECG heart acquisition satisfies the demands above. This design gives a composition of hardware and programming of software about the system in detail which has mainly realized the real-time gathering, the amplifier, the filter, the A/D transformation and the LCD display. Since the AVR includes A/D transformation function and support embedded C language programming, it reduces the peripheral circuit, further more it also decreases the time to design and debug this system.

1. Introduction

Cardiovascular disease is one of the major diseases that threaten human life, and ECG is an important basis for evaluating cardiac function. ECG is a common diagnostic tool commonly used in clinical disease diagnosis. Acquisition system of ECG data is a key component of electrocardiogram[1]. The frequency range of human ECG signal is 0.05 ~ 100Hz, amplitude is about 0 ~ 4mv, the signal is very weak. As the ECG signal is usually mixed other bioelectric signals, combined with 50Hz frequency interference in the main electromagnetic field interference, making the ECG noise background strong, so the measurement condition is more complex. In order to detect the clinical value of the clean ECG signal, often require ECG data acquisition system with high precision, high stability, high input impedance, high common mode rejection ratio, low noise and strong anti-jamming capability. With the widespread use of microcontrollers in the field of medical devices, computer - based ECG signal acquisition, processing and display systems have been widely used.

The amplitude of ECG signal is small, random and strong noise background, with intermittent and need to reflect the local changes in the characteristics (such as ECG QRS wave), the statistical characteristics of the signal is time-varying non-equilibrium signal. Therefore, whether the extraction signal, or the analysis of signal characteristics are very difficult to do. With the extensive application of computers in the medical field, PC-based ECG detection system has got more attention. In this paper, an intelligent ECG signal acquisition, processing and display system based on AVR microcontrollers is designed in combination with the actual needs and research hospital. The system enables real-time data acquisition and display, digital filtering and processing of raw data.

2. Purpose and content

This paper designed an intelligent ECG signal acquisition and display system uses ATMEL's AVR

IOP Conf. Series: Materials Science and Engineering **280** (2017) 012037 doi:10.1088/1757-899X/280/1/012037

microcontroller (ATmega8 chip) as a core device. Under the control of the AVR microcontroller, the system can record the ECG signal of the patient in real time and accurately, and can be clearly and accurately displayed by the hardware circuit[2]. This article discusses the ECG acquisition human, ECG signal amplification, filter processing, using a signal processing chip ATmega8 finished performs A/D conversion, the signal is eventually processed by LCD（GXM12864）.

3 Overall design ideas

3.1 Design

According to the practical problems to be solved, the design idea, design principle and design block diagram of the system are put forward. The hardware structure of the design system is divided into three parts and the flow chart is show in the figure 1:

- ECG signal acquisition part, this part is use of connection multi-parameter patient simulator with the right leg drive circuit;
- The amplification and filtering part of the ECG signal, in which the amplification part is composed of preamplifier and secondary amplification, and the filter circuit is composed of band rejection filter and low pass filter.
- A/D conversion of ECG signal and liquid crystal display part are the core of the design system. Use ATmega8 chip built-in analog-digital conversion function to achieve ECG signal A/D conversion. ECG signal converted into the digital signal can be displayed through the ATmega8 chip and LCD.

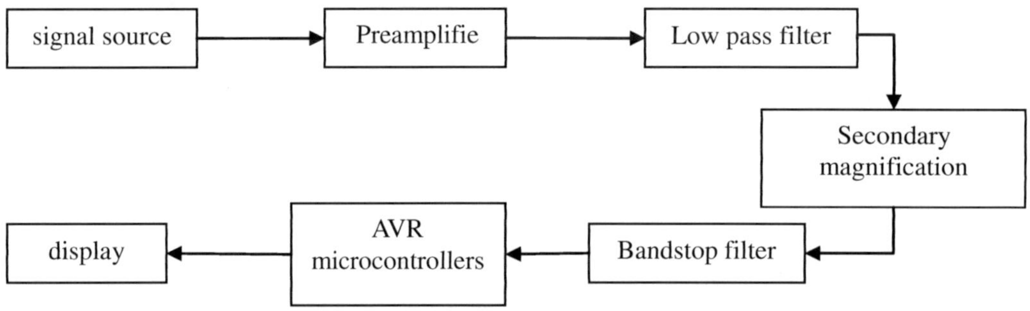

Figure 1. The overall design diagram

3.1.1 Main problem. The main problem of this design system is need to filter out 50Hz frequency interference by using the μA741 notch filter. In addition, ECG signal is real-time changes, so the zoom, filter and display systems should also be real-time.

3.1.2 Hardware circuit design ECG signal acquisition

ECG signal is a low-frequency weak signal, there is a strong interference detection, mainly 50Hz frequency interference. This requires the choice of devices and design of the circuit has a strong anti-interference ability. In this experiment, AD620 amplifiers, operational amplifiers LF365, OP27 and μA741 were selected.

- ECG signal preamplifier and secondary amplification

Preamplifier is the key part of ECG data acquisition. Because the human ECG signal is very weak, the background noise is strong and the signal source impedance is large, and the polarization voltage difference introduced by the electrode is larger than the amplitude of the ECG(A few hundred times greater than the magnitude of ECG). Therefore, the preamplifier is usually required to have a high input impedance, common mode rejection ratio is low, low noise, low drift, non-linearity, the appropriate frequency band and dynamic range performance, the design is generally used differential amplifier circuit. The first stage preamplifier circuit consists of AD620 and LF356. The AD620 bear the amplification function making the ECG signal amplification of more than 30 times[3], through the R4 to adjust the magnification; The LF356 is composed of the right leg drive circuit, to further improve the circuit's anti-jamming capability, which is a feature of ECG detection method. This design uses the instrument amplifier AD620 as a preamplifier. AD620 having a low input bias current, low noise, high-precision, high settling time, and low power consumption characteristics, common mode rejection ratio is up to 130dB, very suitable for use as medical instruments preamplifier. The gain is adjustable (range 1 to 1000 times) and can be determined by the formula G = 1 + 49.4kΩ / R4. To prevent the preamplifier from operating in the full and / or cutoff area, the gain can not be too large. Experiments show that: about 30 times better. Common body signal can be detected through OP27 be used to drive the wire shielding layer to eliminate the distributed capacitance, improve the input impedance and common mode rejection ratio. The common mode signal of human body can be amplified with the "floating" drive circuit used to stimulate the human right leg, thereby reducing or even offset the common mode voltage, in order to achieve a purpose of strong 50Hz frequency interference. OP27, LF356 is mainly used to stabilize the input signal and improve the input impedance, to further improve the common mode rejection ratio. The magnification of the preamplifier circuit is:

$$G = 1 + \frac{49.4k\Omega}{R_4} = 1 + \frac{49.4k\Omega}{1.7k\Omega} = 30 \tag{1}$$

ECG signal comes from amplifier circuit about 1mV or so, preamplifier circuit will amplificate ECG signal for 30 times. Since the circuit need to be A/D conversion after amplified and filtered, the A/D conversion function in AVR microcontroller requires the input signal amplitude for 0 ~ 5V. So the second stage amplifier circuit needs to amplify 100 times, the total magnification is 3000 times, ECG AVR microcontroller is about 3V. The magnification of the second stage amplifier is:

$$G = \frac{R_8}{R_7} + 1 = \frac{4.56K}{470} + 1 = 100 \tag{2}$$

- Electrocardiogram filtering

As the ECG signal is usually mixed with other bio-electrical signals, ECG background noise is strong, the measurement conditions are more complex. In order to detect the clinical value of clean ECG signal without distortion, the ECG data acquisition system has high accuracy, high stability, high input impedance, high common mode rejection ratio, low noise and strong anti-interference ability, etc. The main function of this filter circuit is to filter out the 50Hz frequency interference, so that ECG signal can be clearly displayed.

$$\text{cut-off frequency: } f = \frac{1}{2\pi RC} = \frac{1}{2\pi R_{13}C_5} = \frac{1}{2\pi 1500\times 10^{-6}} = 106H_Z \tag{3}$$

- Design of Interface Circuit between LCD and AVR Microcontroller

The interface circuit is shown in Figure 2: the ECG signal after amplifying and filtering from the input port PC0; PC1, PC2, PC3, PC4, PC5 port of ATmega8 combine with the control port of LCD for input control signal; PC6 port connected to a reset circuit ; PB port is the two-way input and output to control the LCD waveform display; INT0, INT1 are connected with the switches K2, K1 respectively.

The whole circuit work process: After power on, there will be a graphical interface selection. It will conduct A/D conversion through the A/D conversion function built-in ATmega8 when the PC0 port accesses analog signal after acquisition (after amplification, filtering) the chip[4]. The converted digital signal is output to the LCD display via port PB. INT0 and INT1 two interrupts are used to control the selection of the program entry. Capacitors C2, C3, C4 role for the A/D conversion to provide a stable reference element, so that A/D conversion of the data is more accurate, or due to the instability of the reference element, will make A/D conversion data not accurate, thus resulting in distortion of the output waveform, affecting the user's observation of the waveform.

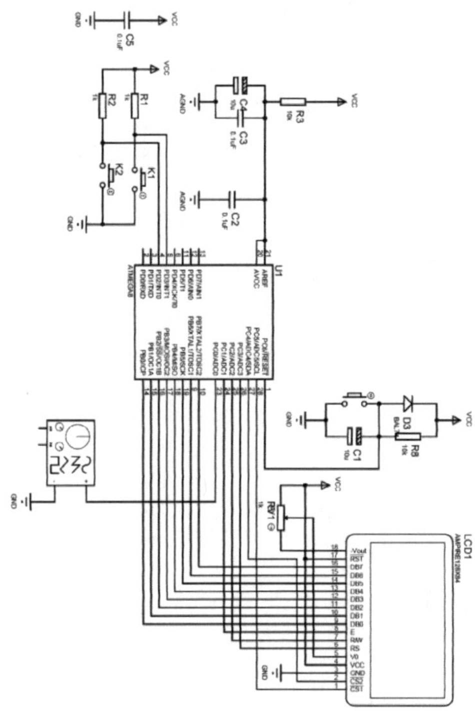

Figure2. Interface Circuit

3.1.3 Software circuit design. The focus of the software design is on the LCD module drive. Since the GXM12864 contains a controlled drive, it is necessary to know the instruction set of the controller and how to use it. In this design, a major part of the collected ECG signal is real-time A/D conversion and display[5]. As the GMX12864LCD itself consists of eight pages, you must press the page, press the column, press the left and right half-screen display control. In order to ensure the continuity of the display waveform, in each display must ensure that the two points between the LCD part to show all. And in order to control the exit of normal display, in the display of the death cycle by adding a control flag to control it can be promptly exit, exit and return to the main program.

As the liquid crystal itself read and read out the time delay, single-chip A/D conversion delay and program execution time delay, the final output of the waveform frequency limit below 140Hz. But because the frequency of ECG signal between 0.05Hz to 100Hz, so fully meet the ECG signal display[6].

ATmega8 has a 10-bit successive comparison of ADC. The ADC is connected to an 8-channel analog multiplexer capable of sampling 8 single-ended voltage inputs with the PORTC port as ADC input pins. See the main program flow chart figure 3. And figure 4 is ECG waveform display results.

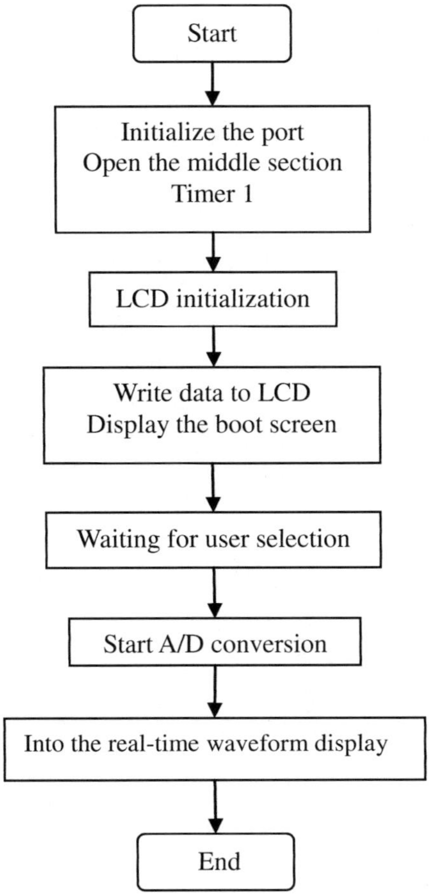

Figure3. Main program flow chart

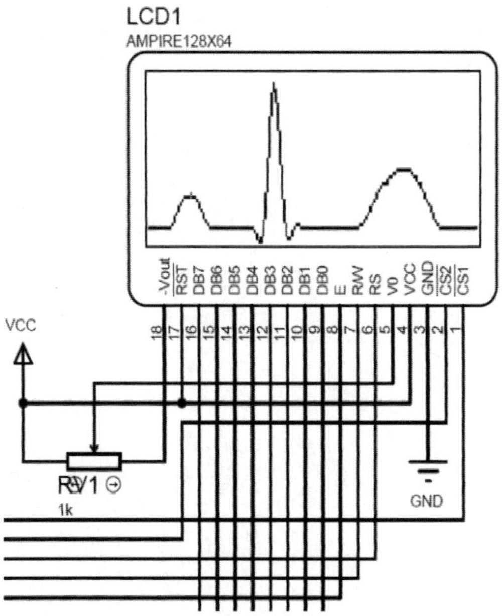

Figure4. ECG waveform display results

4. Conclusion

This article designed an acquisition module based on the characteristics of the ECG signal. The use of pre-differential amplifier circuit configured by AD620 can amplify ECG signal for 30 times. Because ATmega8 built-in A / D conversion function requires the input signal amplitude of 0 ~ 5V, therefore, need to be composed of μA741 as secondary amplifier circuit to amplify ECG signal for 100 times. This ECG signal is amplified 3000 times, access AVR microcontroller ECG signal is about 3V. As the ECG signal is usually mixed with other bio-electrical signals and 50Hz power frequency interference, the use of dual T-band filter and a low-pass filter filter out 50Hz frequency interference and other high-frequency interference signal. Pre-processed ECG signals with AVR microcontroller performs A/D conversion, and through the embedded C language programming and LCD liquid crystal display (GXM12864) interface to achieve ECG waveform display.

Acknowledgement

Provincial scientific research projects: Research on Single - phase Earth Fault Location System of Distribution Network.(project number:20162202020000062)

References

[1] Li Ran. Research on Intelligent Detection and Analysis Method of ECG [D] .Journal of Beijing University of Technology 2007.
[2] Zhang Xudong, Liu Chen, Cui Qiangqiang. Portable ECG measurement system for non - contact electrodes [J]. Chinese Journal of Medical Devices, 2014,03: 168-170 + 185

IOP Conf. Series: Materials Science and Engineering **280** (2017) 012037 doi:10.1088/1757-899X/280/1/012037

[3] Analog Devices Inc.AD620 datasheet.2004.[Online].Avaliable: http://www.analog.com/zh/speciatly-amplifiers/instrumentation-amplifiers/ad620/products/product.html

[4] Dhananjay V. Gadre. PROGRAMMING AND CUSTOMIZING THE AVR MICROCONTROLLER[M]. McGraw-Hill Compamies,2001.

[5] L P Chua,S C M Yu,L Leo,et al. Comparison of flow characteristics of enlarged pump models with different impeller design[J].Heat Mass Transfer,2000,26(3):369-378.

[6] Chi Y M, Cauwenberghs G. Wireless non-contact EEG/ECG eletrodes for body sensor networks[C]//Body Sensor Networks(BSN), 2010 International Conference on.IEEE,2010:297-301.

IOP Conf. Series: Materials Science and Engineering **280** (2017) 012038 doi:10.1088/1757-899X/280/1/012038

A Detection Device for the Signs of Human Life in Accident

LiNing[1], Zhang Ruilan[1,*], Liu Jian[2], Cheng Ruirui[1] and Diao Yuhong[1]

[1]Basic Medical of Beihua University,Jilin,132013; 2. Jilin People's Congress, Jilin,132000

* Email: bhdxzrl@163.com

Abstract: A detection device for the signs of human life in accidents is a device used in emergency situations, such as the crash site. the scene of natural disasters, the battlefield ruins .it designed to detect the life signs of the distress under the injured ambulance vital signs devices. The device can on human vital signs, including pulse, respiration physiological signals to make rapid and accurate response. After some calculations, and after contrast to normal human physiological parameters given warning signals, in order for them to make timely ambulance judgment. In this case the device is required to do gymnastics convenience, ease of movement, power and detection of small flexible easy realization. This device has the maximum protection of the wounded safety significance.

1. Introduction

The detection device for signs of human life in accident is suitable for detecting the vital signs in frontier emergency (such as emergency center, ambulance), accident disaster rescue, and troops battlefield frontier ambulance. At present, many research units and enterprises at home and abroad have set up related projects, put a lot of manpower and material resources research and development of various monitoring equipment, the overall is moving towards multi-parameter, miniaturization, electronic, intelligent and network [1].

For such cases, design a detection equipment of accurate, not easy to pollution, simple and efficient is of great importance. In the accident scene it can be timely measurement of human vital signs (heart rate, pulse, blood pressure, breathing, etc.) and quickly feedback to the medical staff. Not only timely provide a basis diagnostic data for the site rescue but also for the admission treatment, greatly increased first aid efficiency and maximum to save the lives of the wounded health.

2 Purpose and content

Vital signs [2] is used to determine the patient's indications for severity and critical degree. There are heart rate, pulse, blood pressure, breathing, body temperature and so on. Medical staff need to fully understand the meaning of vital signs, but also timely grasp the changes of patient's vital signs in order to take measures for treatment timely and effective. Therefore, in clinical practice it needs some auxiliary detection equipment of vital signs to help doctors diagnose the patient's condition for timely treatment.

Based on this, this paper design and production of a simple, low-cost human vital signs monitoring devices. Atmega8 chips used for the main part, Piezoelectric sensors collecting physiological signal AD620 signal amplification, user-friendly LCD interface output display. At the scene of an accident, the rapid and accurate detection of the wounded vital signs (pulse). Diagnosis of medical staff to provide the required information.

The design uses Atmega8 chip to achieve a dual channel[3] detection device for pulse wave and body temperature life. Using a piezoelectric sensor through a channel to achieve the measurement of human pulse wave, and integrated temperature sensor to detect human body temperature. The signal acquisition from piezoelectric sensors needs filter by AD620, shape and adjust to the microcontroller interface, display output with LCD after processing the programs. The corresponding system is divided into signal extraction module, zoom filter module, single chip and display module, as well as software design and power module. This article focuses on the hardware interface section.

3 Hardware design

3.1. System program demonstration and selection

The four methods[4] of human body physiological signal detection are respectively Photoelectric sensor method, Ultrasonic reflection detection method, Temperature sensor - sensing life signal method and Direct measurement of piezoelectric sensors. Among them photoelectric sensor method is mainly used to measure the number of pulse beat. For the amplitude of vibration, due to differences elasticity of human skin, the error is large, and thus we give up this program in this design. Ultrasonic reflection detection method, in this area the Fourth Military Medical University has developed products. But it bulky and difficult to achieve.

The main form of human radiation is invisible infrared radiation. The higher temperature, the stronger radiation. Infrared rapid temperature detection products is based on the above principles. We designed a pulse signal acquisition device by means of a piezoelectric accelerometer. Through the hardware and software filtering can eliminate the interference of external vibration.

The sensitivity can be determined by the following formula:

$$\frac{d_{33}}{\sqrt{(1/\omega R)^2 + (c_1 + c_2)^2}} \tag{1}$$

Which d_{33} is the piezoelectric coefficient. Since the pulse signal is a low frequency signal, the sensitivity is sufficiently high as long as the input impedance of the amplifier is large enough. The signal output from Piezoelectric sensing is weak, need to be amplified. According to the equivalent circuit of piezoelectric sensing, its output can be either voltage or charge. So the preamplifier also has two circuit forms: the voltage amplifier and the charge amplifier. The input voltage of the preamplifier is:

$$U_{im} = \frac{dF_m wR}{\sqrt{1 + (wR)^2 (C_e + C_i + C_C)^2}} \tag{2}$$

As can be seen from the above equation, the amplitude of input voltage is not only related to the force, but also with the frequency of measured signal, input resistance, capacitance. When $w >> \frac{1}{\sqrt{R(C_e + C_i + C_c}}$,the input voltage of preamplifier is independent of the frequency of the signal. Since the pulse signal is very weak and the frequency is low, the input resistance of amplifier is required great. In addition, the variation of the cable capacitance C also affects the changes of the input voltage of the preamplifier. Therefore, the pulse signal should not use voltage amplifier.

In summary, we use the three method and four method to achieve a detection device of human pulse and body temperature.

3.2. Circuit design

The diagram of system block is shown in Figure 1:

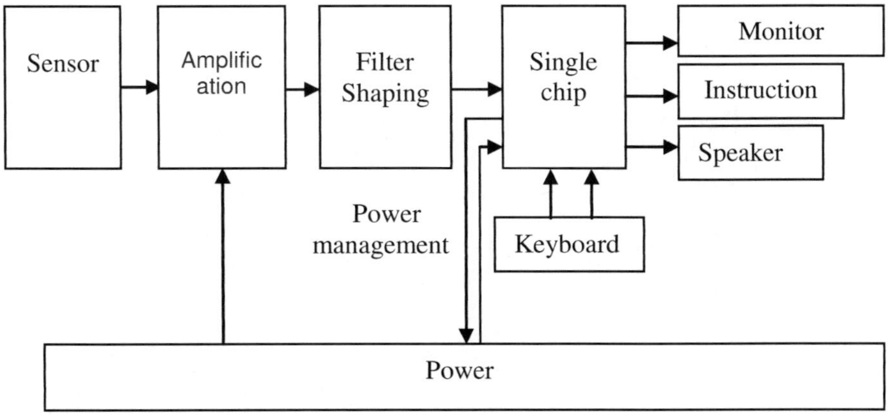

Figure 1. Diagram of System Block

The system consists of power supply part, amplifier part, filter part, keyboard part, display, sound and light alarm part and the microcontroller part.

- Power supply part: Because operational amplifier needs use dual power supply, so use 9V carbon battery. Reduce it to 5V after regulator module 7805 and then use the max232 interface chip to produce positive and negative 10V dual voltage.
- Amplifier part: Using AD620 amplifier, it has high accuracy and high input impedance, and it is very appropriate with piezoelectric sensors. It needs only one external rheostat for adjusting the gain, suitable for a variety of equipment, simple and convenient.
- Amplifier part: The design is mainly to detect the pulse and low demand for the accuracy of the waveform, so use ordinary RC low-pass filter can meet the requirements.
- Keyboard part: With two ordinary buttons, red stop, green start, easy to operate.
- Display: Using LCD1602 display, the information is more rich, user-friendly.
- Sound and light alarm part: Signal is not normal, the red light, speakers high frequency alarm; When the signal is normal, the green light, the speaker regular sound with the pulse beating.
- Microcontroller part: The core of the system. SCM is responsible for the signal processing, calculation, control LCD display, sound and light alarm, accept and deal with key information, management power and so on. Where the power management part of the design is a bright spot.

3.2.1. Power module. The design uses 9V carbon battery, reduced voltage to 5V with use of 7805 regulator module. Because the AD620 operational Amplifier is used, dual power supplies are required. And then use the RS232 interface chip to produce positive and negative 10V dual voltage, you can ensure that the microcontroller part display module driver and amplification filter circuit to work properly. In the actual test, we found that if the output voltage directly connect with 7805 to provide power to the other module, the power will not be fully utilized, so we designed a following power management circuit after improvement, shown in Figure 2:

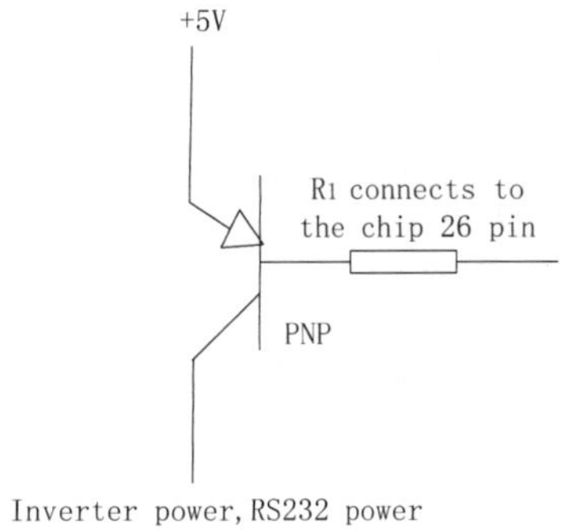

Figure 2. Power management circuit

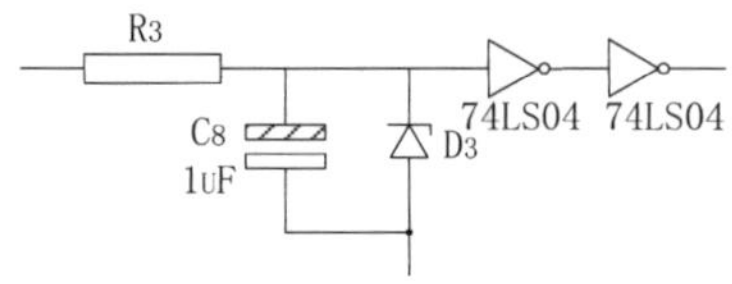

Figure 3. Filter and shaping part

When the system is working normally, the program controls the 26-pin of the microcontroller output a low level, so that the PNP transistor turns on, the amplifier and shaping circuit is energized. The microcontroller output high level when pause or sleep mode, PNP transistor cut-off, amplifier and shaping circuit stop working, and the system is in power-saving state.

When 5V voltage change into ± 10V voltage, the general chip is difficult to meet the requirements. Combined with the usual experience, RS-232 interface chip can achieve this function, and has high efficiency, so use max-232 as the power chip.

*3.2.2. Zoom filter module.*The output signal of basic hardware circuit is weak, need to be amplified. According to the equivalent circuit of piezoelectric sensing [5], its output can be either a voltage or a charge. So the preamplifier also has two circuit forms: which is voltage amplifiers and charge amplifiers. The input voltage amplitude of voltage preamplifier is:

$$U_{im} = \frac{dF_m wR}{\sqrt{1+(wR)^2(C_e+C_i+C_C)}} \tag{3}$$

From the above, the input voltage amplitude is not only related to the force, but also with the frequency of the measured signal, input resistance, capacitance. When

$$w >> \frac{1}{\sqrt{R(C_e+C_i+C_c}} \tag{4}$$

The input voltage of preamplifier is independent of the frequency of the signal measured. Since the pulse signal is very weak and the frequency is low, the input resistance of amplifier is required large. In addition, the variation of the cable capacitance C also affects the change of input voltage. Therefore, the pulse signal should not use voltage amplifier. The output voltage of the charge amplifier is:

$$U_0 = \frac{-Aq}{\sqrt{C_e+C_c+C_i+(1+A)C_f}} \tag{5}$$

When $(1+A)C_f >> (C_e+C_c+C_i)$, The output voltage of the charge amplifier is independent of the cable capacitance and the frequency of measured signal. Therefore, the piezoelectric sensor for measuring the pulse signal uses a charge amplifier [6].

The body's pulse signal change vibration signal into a charge signal by the piezoelectric sensors. And then through the charge amplifier conversion, the amount of charge will be converted to a very weak voltage signal. This voltage signal can change into digital signal by the amplification of voltage amplifier. This also meet the requirements of ADC.

There is negative pressure after filter, and sometimes the peak is too high, we added a 5.6V regulator tube to protect the follow circuit. In order to achieve the better filtering effect, the design plus two-stage reverse shaper behind the regulator tube,as shown in Figure 3.

3.2.3. Microcontroller and display module. LCD display use 1602 module, the module comes with character set and 11 instructions. Considering the reliability of the circuit, input instructions and data by using four bus mode divided twice times.

The requirements of microcontroller is got on A/D conversion and the display module driver. The I/O distribution of hardware part is: The timer / counter receives and input signal; The button and the indicator use two I/O port respective; There are total six I/O port for display using four bus and plusing control signals. There is also a power management control port. Also set up a 10-pin download line interface for facilitate its programming.

4. Design circuit diagram and debugging

System design circuit shown in Figure 4, getting on the proteus software simulation according to the above selection, and then debug the hardware part. Debug waveform as shown in Figure 5.

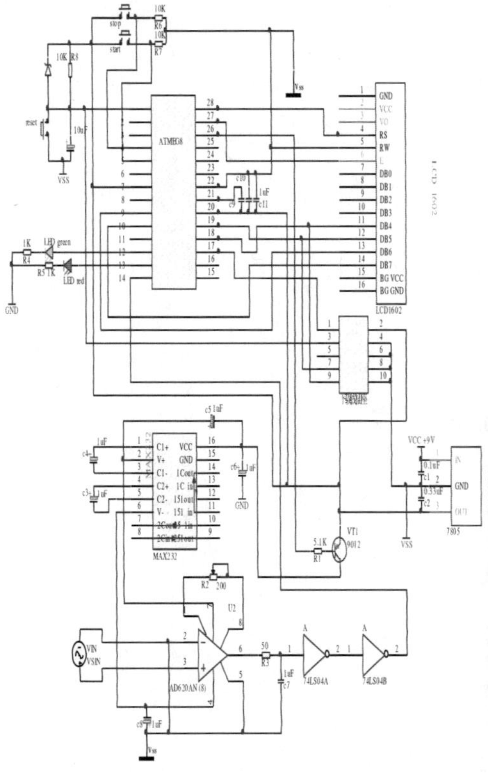

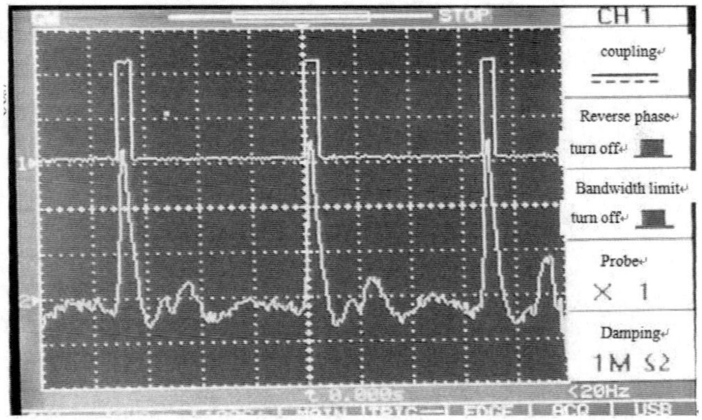

Figure 4. System design circuit diagram **Figure 5.** Pulse graphics after amplification of the signal

When debugging, the components and chip sockets are not inserted into any chip. Starting from the power input, use the multimeter to detect the output of the 7805 in turn. And then detect 16 pin of the max232 socket, 7 pin of the microcontroller socket, 2 pin and 3 pin of the LCD socket, , 14 pin of the 74LS02 socket. Under normal circumstances these pins have a voltage of 5v, and then the power of AD620 appears (output of the max232) ± 10V voltage.

Acknowledgement

School-level key issues: Research and Practice of Ubiquitous Learning System Based on Mobile Terminal

IOP Conf. Series: Materials Science and Engineering **280** (2017) 012038 doi:10.1088/1757-899X/280/1/012038

References

[1] Deng Qinkai. Principle of Modern Medical Instrument Design[M].Beijing: Science Press,2016
[2] Lv Huaijuan, Zhang Chunmei. Disputes and prospects of measurement methods of vital signs [J]. Journal of Nursing Science,2013
[3] Yu Xuefei. Principle and Design of Modern Medical Electronic Instrument(Third edition)[M].Guangzhou: South China University of Technology Press,2013
[4] Zhang Lanxiang. The Technical Principle and Application Improvement of Life Detector [J]. Journal of the Armed Police Academy,2017
[5] Wang Yuyin. Digital Circuit Logic Design (Second edition)[M].Beijing: Higher Education Press,2007
[6] Zhou Changsen. Electronic Circuit Computer Simulation Technology[M] Jinan: Shandong Science and Technology Press,2009.

IOP Conf. Series: Materials Science and Engineering **280** (2017) 012039 doi:10.1088/1757-899X/280/1/012039

Meta Modelling of Submerged-Arc Welding Design based on Fuzzy Algorithm

Chang-Yong Song[1], Jonghwan Park[1], Dugab Goh[1], Woo-Chang Park[1], Chang-Ha Lee[1], Mun Yong Kim[2] and Jinseo Kang[2]

[1]Dept. of Naval Architecture & Ocean Engineering, Mokpo National University, Jeonnam, Korea
[2]Myung Il Co., Ltd., Jeonbuk, Korea

*cysong@mokpo.ac.kr

Abstract. Fuzzy algorithm based meta-model is proposed for approximating submerged-arc weld design factors such as weld speed and weld output. Orthogonal array design based on the submerged-arc weld numerical analysis is applied to the proposed approach. The nonlinear finite element analysis is carried out to simulate the submerged-arc weld numerical analysis using thermo-mechanical and temperature-dependent material properties for general mild steel. The proposed meta-model based on fuzzy algorithm design is generated with triangle membership functions and fuzzy if-then rules using training data obtained from the Taguchi orthogonal array design data. The aim of proposed approach is to develop a fuzzy meta-model to effectively approximate the optimized submerged-arc weld factors. To validate the meta-model, the results obtained from the fuzzy meta-model are compared to the best cases from the Taguchi orthogonal array.

1. Introduction

In this study, the new approach combining both Taguchi method and fuzzy logic is proposed for estimating design parameters of the weld process. Fuzzy logic is an intelligent soft computing technique that is able to describe a crisp conclusion using more realistic phenomenon under the environment of uncertainty [1-3]. This method requires a knowledge or training data set that includes not only some quantitative data such as experiment results or engineering database but also some experienced facts or rules. Therefore, a reliable approximate reasoning of fuzzy logic largely depends on what kind of knowledge data set is applied. The Taguchi method is usefully applied to generate knowledge data set because the robustness is considered in S/N (Signal-to-Noise) ratio results used to measure the deviation of the performance characteristics from the object value. Using results obtained from the Taguchi method, control parameters and response are defined as fuzzy membership functions, and then fuzzy rules are derived through the fuzzy reasoning process. Using approximate reasoning suggested fuzzy logic in this paper it is possible to infer some useful information with the consideration of the robustness of knowledge data set and the uncertainty. The finite element analysis of a fine-grain steel plate is first carried out to simulate the submerged-arc welding process using thermo-mechanical and temperature-dependent material properties. Simulation results on thermal-mechanical coupled physical phenomena are compared with experiment. Taguchi method is then applied to evaluate the relationships between parameters and responses in the submerged-arc welding process and to generate some training data for fuzzy application. Fuzzy membership functions and fuzzy rules are derived from the results of Taguchi method. The results of Taguchi method based fuzzy algorithm design would be usefully applied to approximate optimization that could be an efficient methodology to solve design issue with high computational cost in various engineering design [4-6]. The present study facilitates a quantitative decision of welding parameters such as welding speed and welding output that are generally expressed by qualitative linguistic terms in the welding process.

2. Numerical analysis for weld

The numerical analysis for weld is carried out to provide the Taguchi design to experimental data. The thermal-mechanical characteristics of weld are also reviewed from analysis results, and the numerical analyses are performed out according to the level variation of weld design factors in Taguchi design.

The two plate halves, 25(thickness)x500(transverse width)x1000(longitudinal height) mm, made of mild steel, are butt-welded together, wherein the weld electrode moves along the longitudinal top edge of the steel plate. The plate is mechanically fixed at both ends. The temperature-dependent material properties used in the numerical analysis are referred to the study of Andersson [7]. The symmetry boundary condition along the longitudinal plane is spanned by the seam and through thickness axis. The cross-section of one of the plate halves is assumed to be the generalized 2-dimensional plane strain condition to minimize the computing time. The finite analysis models of base metal and filler are made as shown in Figure 1 [8].

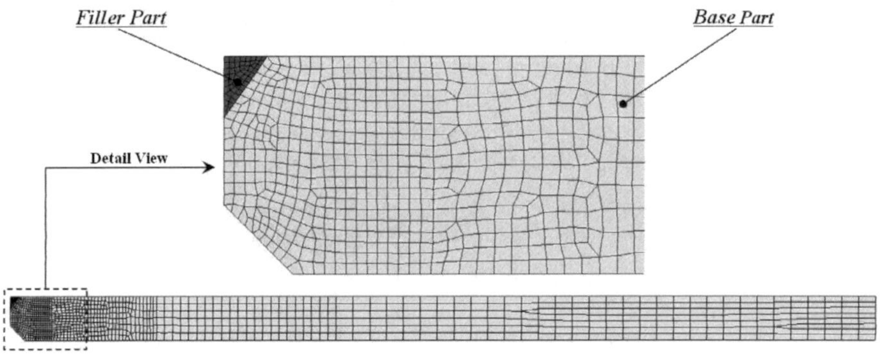

Figure 1. Finite analysis model of weld simulation [8]

The element type of filler is the heat transfer plane element transferring the thermal-structural stiffness and the thermal energy on the welding area. Melting point temperatures are defined on the nodes of the created filler elements as long as the latter ones remain in the weld pool. In case that the heat source moves on, temperature boundary conditions are eliminated and the filler elements are cooled down. The deformable contact condition is applied to the base metal elements and the filler. The heat transfer coefficient is calculated to transfer the filler heat input to the base metal elements. The weld heat source equation is used from Goldak's double ellipsoidal shaped model [9] that is generally well known for the deep penetration weld analysis.

The weld temperature is applied to the filler element, and the ambient temperature and the heat transfer coefficient are considered in the base metal. For the weld simulation based on the finite element analysis, MSC.MARC [10], a general nonlinear finite element analysis code, is used. The detailed temperature contour results are presented in Figure 2. Residual stress and deformation contour results are also shown in Figure 3.

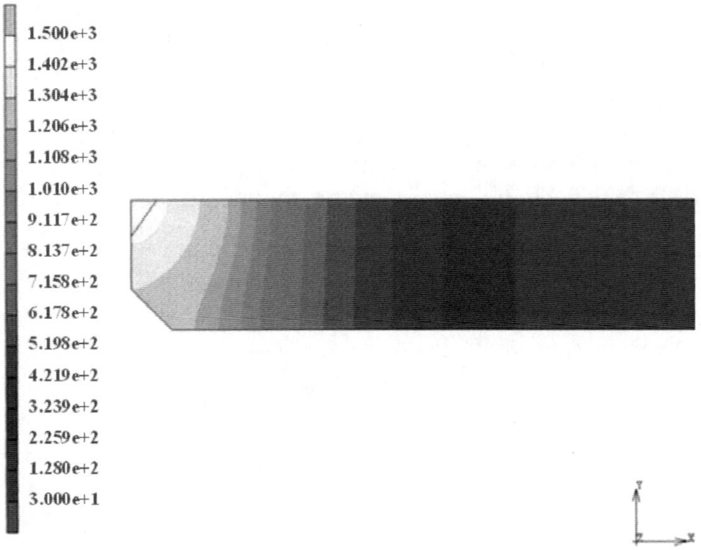

Figure 2. Weld temperature results (unit: °C)

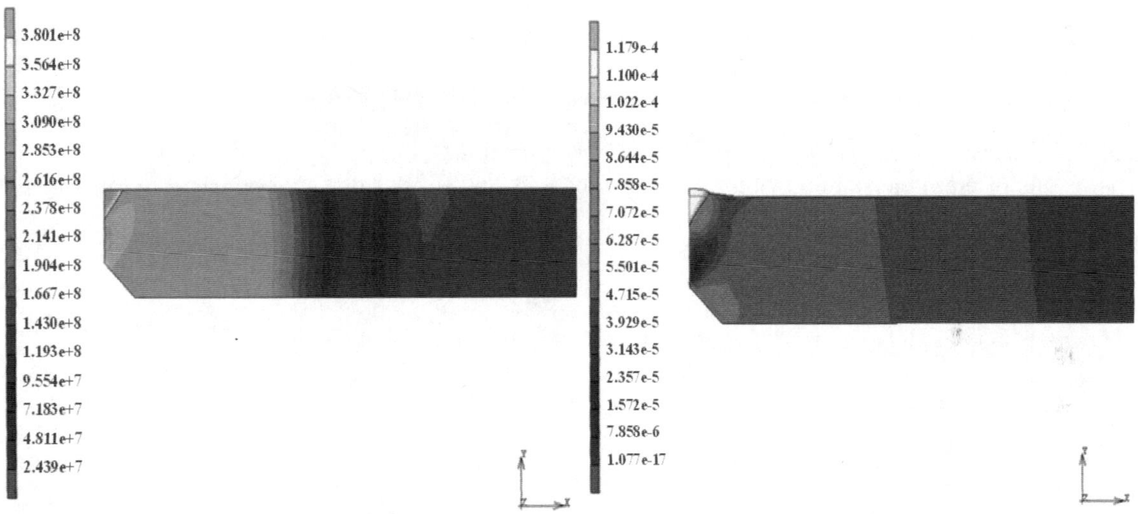

Figure 3. Weld stress (left; unit: m) and deformation (right; unit: m) results

3. Taguchi and fuzzy algorithm design

3.1. Taguchi design

In the Taguchi design, weld output (kW) is considered as control factor with 4 levels such 16, 22, 28, and 34, weld speed (mm/s) is also defined as control factor with 4 levels such as 10, 15, 20, and 25, and the noise factor of 3 levels such as 15, 30, and 45 is defined by ambient temperature (℃). The maximum weld stress is regarded as the response for the 'smaller-the-better' in the Taguchi design. The weld simulation is carried out according to the L16 orthogonal array that the weld numerical analyses of 48 times are performed. From the results of Taguchi design, it is found that the combination of the lowest level weld output and the highest weld speed is the best case to minimize the weld stress.

3.2. Fuzzy algorithm design

Using the fuzzy algorithm design, the weld factors could be considered realistically as linguistic term and on the continuous design space. In this study, the Taguchi design results are applied to the fuzzy algorithm design to derive the fuzzy membership functions and inference rule. The procedures deciding membership functions and decision rules consist of the five stages. In the first stage, the Taguchi design results are clustered and fuzzified. In the second and third stages, initial membership functions and decision table are constructed. In the fourth stage, the initial decision table is simplified. Finally the membership functions are rebuilt in the simplification process with the triangle fuzzy membership function shown in Figure 4 [8], and then the decision rules are obtained using the fuzzy if-then rule.

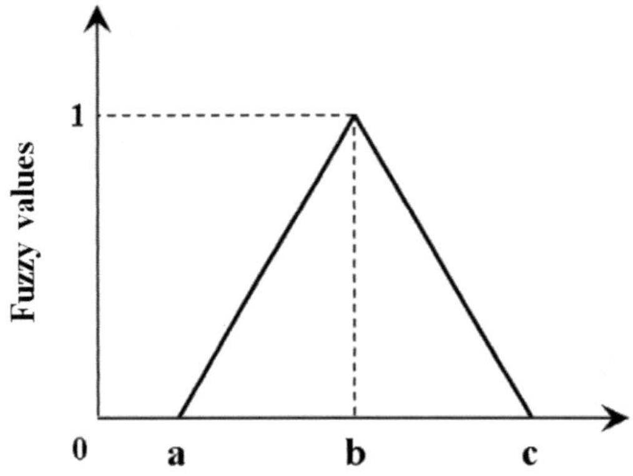

Figure 4. Triangle fuzzy membership function [8]

3.3. Comparison of Taguchi design and fuzzy algorithm results

The results of the proposed fuzzy membership functions and fuzzy rules are compared to the Taguchi design as shown Figure 5.

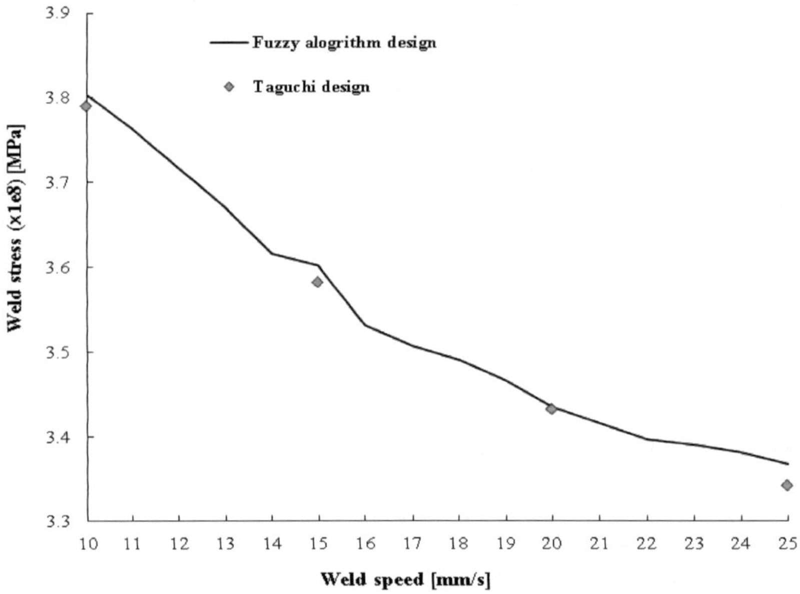

Figure 5. Comparison results for fuzzy algorithm and Taguchi design

In the Figure 5, the meta-model from fuzzy algorithm design is generated by the defuzzification equation [11]. It is noted that the meta-model from fuzzy algorithm design represents the high accuracy with less computational costs.

4. Conclusion

In this study, the fuzzy algorithm based meta-model was proposed for approximating submerged-arc weld design factors such as weld speed and weld output using the Taguchi design results. The Taguchi orthogonal array design based on the finite element analysis of weld simulation was applied to the proposed approach. The nonlinear finite element analysis was carried out to simulate the weld numerical analysis using the thermo-mechanical and temperature-dependent material properties for the general mild steel. The proposed meta-model based on fuzzy algorithm design was generated with the triangle membership functions and the fuzzy if-then rules using the training data obtained from the Taguchi orthogonal array design data. To validate the meta-model, the results obtained from the fuzzy meta-model are compared to the best cases from the Taguchi orthogonal array. It was found that the meta-model from fuzzy algorithm design represented the high accuracy with less computational costs.

Acknowledgments

This research was supported by MOTIE & KIAT through the promoting Regional Specialized Industry, Training of Professional Manpower Utilizing Retired Workforce of Ship & Offshore Industry(Industrial Manpower Capacity Enhancement Project) by MOTIE, the Fostering the Convergence of Industry and Academia by MOTIE, and Regional Innovation Center for Southwestern Mid-sized Shipbuilding Industry Project in the year of 2017, Mokpo National University.

References

[1] Jang JSR, Sun CT, Mizutani E, Neuro-Fuzzy and Soft Computing, Prentice-Hall, 1997.

[2] Lee J, Kim, S, Applications of Soft Computing Techniques in Response Surface Based Approximate Optimization. 2001, *KSME International Journal*, 15(8), 1132-1142.

[3] Lee J, Kim S, Kang S, Evolutionary Fuzzy Modelling in Global Approximate Optimization. 2002, *International Journal of Vehicle Design*, 28(3), 81-97.

[4] Kim SW, Kang K, Yoon K, Choi DH, Design optimization of an angular contact ball bearing for the main shaft of a grinder. 2016, Mech. Machine Theory, 104, 287–302.

[5] Song CY, Choi HY, Lee J, Approximate multiobjective optimization using conservative and feasible moving least squares method: application to automotive knuckle design. 2014, Structural and Multidisciplinary Optimization, 49(5), 851–861.

[6] Korta JA, Mundo D, Multiobjective micro-geometry optimization of gear tooth supported by response surface methodology. 2017, Machine Theory, 109, 278–295.

[7] Andersson BAB, Thermal Stresses in a Submerged-Arc Welded Joint Considering Phase Transformations. 1978, *ASME Journal of Engineering Materials and Technology*, 100, 356-362.

[8] Lee J, Song CY, Estimation of Submerged-arc Welding Design Parameters using Taguchi Method and Fuzzy Logic. 2013, *Proceedings of the Institution of Mechanical Engineers, Part B: Journal of Engineering Manufacture*, 227, 532-542.

[9] Goldak J, Chakravarti A, Bobby M, A New Finite Element Model for Welding Heat Source. 1984, *Metallurgical Transactions B*, 15, 299-305.

[10] MSC.Marc Theory Manual, MSC Software, 2010.

[11] Hong TP, Lee CY, Induction of Fuzzy Rules and Membership Functions from Training Examples. 1996, *Fuzzy Sets and Systems*, 84, 33-47.

IOP Conf. Series: Materials Science and Engineering **280** (2017) 012040 doi:10.1088/1757-899X/280/1/012040

Synthesis and Evaluation of A High Precision 3D-Printed Ti6Al4V Compliant Parallel Manipulator

Minh Tuan Pham[1, 2, 3, *], Tat Joo Teo[2], Song Huat Yeo[3], Pan Wang[4], and Mui Ling Sharon Nai[4]

[1]Singapore Centre for 3D Printing, Nanyang Technological University, Singapore 639798, Singapore
[2]Mechatronics Group, Singapore Institute of Manufacturing Technology, Singapore 138635, Singapore
[3]School of Mechanical and Aerospace Engineering, Nanyang Technological University, Singapore 639798, Singapore
[4]3D Additive Manufacturing Team, Singapore Institute of Manufacturing Technology, Singapore 637662, Singapore

*E-mail: PHAMMINH001@e.ntu.edu.sg

Abstract. A novel 3D printed compliant parallel manipulator (CPM) with $\theta_X - \theta_Y - Z$ motions is presented in this paper. This CPM is synthesized using the beam-based method, a new structural optimization approach, to achieve optimized stiffness properties with targeted dynamic behavior. The CPM performs high non-actuating stiffness based on the predicted stiffness ratios of about 3600 for translations and 570 for rotations, while the dynamic response is fast with the targeted first resonant mode of 100Hz. A prototype of the synthesized CPM is fabricated using the electron beam melting (EBM) technology with Ti6Al4V material. Driven by three voice-coil (VC) motors, the CPM demonstrated a positioning resolution of 50nm along the Z axis and an angular resolution of ~0.3" about the X and Y axes, the positioning accuracy is also good with the measured values of ±25.2nm and ±0.17" for the translation and rotations respectively. Experimental investigation also shows that this large workspace CPM has a first resonant mode of 98Hz and the stiffness behavior matches the prediction with the highest deviation of 11.2%. Most importantly, the full workspace of $10^{o} \times 10^{o} \times$ 7mm of the proposed CPM can be achieved, that demonstrates 3D printed compliant mechanisms can perform large elastic deformation. The obtained results show that CPMs printed by EBM technology have predictable mechanical characteristics and are applicable in precise positioning systems.

1. Introduction

Compliant parallel manipulator (CPM), which is developed based on the merits of elastic deformation and closed-loop parallel architecture, plays an important role in high precision motion systems. This is because elastic deformation offers a low-cost solution to deliver frictionless repeatable motions [1] while the closed-loop parallel architecture makes the CPM less sensitive to external disturbances and achieves higher non-actuating stiffness compared to the serial counterpart [2]. Based on the past literatures, most of the developed multiple degrees-of-freedom (DOF) CPMs were limited to few hundred micrometers in translation axes and less than a degree in the rotation axes [3–9]. With the assembly of next-generation optical fiber, optics and silicon photonics devices demanding for high precision motion systems with larger workspace and higher positioning resolution, CPM needs to meet these requirements to extend its application into the automation and production of those emerging technologies. However, only a handful of multi-DOF CPMs managed to achieve larger workspace in recent literatures. Examples include the 3-DOF spatial motion ($\theta_X - \theta_Y - Z$) CPMs that demonstrated workspace of 5° × 5° × 5mm [10, 11] and 8° × 8° × 5.5mm [12], the 3-DOF planar-motion ($X - Y - \theta_Z$) CPM that demonstrated a workspace of 2mm × 2mm × 2° [13]. Yet, having larger workspace also lead to poor dynamic performance. In particular, the developed 3-DOF spatial-motion ($\theta_X - \theta_Y - Z$) in [11] only has a low first resonant mode of 15Hz; and 84.4Hz is the dynamic response that one of the largest workspace CPMs [12] has been achieved.

Electron Beam Melting (EBM), a powder-based 3D printing technology that is able to build nearly full-dense metal parts with good mechanical properties [14], is used to fabricate the synthesized CPM with Ti6Al4V material so as to reduce or avoid any form of undesired assembly error. Recently, a 3D printed 3-DOF CPM has been developed with a limited work range of 6° × 6° × 4mm [15]. In this work, the beam-based method [12] is employed to synthesize a new design of 3-DOF ($\theta_X - \theta_Y - Z$) CPM with large workspace of more than 8° and 5.5mm and the targeted first resonant mode of 100Hz. The factor of 1.27 proposed in [15] is used to correct the actual thickness of EBM printed thin features. Experiments are conducted to investigate the mechanical characteristics of the synthesized CPM, especially the large elastic deformation of the 3D printed structure. Subsequently, three voice-coil (VC) motors are used to drive the CPM and experimental investigations are conducted to evaluate its positioning performance.

The rest of this paper is organized as: the synthesis of the 3-DOF ($\theta_X - \theta_Y - Z$) CPM is presented in Section 2 and the performance evaluation of the 3D-printed prototype is discussed in Section 3. Lastly, several conclusions are summarized in Section 4.

2. Synthesis of the 3-DOF ($\theta_X - \theta_Y - Z$) CPM

2.1. *Problem formulation*

The beam-based method presented in [12] is employed to design a 3-DOF spatial motion ($\theta_X - \theta_Y - Z$) CPM, which requires to have a workspace of at least 8° × 8° × 5.5mm and a targeted first resonant mode of 100Hz. Subsequently, the CPM is fabricated by EBM technology and a common titanium alloy for EBM, Ti6Al4V, is chosen as the fabricating material. The mechanical properties of Ti6Al4V are as follows: elastic modulus of 111GPa, Poisson ratio of 0.34 and density of 4.5g/cm^3. To obtain a 3-DOF motion, a symmetrical parallel configuration with three compliant limbs is adopted and the design domain of a limb is chosen as 50 × 50 × 50 mm^3 as shown in figure 1a. The construction of each compliant limb is illustrated in figure 1b. Elastic deformation of two reflecting curved-and-twisted (C-T) beams is used to generate the desired motions for the entire CPM. The C-T beam is created by sweeping a rectangular area through a cubic Bezier curve with different orientations of two ends. Finite element method (FEM) is used to model the entire CPM structure and the C-T beams are discretized by a number of 2-node beam elements. As the CPM will then be fabricated by EBM technology, the coefficient factor of 1.27 [15] is implemented in the synthesis process to correct the thickness of the printed prototype.

2.2. *Synthesis process*

The synthesis process is separated into two steps; the stiffness optimization and the dynamic optimization as shown in figure 2. The objective of the synthesis is to optimize the stiffness characteristics of the CPM with desired dynamic property. Hence, the stiffness optimization process is first carried out to define the most suitable structure for the C-T beams in order to achieve the largest workspace and the lowest actuating stiffness as well as the highest non-actuating stiffness. The fitness (objective) function for the stiffness optimization has been formulated using the works done by external loads. Supposing a load vector, P, consists of three forces (F_x, F_y, F_z) and three moments (M_x, M_y, M_z) along/about the X, Y and Z axes respectively, is applied to the CPM at the end effector's center. Six displacements of the end effector (Δ_x, Δ_y, Δ_z, θ_x, θ_y and θ_z) caused by P are denoted by the vector U as illustrated in figure 3.

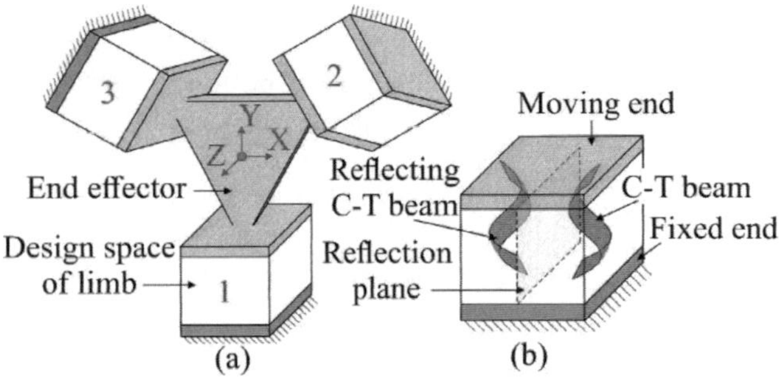

Figure 1. (a) Symmetrical 3-limb configuration of the CPM and (b) construction of a limb.

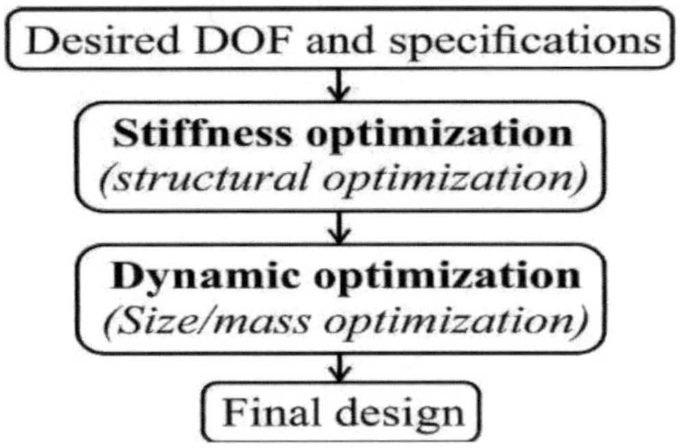

Figure 2. Synthesis steps of the beam-based method.

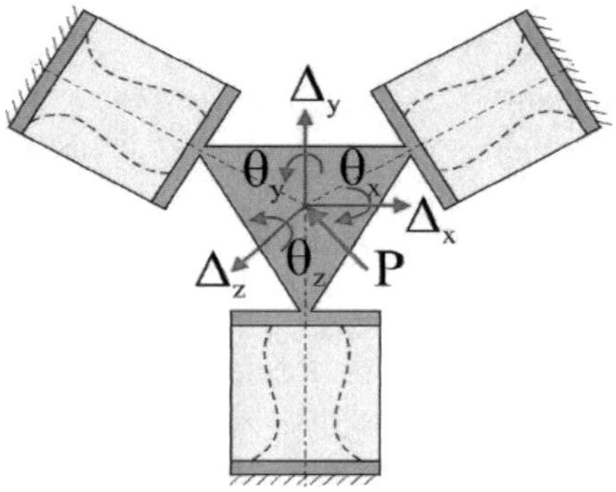

Figure 3. Load-displacement relation.

By letting W_i be the work done by P_i and U_i, their relationship is expressed as

$$W_i = \int_0^{P_i} U_i \, dP_i = \int_0^{P_i} \frac{P_i}{K_i} dP_i = \frac{1}{2}\frac{P_i^2}{K_i}, \text{ where } P_i = K_i \cdot U_i \quad (1)$$

Here, K is the 6 × 6 matrix represents the stiffness property of the entire CPM and K_i indicates the diagonal component of K that corresponds with P_i and U_i. By referring to [12] and equation (1), the fitness function for the stiffness optimization process is derived by the ratios between the works done of undesired displacements, i.e., $W_{\Delta x}$, $W_{\Delta y}$, $W_{\theta z}$, and the works done of desired displacements, i.e., $W_{\theta x}$, $W_{\theta y}$, $W_{\Delta z}$, written as

$$\text{minimize } f = \frac{W_{\Delta x}^3 \cdot W_{\Delta y}^3 \cdot W_{\theta z}^3}{W_{\theta x}^3 \cdot W_{\theta y}^3 \cdot W_{\Delta z}^3} = \kappa \cdot \frac{K_{\theta x}^3 \cdot K_{\theta y}^3 \cdot K_{\Delta z}^3}{K_{\Delta x}^3 \cdot K_{\Delta y}^3 \cdot K_{\theta z}^3}, \text{ where } \kappa = \frac{F_x^6 \cdot F_y^6 \cdot M_z^6}{M_x^6 \cdot M_y^6 \cdot F_z^6} \quad (2)$$

Note that f is dimensionless and $\kappa = 1$ when P is a unit load vector. Equation (2) is solved via the genetic algorithm (GA) solver from MATLAB and figure 4a shows the CPM structure after the stiffness optimization. It is observed that the C-T beams became flat and two reflecting C-T beams cut each other at the free ends to generate a V-shape flexure for each compliant limb. This structure is able to provide the largest workspace and good decoupled motions for the CPM.

Next, the dynamic optimization process is done to distribute the masses within the CPM to achieve the targeted first resonant mode. The fitness function of the dynamic optimization consists of two equations as shown in equation (3). In the first equation, the first resonant frequency, F_1, is guaranteed to match the desired value of 100Hz by minimizing their difference. On the other hand, the large work range of the CPM can be obtained via maximizing the total desired works done as expressed in the second equation.

$$\begin{cases} \text{minimize}\left(\left|F_1 - 100\right|\right) \\ \text{maximize}\left(W_{\theta x} + W_{\theta y} + W_{\Delta z}\right) \end{cases} \tag{3}$$

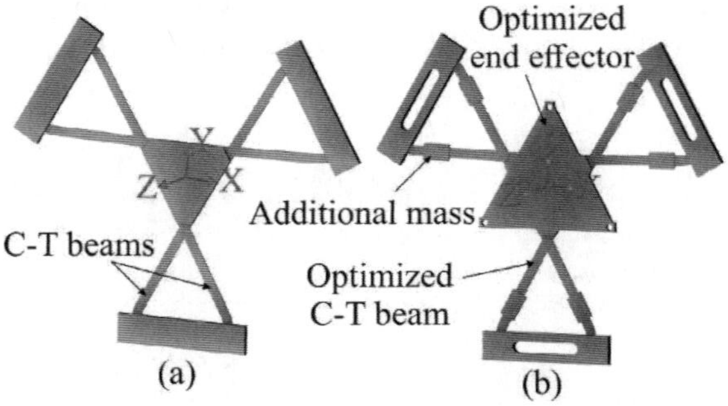

Figure 4. Results of the (a) stiffness optimization and (b) dynamic optimization.

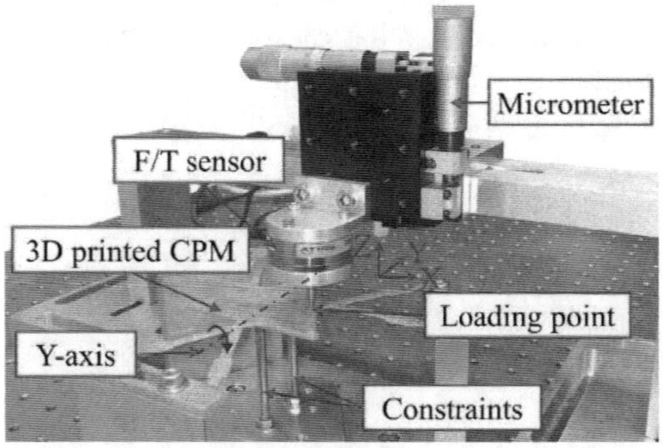

Figure 5. Setup for evaluating the compliance about the Y-axis.

The dynamic optimization problem is also solved by GA solver as the stiffness optimization. The final design of the CPM has a predicted first resonant mode of 99Hz, which is very close to the targeted value. The structure of the final design is shown in figure 4b. The compliant matrix, C, that represents the stiffness characteristics of the CPM is written as

$$C = \text{diag}\left[5.53e-8 \quad 5.53e-8 \quad 1.99e-4 \quad 1.32e-1 \quad 1.32e-1 \quad 2.33e-4\right] \tag{4}$$

Equation (4) shows that the CPM has better decoupled motion property as compared to the existing designs [11, 12, 15] since all non-diagonal components are zeros. The stiffness characteristics of the final CPM are shown by two ratios, expressed as

$$\frac{K_{\Delta x}}{K_{\Delta z}} = \frac{K_{\Delta y}}{K_{\Delta z}} = \frac{C_3}{C_1} = \frac{C_3}{C_2} \approx 3599 \ ; \ \frac{K_{\theta z}}{K_{\theta x}} = \frac{K_{\theta z}}{K_{\theta y}} = \frac{C_4}{C_6} = \frac{C_5}{C_6} \approx 567 \tag{5}$$

Here, C_i represents the ith diagonal component within the compliant matrix. Equation (5) shows that the non-actuating translational stiffness along the X and Y axes is higher than the actuating translational stiffness along the Z-axis by 3599 times, the non-actuating rotational stiffness about the Z-axis is higher than the actuating rotational stiffness about the X and Y axes by 567 times. Such large ratios suggest that the CPM is very stiff in non-actuating directions and other resonant modes will be much higher than the first resonant mode. It is seen that the obtain lowest ratio of 567 is significantly improved as compared to the corresponding values of the previous designs [12, 15]. In addition, the stress analysis via ANSYS is also carried out to evaluate the workspace of the CPM. With the Ti6Al4V material providing a yield strength of 850MPa, the synthesized CPM can deliver a large workspace up to $\pm 5^{\circ} \times \pm 5^{\circ} \times \pm 3.5$mm, which satisfies the desired requirements and much larger than the existing CPMs [11, 12, 15].

3. Experimental investigation and results

3.1. Stiffness evaluation

A prototype of the CPM was built by EBM method with Ti6Al4V material and experiments were carried out to evaluate the stiffness characteristics. Figure 5 shows the experimental setup for measuring the compliance about the Y-axis. It is seen that two rigid rods with sharp tips were placed below the end effector constraining the motion along the Z-axis as well as creating the rotation axis for the CPM. The input displacement was applied by a micrometer and the actuating force was measured by a force/torque (F/T) sensor (ATI Mini-40). The bending moment and rotation angle were defined based on the distance between the loading point and the Y-axis. The similar setup was adopted for evaluating the compliance about the X-axis. Figures 6b and c plot the measured compliance about the X and Y axes respectively. Results are compiled from five separate measurements with fourteen data points in each run. The experimental setup for evaluating the translational compliance along the Z-axis is similar except that the two constraints along the Z-axis were removed and the loading point was shifted to the end effector's center. The experiment was also conducted with five measurement runs and the results are plotted in figure 6a.

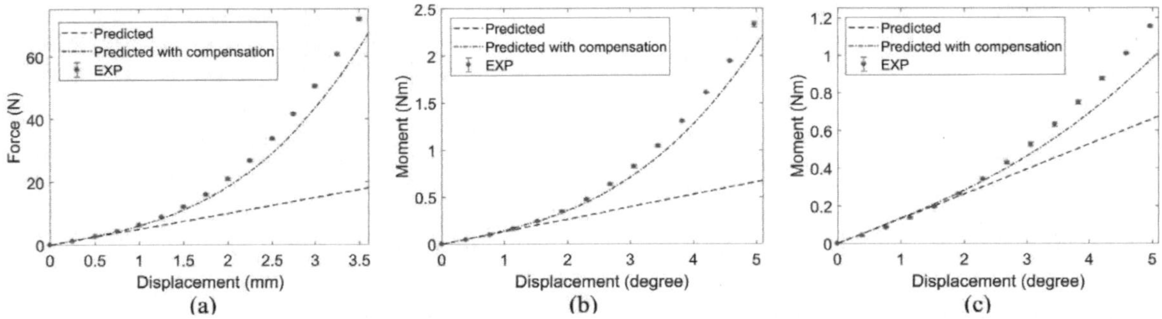

Figure 6. Measured compliance (a) along the Z-axis, (b) about the X-axis and (c) about the Y-axis.

In figure 6, the predicted data is given by equation (4). It is observed that the predicted data agrees with the measured results within a small deformation range, e.g., up to $<2^{\circ}$ for the rotation and <1mm for the translation. Yet at larger deformation range, the compliance of the CPM becomes nonlinear and higher than the predicted data, which are linear because the beam-based method uses FEM to determine the stiffness. The nonlinear stiffness characteristic that occurs during the large deformation is due to the over-constrained design of the synthesized CPM. Such design not only requires the C-T beams to bend but they also need to be elongated/stretched during large deformation. For better prediction of such nonlinear stiffness characteristic, the elongation of the C-T beams was calculated by the pseudo-rigid-body model [12, 16] and added to the initial predicted data. As a result, the predicted data with such compensation is also plotted in figure 6. After considering the elongation of the beams, the predicted data with compensation agrees with the experimental results. For the rotations about the X and Y axes, the average deviations between the prediction and measured results are 8% and 11.2% respectively. For the linear displacement along the Z-axis, the average deviation is 9.5%.

3.2. Dynamic evaluation

The experiment for measuring the dynamic behavior of the CPM was carried out using a standard modal analysis equipment. As the first resonant mode of the CPM is along the Z-axis, an impact hammer was used to apply a knock on the end effector perpendicularly. An accelerometer attached on the opposite side of the end effector was used to measure the acceleration in the same direction. Their outputs were then processed by the signal acquisition device. The first mode of the 3D printed CPM after analysis is plotted in figure 7. The measured resonant frequency is 98Hz, which matches well with the prediction of 99Hz with a small deviation of 1%.

3.3. Positioning evaluation

Three VC motors were used to actuate the 3D printed CPM, each VC motor has a direct feedback via a high-resolution encoder (MicroE System MII5000/6000 series). With all three limbs of the CPM having direct positioning feedback, a joint-space control by having a proportional-integral-derivative (PID) controller to control individual limb is implemented on the CPM. A three-axes laser interferometer (SIOS SP 2000 TR) was employed to monitor the positioning performance of the system. Figure 8 shows the experimental setup for measuring the positioning accuracy of the 3D

IOP Conf. Series: Materials Science and Engineering **280** (2017) 012040 doi:10.1088/1757-899X/280/1/012040

printed manipulator. The CPM was placed at a distance away from the laser interferometer and the laser beams were reflected back to the interferometer by a mirror attached to the end effector.

At this stage of research, the motion of the end effector is considered open-loop since there is no direct positioning feedback control implemented on it. Using just the joint-space control, figure 9a shows a 50nm step resolution of the end effector along the Z-axis that was measured directly from the laser interferometer. It is observed that the accuracy and the maximum positioning error are ±25.2nm and 14.2nm respectively.

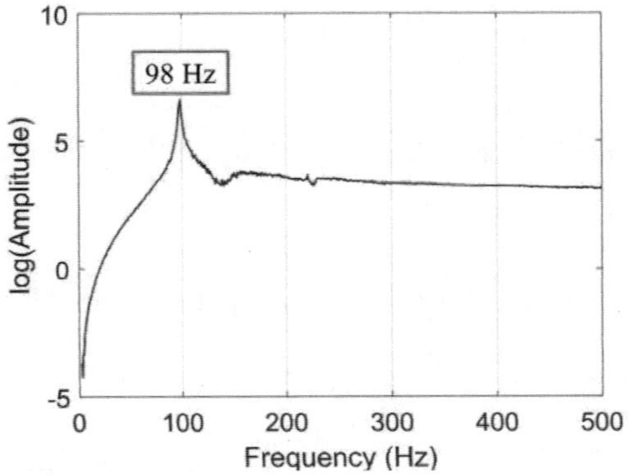

Figure 7. The first resonant frequency of the 3D printed CPM.

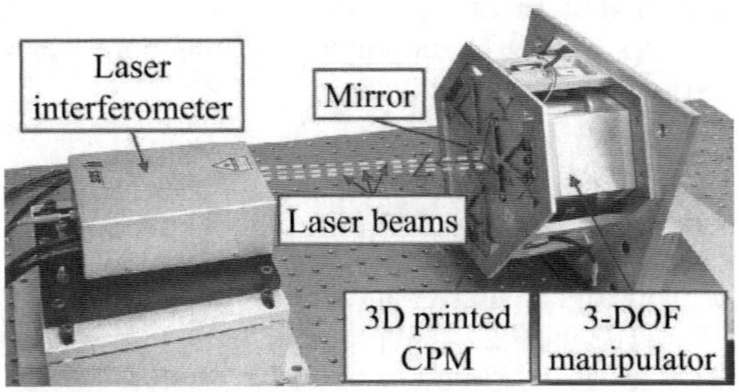

Figure 8. Setup for measuring the positioning accuracy of the CPM.

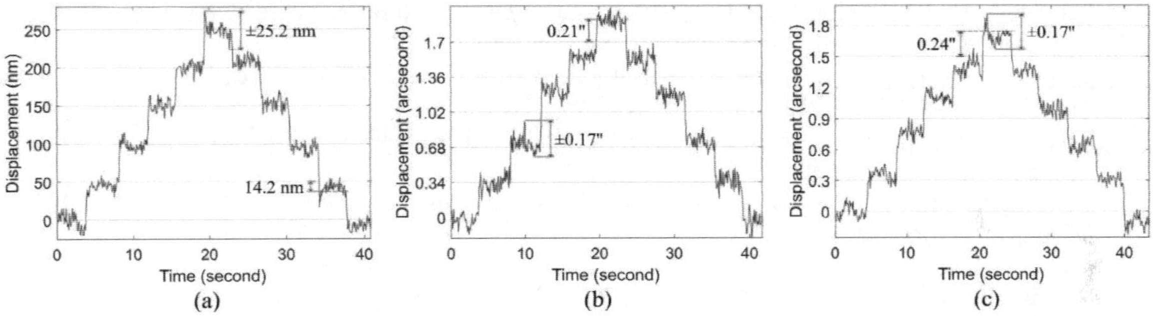

Figure 9. Measured step displacement (a) along the Z-axis, (b) about the X-axis and (c) about the Y-axis.

By controlling each VC motor to achieve a step resolution of 50nm and having a fixed distance between each VC motor, the end effector of the 3D printed CPM was able to achieve an angular resolution of 0.34" and 0.3" about the X and Y axes respectively. Figures 9b and c show the measured data of the step angular displacements about the X and Y axes respectively. Both results show that the 3D printed CPM achieved an accuracy of ±0.17" for the angular motions and maximum positioning errors of 0.21" and 0.24" about the X- and Y-axis respectively.

4. Conclusion

This paper presents a high precision 3D printed Ti6Al4V CPM with 3-DOF ($\theta_X - \theta_Y - Z$) out-of-plane motion. The CPM was designed by a novel beam-based method to achieve optimized stiffness characteristics with targeted dynamic performance. The synthesized CPM delivers a large work range of $\pm 5^{\circ} \times \pm 5^{\circ} \times \pm 3.5$mm with good stiffness ratios of ≥ 567 and fast dynamic response of 98Hz. The 3D printed CPM was fabricated by EBM technology to eliminate assembly errors and several experiments were carried out to evaluate its performance. Experimental results show that the 3D printed CPM provides large elastic deformation and the actuating compliance also matches the prediction with the highest deviation of 11.2% over the full workspace. Most importantly, the 3D printed CPM demonstrated predictable dynamic response since the deviation between the measured first resonant mode and the predicted value is only 1%. The experimental results suggest that the 3D printed Ti6Al4V CPM can produce predictable stiffness and dynamic characteristics. In addition, the CPM can provide repeatable motions with the end effector achieving a positioning resolution of 0.34" × 0.3" × 50nm and accuracy of ±0.17" × ±0.17" × ±25.2nm. The obtained results demonstrate that the 3D printed CPMs can be used in precise positioning systems with predictable mechanical properties and large workspace. The future work is to focus on exploiting advantages of 3D printing technology to fabricate compliant mechanisms with good performance.

Acknowledgment
This research was supported by Singapore Centre for 3D Printing (SC3DP) and National Research Foundation (NRF) Singapore.

References

[1] Howell L L, Magleby S P and Olsen B M 2013 *Handbook of compliant mechanisms* (John Wiley & Sons)
[2] Teo T, Yang G and Chen I M 2014 *Compliant Manipulators* (Springer London) book section 102, pp 2229-2300
[3] Chao D, Zong G and Liu R 2005 *IEEE/ASME Int. Conf. on Advanced Intelligent Mechatronics* (Monterey, CA, USA) pp 765-770
[4] Yangmin L and Qingsong X 2009 *IEEE Transactions on Robotics* **25** 645-657
[5] Kim H and Gweon D G 2012 *Review of Scientific Instruments* **83** 085102
[6] Kim H Y, Ahn D H and Gweon D G 2012 *Review of Scientific Instruments* **83** 055114
[7] Hui T and Yangmin L 2013 *IEEE Transactions on Robotics* **29** 650-662
[8] Bhagat U, Shirinzadeh B, Clark L, Chea P, Qin Y, Tian Y and Zhang D 2014 *Mechanism and Machine Theory* **74** 173-187
[9] Xiao X and Yangmin L 2014 *IEEE/ASME Int. Conf. on Advanced Intelligent Mechatronics* (Besacon, France) pp 1480-1485
[10] Yang G, Teo T J, Chen I M and Lin W 2011 *IEEE Int. Conf. on Robotics and Automation*, 2011 (Shanghai, China) pp 2751-2756
[11] Teo T J, Yang G and Chen I M 2014 *Precision Engineering* **38** 861-871
[12] Pham M T, Teo T J and Yeo S H 2017 *Precision Engineering* **47** 131-139
[13] Lum G Z, Teo T J, Yang G, Yeo S H and Sitti M 2013 *IEEE/ASME Int. Conf. on Advanced Intelligent Mechatronics* (Wollongong, NSW, Australia) pp 247-254
[14] Wang P, Tan X, Nai M L S, Tor S B and Wei J 2016 *Materials & Design* **95** 287-295
[15] Pham M T, Teo T J, Yeo S H, Wang P and Nai M L S 2017 *IEEE/ASME Transactions on Mechatronics* **22** 2359-2368
[16] Teo T J, Chen I M, Yang G and Lin W 2007 *IEEE Int. Conf. on Robotics and Automation* (Roma, Italia) pp 1419-1424

IOP Conf. Series: Materials Science and Engineering **280** (2017) 012041 doi:10.1088/1757-899X/280/1/012041

Development of Multi-Layered Floating Floor for Cabin Noise Reduction

Jee-Hun Song[1,*], Suk-Yoon Hong[2] and Hyun-Wung Kwon[3]

[1] Department of Naval Architecture and Ocean Engineering, Chonnam National University, San 96-1, Dunduck-dong, Yeosu, Chonnam, 550-749, Korea
[2] Department of Naval Architecture and Ocean Engineering, Seoul National University, San 56-1, Silim-dong, Kwanak-gu, Seoul, 151-742, Korea
[3] Department of Naval Architecture and Ocean Engineering, Koje College, 91, Majeon 1-gil, Geoje, Gyeongnam, 656-701, Korea

* jhs@jnu.ac.kr

Abstract. Recently, regulations pertaining to the noise and vibration environment of ship cabins have been strengthened. In this paper, a numerical model is developed for multi-layered floating floor to predict the structure-borne noise in ship cabins. The theoretical model consists of multi-panel structures lined with high-density mineral wool. The predicted results for structure-borne noise when multi-layered floating floor is used are compared to the measurements made of a mock-up. A comparison of the predicted results and the experimental one shows that the developed model could be an effective tool for predicting structure-borne noise in ship cabins.

1. Introduction

Recently, ship cabins noise has become a big issue in ship yards as ship's owner and operators are demanding quitter vessels and relevant noise regulations are getting more stringent. As a rule, however, a ship consists of complex steel structures where structure-borne noise is strong and there are a number of noise sources, such as main engine, diesel generator and auxiliary machinery. From the ship cabins noise tests, it was found that the combined noise level of a cabin was dominated by the radiated noise from the floor structure, which is a combined of a stiffened steel deck and deck covering [1].

Floor impact noise of floating floors has been investigated since the 1950s, and much effort has been directed toward the measurement of such floors. Cremer [2] analyzed the impact noise insulation of floating floors. Istvan [3] reported the characteristics of floor impact noise using a hard floor with an elastic surface material. Yago and Inoue [4] reported experimental results about the characteristics of floor impact noise. Buiten [5] and Jensen [6] reported experimental results of various types of floating floors for ships.

Until now, most research on floating floors has been conducted without a proper understanding of how such an installation works for sound insulation. In particular, research on floating floors with an inserted viscoelastic layer has never been properly conducted.

The aim of this work is to develop an adequate model for multi-layered floating floor that can be used to predict structure-borne noise in ship cabins. To verify the accuracy and validity of the model, the model predictions are compared with measurements made of a mock-up that simulates a typical ship cabin structure.

2. Analysis of multi-layered floating floor

As shown in Figure 1, multi-layered floating floor was investigated. To analyze this floor, we consider two regions: the first is the three-layered sandwich plate, which consists of two isotropic face plates sandwiching a viscoelastic core. The GHM method [7] is used to account for the frequency-dependent complex shear modulus of the viscoelastic layer. The second region is the single panel with high-density mineral wool. The Biot [8, 9] and Bolton [10] models can be used to describe the wave propagation in the wool. We assumed that the transverse displacement is compatible between the regions.

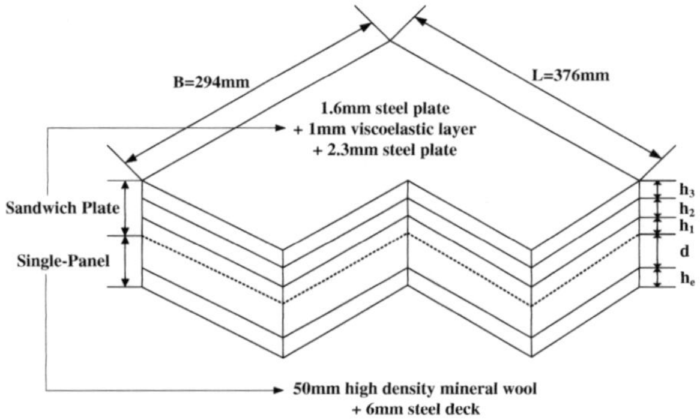

Figure 1. The multi-layered floating floor model.

2.1. *The first region: Three-layered sandwich plate*

The model is formulated according to Kirchoff's theory [11] for plates. The face plates are assumed to exhibit bending, in-plane shear, and extensional stiffness, and their lateral shear deformation is neglected. The viscoelastic layer is assumed to exhibit transverse shear stiffness alone. Furthermore, perfect continuity is assumed at the interface, with no slip occurring during the plate bending. Finally, every point of the three-layer cross section undergoes the same transverse displacement.

Under this set of assumptions, We can write the total kinetic energy T and the potential energy U of the sandwich plates as follows:

$$T=\frac{1}{2}\int_A\left[\rho h\left(\frac{\partial w}{\partial t}\right)^2+\sum_{i=1,3}\rho_i h_i\left(\left(\frac{\partial u_i}{\partial t}\right)^2+\left(\frac{\partial v_i}{\partial t}\right)^2\right)\right]dA \tag{1}$$

$$U=\frac{1}{2}\int_A\left[\begin{array}{l}\sum\limits_{i=1,3}N_{x,i}\dfrac{\partial u_i}{\partial x}+\sum\limits_{i=1,3}N_{y,i}\dfrac{\partial v_i}{\partial y}+\sum\limits_{i=1,3}N_{xy,i}\left(\dfrac{\partial u_i}{\partial y}+\dfrac{\partial v_i}{\partial x}\right)\\ -\sum\limits_{i=1,3}M_{x,i}\dfrac{\partial^2 w}{\partial x^2}-\sum\limits_{i=1,3}M_{y,i}\dfrac{\partial^2 w}{\partial y^2}-2\sum\limits_{i=1,3}M_{xy,i}\dfrac{\partial^2 w}{\partial x\partial y}+Q_{xz}\gamma_{xz}+Q_{yz}\gamma_{yz}\end{array}\right]dA \tag{2}$$

ρh and G are the mass per unit area for the entire total sandwich plate and the shear modulus of the viscoelastic layer, respectively. In order to discretize the system and generate the mass and stiffness matrices, the assumed mode method [12] was used to analyze the sandwich plate. Substituting the displacements for sandwich plate into Eqs. 1-2 yields

$$[\mathbf{M_e}]\{\ddot{\mathbf{x}}\}+[\mathbf{K_e}]\{\mathbf{x}\}+[\mathbf{K_v}]\{\mathbf{x}\}=\{\mathbf{F}\} \tag{3}$$

where $[\mathbf{M_e}]$ is the mass matrix, $[\mathbf{K_e}]$ is the stiffness matrices from the face plates and $[\mathbf{K_v}]=G^*[\overline{\mathbf{K}}]$ is the stiffness matrix from the viscoelastic layer.

2.2. The second region: Single-panel

The single-panel model is based on Biot's theory [8, 9] which has been adapted to describe wave propagation in high-density mineral wool. The elastic panel is assumed only to exhibit transverse displacement. The high-density mineral wool is assumed to experience solid and fluid longitudinal waves and a solid transverse shear wave. The solid and fluid stresses are defined to be forces acting on the frame:

$$\sigma_i = 2N e_i + A e_s + Q\varepsilon \tag{4}$$

$$s = R\varepsilon + Q e_s \tag{5}$$

$i = x, y, z$, where $e_s = \nabla \cdot \overline{u}$ and $\varepsilon = \nabla \cdot \overline{U}$ are the solid and fluid volumetric strain, respectively. $N = E_s / 2(1+\nu)$ and $A = \nu E_s /(1+\nu)(1-2\nu)$ are the shear modulus and Lame constant, respectively. Also, the quantity $Q = (1-Y)E_f$ is the potential coupling coefficient [13]. $R = YE_f$ is a constant that depends on the fluid stress and strain [13]. Y is the porosity.

The longitudinal and transverse shear wave strains in the solid were assumed to be:

$$e_s = e^{-jk_x x}(C_1 e^{-jk_{1y} y} + C_2 e^{jk_{1y} y} + C_3 e^{-jk_{2y} y} + C_4 e^{jk_{2y} y}) \tag{6}$$

$$\Omega_{s,z} = e^{-jk_x x}(C_5 e^{-jk_{ty} y} + C_6 e^{jk_{ty} y}) \tag{7}$$

From the Helmholtz decomposition, we can obtain the x - and y - components of the displacement of the solid and fluid, respectively:

$$u_x = jk_x e^{-jk_x x}(\frac{C_1}{k_{l1}^2} e^{-jk_{1y} y} + \frac{C_2}{k_{l1}^2} e^{jk_{1y} y} + \frac{C_3}{k_{l2}^2} e^{-jk_{2y} y} + \frac{C_4}{k_{l2}^2} e^{jk_{2y} y}) - j\frac{k_{ty}}{k_t^2} e^{-jk_x x}(C_5 e^{-jk_{ty} y} - C_6 e^{jk_{ty} y}) \tag{8}$$

$$u_y = je^{-jk_x x}(\frac{k_{1y} C_1}{k_{l1}^2} e^{-jk_{1y} y} - \frac{k_{1y} C_2}{k_{l1}^2} e^{jk_{1y} y} + \frac{k_{2y} C_3}{k_{l2}^2} e^{-jk_{2y} y} - \frac{k_{2y} C_4}{k_{l2}^2} e^{jk_{2y} y}) - j\frac{k_x}{k_t^2} e^{-jk_x x}(C_5 e^{-jk_{ty} y} + C_6 e^{jk_{ty} y}) \tag{9}$$

$$U_x = jk_x e^{-jk_x x}(\frac{b_1 C_1}{k_{l1}^2} e^{-jk_{1y} y} + \frac{b_1 C_2}{k_{l1}^2} e^{jk_{1y} y} + \frac{b_2 C_3}{k_{l2}^2} e^{-jk_{2y} y} + \frac{b_2 C_4}{k_{l2}^2} e^{jk_{2y} y}) + j\frac{\rho_{12}^*}{\rho_{22}^*}\frac{k_{ty}}{k_t^2} e^{-jk_x x}(C_5 e^{-jk_{ty} y} - C_6 e^{jk_{ty} y}) \tag{10}$$

$$U_y = je^{-jk_x x}(\frac{b_1 k_{1y} C_1}{k_{l1}^2} e^{-jk_{1y} y} - \frac{b_1 k_{1y} C_2}{k_{l1}^2} e^{jk_{1y} y} + \frac{b_2 k_{2y} C_3}{k_{l2}^2} e^{-jk_{2y} y} - \frac{b_2 k_{2y} C_4}{k_{l2}^2} e^{jk_{2y} y}) - j\frac{\rho_{12}^*}{\rho_{22}^*}\frac{k_x}{k_t^2} e^{-jk_x x}(C_5 e^{-jk_{ty} y} + C_6 e^{jk_{ty} y}) \tag{11}$$

By substituting Eqs. 6-7 and 8-11 into Eqs. 4-5, we can obtain the normal solid stress σ_y, the shear stress τ_{xy}, and the fluid stress s. The unknown cons-tants C_1- C_6 in Eqs. 6-7 and 8-11 are determined by the boundary conditions at the interfaces of the high-density mineral wool.

3. Experiment Validation

To verify the accuracy and validity of the developed model, the predicted results have been compared with the corresponding experimental results, which comprise measurements of mock-up built that simulates a typical ship cabin structure.

The length and width of the sandwich plate are 3.76 and 2.94 m, respectively. The viscoelastic material modulus and loss factor were obtained from the product information provided by 3M Scotchdamp ISD-11217. A standard tapping machine was used to excite bending motion of the floating floor. The structure-borne noise transmitted by the tapping machine to the lower receiving room is measured in one-third-octave bands, from 31.5Hz to 8,000Hz. We measure structure-borne noise in dB defined as 20 log (rms acceleration (m/s^2) / ref. acceleration ($10^{-5} m/s^2$)).

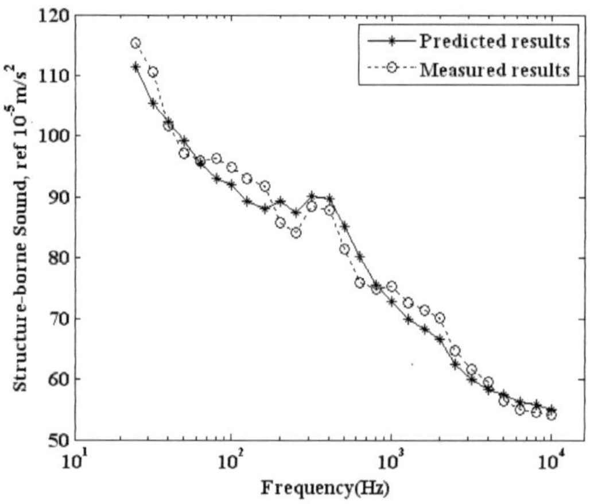

(a) 25mm MW

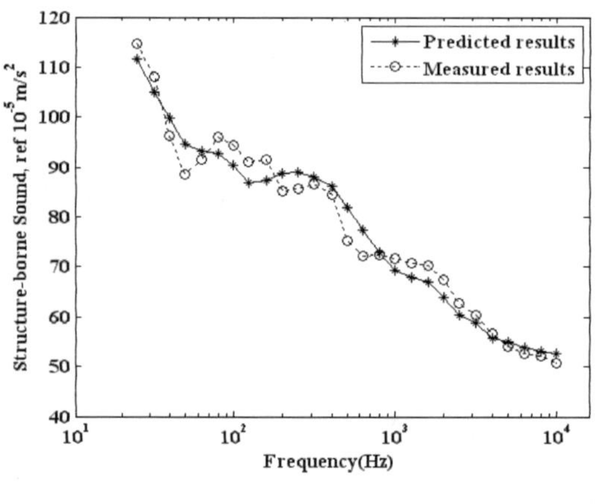

(b) 50mm MW

Figure 2. The effect of high density mineral wool thickness in floating floor with an inserted viscoelastic layer structure-borne noise

In Figures 2-3, we compared structure-borne noise measurements to predictions for floating floor with an inserted viscoelastic layer. Comparison shows that the results simulated by the developed floating floor predict the frequency dependency of experimental structure-borne noise relatively well.

In Figure 2, we show the effects of the high-density mineral wool (MW) thickness on structure-borne noise. The floating floor structure is: 6mm deck + MW + 1.6mm steel plate + 1mm viscoelastic layer + 1.6mm steel plate, in which MW thickness varies from 25mm to 50mm. It is found that as the thickness of MW increases from 25mm to 50mm, the structure-borne noise decreases by approximately 4-5 dB.

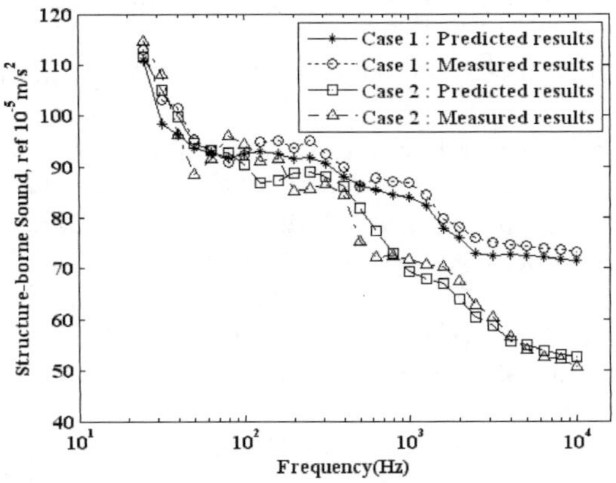

Figure 3. The effect of viscoelastic layer in floating floor structure-borne noise.

In the next example, Figure 3 illustrates the importance of the viscoelastic layer for damping a vibration in the medium-to-high frequency range. The floating floor structure is: Case 1 : 6mm deck + 50mm MW + 3.2mm steel plate, Case 2 : 6mm deck + 50mm MW + 1.6mm steel plate + 1mm viscoelastic layer + 1.6mm steel plate. By using a viscoelastic layer, the structure-borne noise is reduced by approximately 10-15 dB at medium-to-high frequencies. It provides also a more economically and technically effective solution to noise and vibration transmission problems.

4. Conclusions

In this paper, a model is developed of multi-layered floating floor to predict structure-borne noise in ship cabins. To verify the accuracy and validity of the developed model, the predicted results are compared with corresponding experimental results, which comprised measurements of a mock-up of a typical ship cabin structure. Various floating floor structures are studied, and the effects of variations of the thickness and viscoelastic layer. On the whole, the structure-borne noise can be controlled to some extent by varying any of the three prescribed parameters. Moreover, by using a viscoelastic layer, structure-borne noise is reduced by approximately 10-15 dB at medium-to-high frequencies. The developed model could be an effective tool for predicting structure-borne noise in ship cabins.

Acknowledgments

This work was supported by the following: the Basic Science Research Program through the National Research Foundation of Korea (NRF) funded by the Ministry of Education, Science and Technology (2015R1D1A1A01060387, 2016R1D1A1A09918294).

References

[1] Joo W.-H, Kim S.-H, Kim D.-H, Bae J.-G. and Hong S.-Y. 2008 Quantitative Evaluation of Airborne Sound Insulation in Ship's Accommodation using Large Scale Noise Test Facilities *Noise Control Eng. J.* **56(1)** 45-51

IOP Conf. Series: Materials Science and Engineering **280** (2017) 012041 doi:10.1088/1757-899X/280/1/012041

[2] Cremer L 1952 Theorie des kolphschalles bei decken mit schwimmenden estrich *Acustica* **2(4)** 167-178
[3] Istvan L V 1971 Impact noise isolation of composite floors *Journal of the Acoustical Society of America* **50** 1043-1105
[4] Yago S, Inoue K 1983 An experimental study on the vibration characteristics of floor impact noise *Journal of the Acoustical Society of Japan* **332** 83-93
[5] Buiten J 1971 *Acoustical investigations of asphaltic floating floors applied on a steel deck Report 152S, TNO, Delft*
[6] Jensen J O 1975 *Noise in accommodation spaces onboard ships The Technical University of Denmark*
[7] McTavish D J, Hughes P C 1993 Modeling of Linear Viscoelastic Space Structures *J. Vib. Acoust* **115(1)** 103-110
[8] Biot M A 1956 Theory of propagation of elastic waves in a fluid-saturated por-ous solid: I. low-frequency range *Journal of the Acoustical Society of America* **28** 168-178
[9] Biot M A 1956 Theory of propagation of elastic waves in a fluid-saturated por-ous solid: II. higher frequency range *Journal of the Acoustical Society of America* **28** 179-191
[10] Bolton J S, Shiau N M, Kang Y J 1996 Sound transmission through multi-panel structures lined with elastic porous materials *Journal of Sound and Vibration* **191** 317-347
[11] Rao Y V K S, Nakra B C 1974 Vibration of unsymmetrical sandwich beams and plates with viscoelastic cores *Journal of Sound and Vibration* **34(3)** 309-326
[12] Inman D J 1994 *Engineering vibration Practice Hall Englewood Cliffs New Jersey*
[13] Allard J F 1993 *Propagation of sound in porous media Elsevier Applied Science London and New York*

IOP Conf. Series: Materials Science and Engineering **280** (2017) 012042 doi:10.1088/1757-899X/280/1/012042

Software-In-the-Loop based Modeling and Simulation of Unmanned Semi-submersible Vehicle for Performance Verification of Autonomous Navigation

Kwangkook Lee[1,*], Mijin Jeong[1] and Dong Hun Kim[2]

[1]Dep't of Naval Architecture & Ocean System Engineering, Kyungnam University, Changwon, South Korea
[2]Dep't of Electrical Engineering, Kyungnam University, Changwon, South Korea

[*]kklee@kyungnam.ac.kr

Abstract. Since an unmanned semi-submersible is mainly used for the purpose of carrying out dangerous missions in the sea, it is possible to work in a region where it is difficult to access due to safety reasons. In this study, an USV hull design was determined using Myring hull profile, and reinforcement work was performed by designing and implementing inner stiffener member for 3D printing. In order to simulate a sea state 5.0 or more at sea, which is difficult to implement in practice, a regular and irregular wave equation was implemented in Matlab / Simulink. We performed modeling and simulation of semi - submersible simulation based on DMWorks considering the rolling motion in wave. To verify and improve unpredicted errors, we implemented a numeric and physical simulation model of the USV based on software-in-the-loop (SIL) method. This simulation allows shipbuilders to participate in new value-added markets such as engineering, procurement, construction, installation, commissioning, operation, and maintenance for the USV.

1. Introduction

In the Maritime Classification Code of the World Meteorological Organization (WMO), sea level 5.0 is the average sea level at 2.5 m to 4.0 m above sea level. If the sea level is above 5.0, offshore operation will be interrupted.

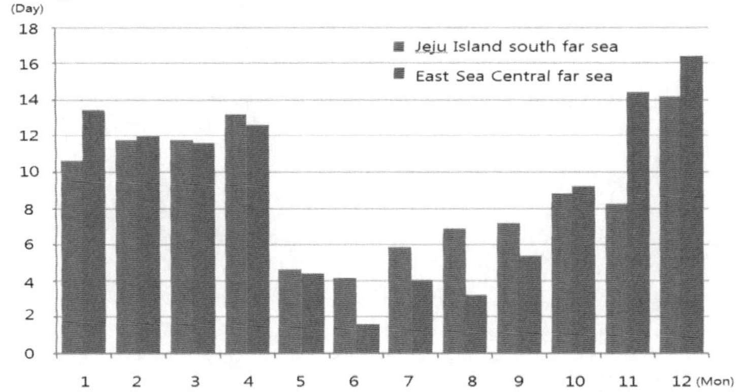

Figure 1. 2011~2015 Monthly storm weather days

The storm warning is announced when the wind speed of 14m/s or more on the sea lasts more than 3 hours or when the significant wave height is expected to be more than 3m. The number of days for the 2011~2015 monthly storm surge is 108 days based on the South Sea and East Sea in Korea as shown in Figure 1. At sea level of 30% in 1 year, it is 5.0, so there is a limit to safety navigation on board, which is increasing the need for unmanned maritime monitoring and security / surveillance missions.

As demonstrated in Figure 2, unmanned vessel is mainly used in coastal waters and polluted waters, which are difficult to access from the safety aspect of the missions carried out by the manned vessel. Long-term operations such as marine surveillance and marine surveys can minimize operational costs and crew fatigue accumulation. It is also used primarily for the purpose of carrying out dangerous missions such as search and removal of mines, infiltration operation, and operation in the tough sea. Unmanned ships are mainly used for military purposes, but recently demand for public purposes is increasing. There is still little demand for civilian use and it is mainly used for scientific purposes [1].

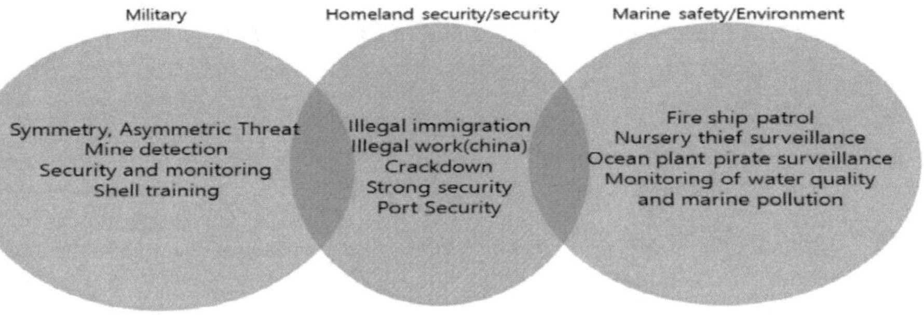

Figure 2. Application area of unmanned ship [1]

Unmanned vehicles that can be operated at sea are unmanned airplanes and unmanned vessel. Unmanned ships are more capable of stealthing than unmanned airplanes, and unmanned ships can be classified as unmanned surface vehicle (USV), unmanned semi-submersible vehicle (USSV) and unmanned underwater vehicle (UUV). UUV is superior to USV in terms of stabilization of the hull, but the moving speed is slower than USV. The USV has a communication environment because the communication device and the external camera are located on the water surface. Therefore, to compensate for the disadvantages of USV and UUV, the unmanned semi-submersible form is an optimized platform for performing missions in rough marine environments as illustrated in Figure 3[1].

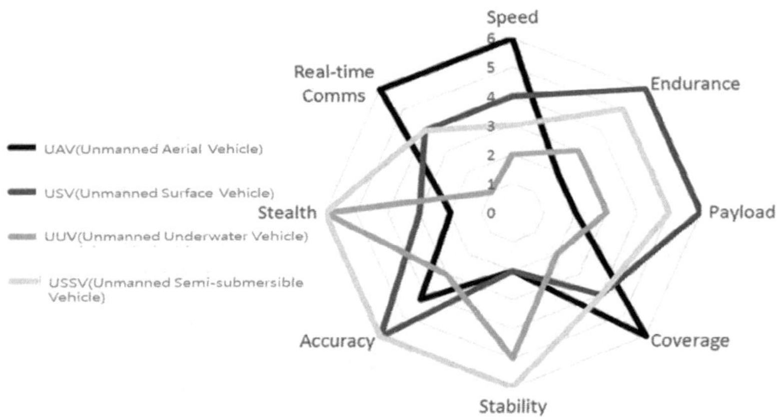

Figure 3. Comparison of characteristics between unmanned semi-submersible and other unmanned vehicles

In this study, we design and implement a platform for autonomous operation of unmanned semi-submersible to operate in rough sea environment above 5.0 sea level. Therefore, it is aimed to establish the environment where the hull design of the unmanned submersible and the marine environment implemented by Matlab/simulink can be visualized in three dimensions on the DMWorks(Digital Manufacturing Works), and to verify the dynamic behavior of the unmanned semi-submersible in a virtual environment.

2. Unmanned semi-submersible hull design

2.1. Basic design of Myring hull based unmanned submersible

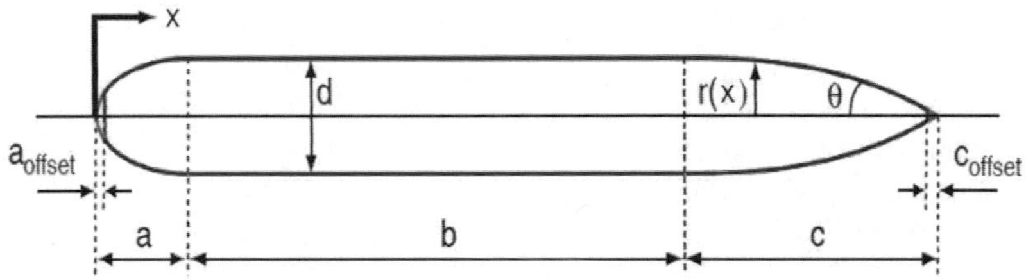

Figure 4. Myring hull profile

The semi-submersible designed in this study was selected as a cylinder type to minimize the resistance of the fluid, and the hull type was designed by comparing it with the cylindrical type of the Myring hull type. In Figure 4, The Myring hull form is respectively determined by the following equation [2] [3] [4].

$$\text{Stem section}: \ r(x) = \frac{1}{2}d\left[1-\left(\frac{x-a}{a}\right)^2\right]^{\frac{1}{n}} \tag{1}$$

$$\text{Middle section}: \ r(x) = d \tag{2}$$

$$\text{Stern section}: \ r(x) = \frac{1}{2}d - \left[\frac{3d}{2c^2} - \frac{tan\theta}{c}\right](x-a-b)^2 + \left[\frac{d}{c^3} - \frac{tan\theta}{c^2}\right](x-a-b)^3 \tag{3}$$

As a result of the comparison, the average error rate of the stern was 3.08% as shown in Figure 5. Finally, the total length was determined as 1000 mm, the width was 350 mm, and the height was 300 mm.

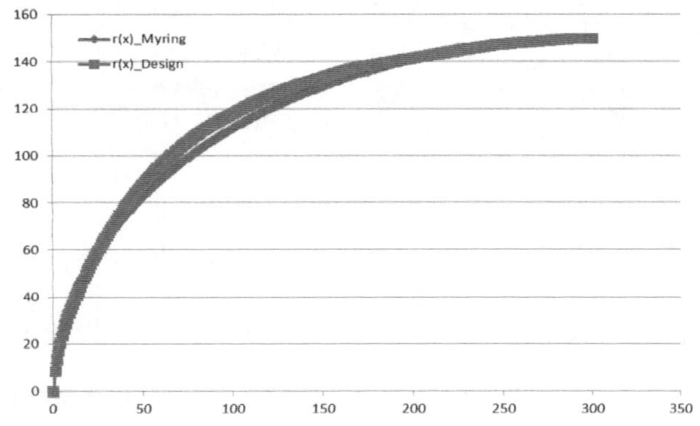

Figure 5. Myring calculation and design ship stem comparison

In order to make the prototype of the unmanned semi-submersible, the main specifications were determined through the hydrostatic calculation and the hull form design with AVEVA Marine by referring to the main specifications and the hull form of the unmanned surface vehicle, unmanned semi-submersible and unmanned underwater vehicle[5] [6].

IOP Conf. Series: Materials Science and Engineering **280** (2017) 012042 doi:10.1088/1757-899X/280/1/012042

Table 1. Main characteristics of unmanned submersible and hydrostatic calculation result

Division	Unit(mm)
LOA	1000
B	350
D	300
Draft	150
Displacement Volume	29454281.98
LCB	514.182
VCB	89.153
Block Coefficient, Cb	0.6545
Midsection Coefficient	0.7835
Prismatic Coefficient, Cp	0.8354
Water Plane Area	262949.156
LCF	502.952
Water Plane Coefficient	0.8765
KMT	147.883

2.2. 3D modeling and 3D printing of unmanned semi-submersible

3D modeling was performed for 3D printing of inside and outside of the designed unmanned semi -submersible hull form. As shown in Figure 6 (a), (b) a connection path for mounting a GPS receiver and an obstacle recognition sensor are installed on the outside.

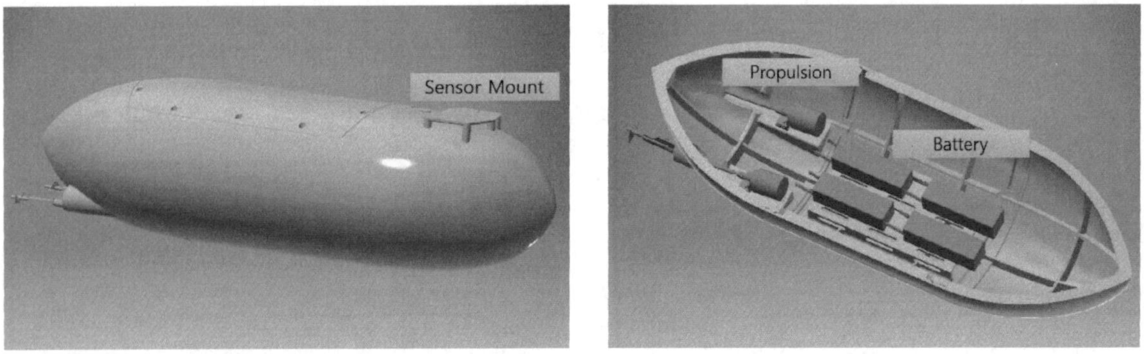

(a) Exterior model (b) Interior model

Figure 6. 3D model of unmanned semi-submersible

3. Modeling and Simulation of DMWorks based unmanned semi-submersible

We simulated the wave using the wave spectrum and implemented it in Matlab / Simulink to simulate the rolling motion of the unmanned semi-submersible which can communicate with the central server in the rough sea environment and can operate easily [7].

Table 2. Wave Definition [7]

Division	Wave definition
Regular wave	$\zeta(x, y, t) =$ $\zeta \sin[(\omega t + \epsilon - kx\cos(\chi) - ky\sin(\chi)]$ where $\omega = \frac{2\pi}{T}(rad/s)$: Wave frequency $k = \frac{\omega^2}{g} = \frac{2\pi}{\lambda}(rad/m)$: Wave number $\lambda = \frac{g}{2\pi}T^2(m)$: Wave length $c = \sqrt{\frac{g\lambda}{2\pi}} = \frac{\lambda}{T}(m/sec)$: Phase velocity $h > \frac{\lambda}{2}(H : water\ depth(m))$ $\epsilon(rad/s)$: Phase of wave $\chi(rad)$: Direction of wave
Irregular wave	$\zeta(x, y, t) =$ $\sum_{i=1}^{n} \zeta(i)\sin(\omega(i)t + \epsilon(i) - k(i)((x\cos(\chi(i) + y\sin(\chi(i))$ (n : The of the harmonic wave components)

Based on Table 2, a wave block model was created as shown in Figure 7

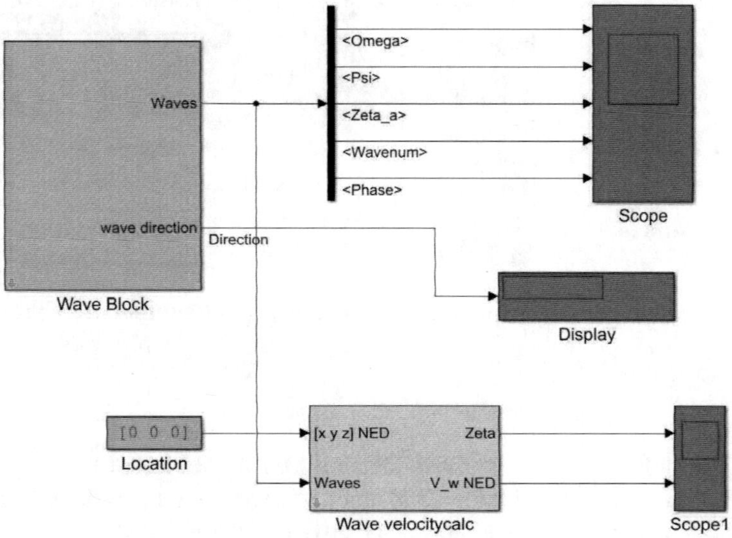

Figure 7. Matlab/Simulink wave block model

To simulate real sea area test or model test in marine environment of 5.0 sea level or more, we constructed a rolling motion simulating platform environment of prototype based on DMworks with wave parameters of unmanned submersible as shown in Figure 8.

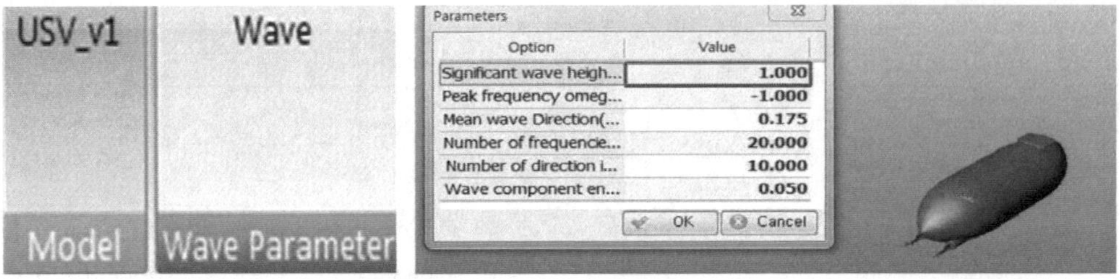

Figure 8. DMWorks wave parameter control implementation concept

In order to test the simulation platform in a real-time environment, we propose an autonomous navigation simulation architecture of an unmanned semi-submersible that can operate in a rough environment by linking National Instrument Veristand, LabView Real-Time, and Phar Lap ETS with DMWorks as shown in Figure 9.

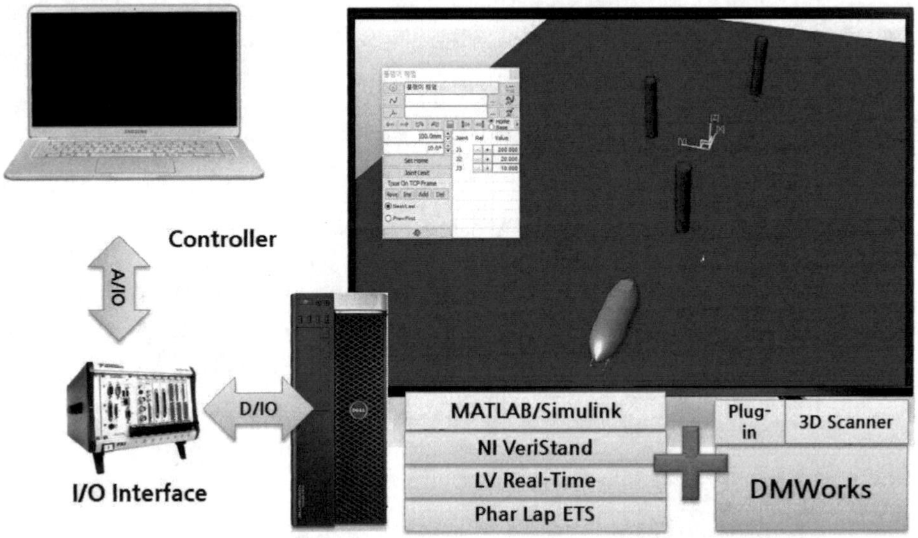

Figure 9. Real-time simulation architecture of unmanned semi-submersible based on DMWorks

4. Conclusion

In this study, an USV hull form was determined using Myring Hull profile, and reinforcement work was performed by designing and implementing inner stiffener member for 3D printing. In order to simulate a sea state 5.0 or more at sea, which is difficult to implement in practice, a regular and irregular wave equation was implemented in Matlab / Simulink. We performed modeling of semi - submersible simulation based on DMWorks considering the rolling motion in wave.

In the near future, we plan to conduct real-time testing based on hardware-in-the-loop and to design and implement a motion compensator to minimize the rolling motion.

Acknowledgement

This research was supported by Basic Science Research Program and Leaders in INdustry-university Cooperation + Project through the National Research Foundation of Korea(NRF) funded by the Ministry of Education, Science and Technology(NRF-2017R1A2B4011329)

IOP Conf. Series: Materials Science and Engineering **280** (2017) 012042 doi:10.1088/1757-899X/280/1/012042

References

[1] J. H. Kim, S. Y. Kim (2015). "Development Trends and Industry Status of Unmanned Ship Technology", Korea Institute of Industrial Technology Evaluation and Management, pp 83-95

[2] F. Y. Lee, B. H. Jun, H. W. Shim, P. M. Lee(2008). "Design of Autonomous Underwater Vehicle 'ISiMI100' to Survey Undersea", Joint conference on Korean Association of Ocean Science and Technology Societies. P2295~2299

[3] G. M. Lee, J. Y. Park, P. M Lee, B. H. Jun, K. H. Kim, C. M. Lee, B. Kim, H. Baek , H. W. Shim, J. I. Kim, S. T. Kwon, F. Y. Lee, and H. S. Jeong(2012). ' Design of 6,000m Autonomous Underwater Vehicle. IsiMi6000', Joint conference on Korean Association of Ocean Science and Technology Societies. pp1719~1722

[4] Chao Chuzhe, H. G. Soon, E. Y. Hong, Mandar Chitre (2011). "Modular modeling of autonomous underwater vehicle", Waikoloa, HI, USA

[5] Korea Ocean Research Development Institute, "Development and utilization of unmanned system for marine observation field" , [Internet] Available: http://kiost.ac.kr/cop/bbs/BBSMSTR_000000000331/selectBoardArticle.do;jsessionid=DD469DEA9533700CD46FD25051A9CD6D?nttId=16277&pageIndex=3&searchCnd=&searchWrd=

[6] ITTC(1978) Performance, Propulsion 1978 ITTC Performance Prediction Method

[7] S. C. Shin(2015). 'Simulation modeling of wave, wind, current, and marine environments for PMS HILS'. Bulletin of the Society of naval architects of Korea, 52 (4), pp.11~17

IOP Conf. Series: Materials Science and Engineering **280** (2017) 012043 doi:10.1088/1757-899X/280/1/012043

Small Sized Drone Fall Recover Mechanism Design

Tzu-Heng LIU[1], Fang-Lin CHAO[1,*] and Jhen-Yuan LIOU[1]

[1]Department of Industrial Design, Chaoyang University of Technology, Taichung, 436, Taiwan, R.O.C.

*flin@cyut.edu.tw

Abstract: Drones uses four motors to rotate clockwise, counter-clockwise, or change in rotational speed to change its status of motion. The problem of Unmanned Aerial Vehicle turnover causes personal loses and harm local environment. Designs of devices that can let falling drones recover are discussed. The models attempt to change the orientation, so that the drone may be able to improve to the point where it can take off again. The design flow included looking for functional elements, using simplify model to estimate primary functional characteristics, and find the appropriate design parameters. For reducing the complexity, we adopted the simple rotate mechanism with rotating arms to change the fuselage angle and reduce the dependence on the extra-components. A rough model was built to verify structure, and then the concept drawing and prototype were constructed. We made the prototype through the integration of mechanical part and the electronic control circuit. The electronic control module that selected is Arduino-mini pro. Through the Bluetooth modules, user can start the rebound mechanism by the motor control signal. Protections frames are added around each propeller to improve the body rotate problem. Limited by current size of Arduino module, motor and rebound mechanism make the main chassis more massive than the commercial product. However, built-in sensor and circuit miniaturization will improve it in future.

1. Introduction

Drones have increasingly broad applications. It uses four motors to rotate clockwise, counter-clockwise, or change in rotational speed to change its status of motion. It can demonstrate styles of flying such as (1) perpendicular motion, (2) pitch motion, (3) roll motion, (4) yaw motion, (5) front and back motion, and (6) Sideways move [1]. The problem of drones/ Unmanned Aerial Vehicle (UAV) falling causes personal loses and disturbing local environment [2]. One may often hear of incidents where drones used for aerial photography and filming or other purposes lose control and fall, thus endangering public safety.

Currently some patented applications already exist that functions like the active safety systems of cars, which intervene once the driver's actions endanger safety: For instance, Taiwan Patent no. M518111 [3] (active obstacle avoidance system of drone) is similar to a car's active safety system that changes the driver's manner and amount of control at times. Processing unit features receive inputs from multiple distance sensors, decode control signals, and simultaneously read the measurement results from the sensor. Using this information, it can then use the rules for obstacle avoidance and modify the controls at appropriate times.

Drones have also been used in field data collection, in Patent No. M510456 [4], a drone is designed to collect water samples. It uses GPS navigation or remote control to fly to a destination, hover above the water surface, and uses weight and buoyancy to change the angle and orientation of the automatic water-sampling cup, and proceed to collect water samples near a pollution source. During field data collection, the drone may be affected by the terrain or above ground objects, and fall and turn upside-down [4]. The propellers of the drone cannot freely rotate while accidentally falls. It is unable to recover from falling by itself. In US3273835A [5], a practical system for monitoring in-flight conditions aboard a target drone which will respond to power failures on the drone electrical system or to drone destruct signals. System ejects the instrument package and initiates a transmission of homing beacon signals to ensure the maintainability.

In this study, we design the mechanism that can let falling drones recover. The designs attempt to change the orientation so that the drone can back to the point where it can take off again. These compositions may help increase the probability of a drone involved in an accident to return.

2. Design approach

Drones use built-in micro-controllers to control the motors, thus changing their mode of locomotion. The flipping that occurs after a fall is related to jumping motion. UC Berkeley has designed a small robot that can leap into the air and then spring off a wall [6], or perform multiple vertical jumps in a row. Spring stored energy within long duration, later released in a short time. For reducing the complexity, we adopted the simple rotate mechanism which uses residual power.

2.1. Design process

The design flow shows in Figure 1; first look for related functional elements then simplify the functional characteristics of simplified model. Through the simulation, we can find the appropriate design parameters. A rough model was built to verify structure status, and then the concept drawing and prototype were constructed by designer.

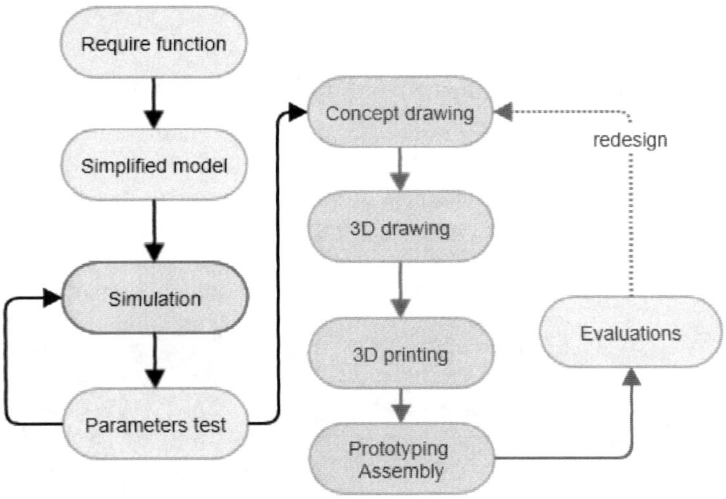

Figure 1. Design flowchart

2.2. Rollover

Micro-UAVs usually have four blades, controlled by speed changes to ensure a stable posture. When performing the rollover, it is necessary to control the small support foot of the self-starting device to change the fuselage angle. The self-starting device shall be low weight so that the structure which attempts to change the body angle will not cause a significant increase in the total weight. For reducing complexity, the simple rotate mechanism was adopted with rotating arm to adjust the fuselage angle and reduce the dependence on the extra-components.

Firstly, the Working Model 2D has utilized to simulation different conditions of rotate mechanism given uniform mass distribution. When the fulcrum position moved to the center of gravity, the object does not move horizontally, and the main chassis can be reversed in short time (Figure 2). Refer to Figure 3 of qualitative observation; the length of the support shaft is increased to make sure that the rollover occurs. When the fulcrum position moves to the center of gravity, rollover also occurs earlier.

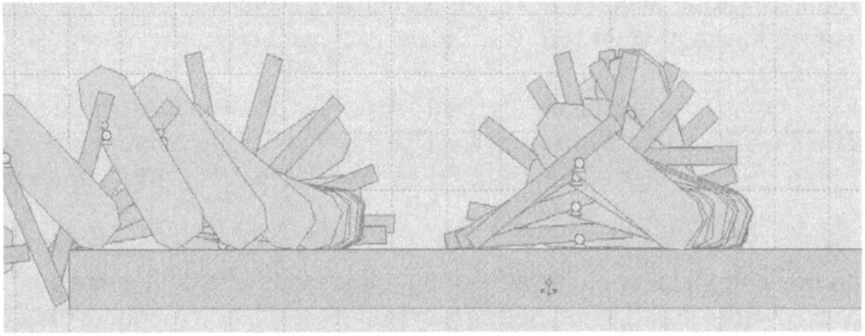

Figure 2. The change of the shaft position on the flipping

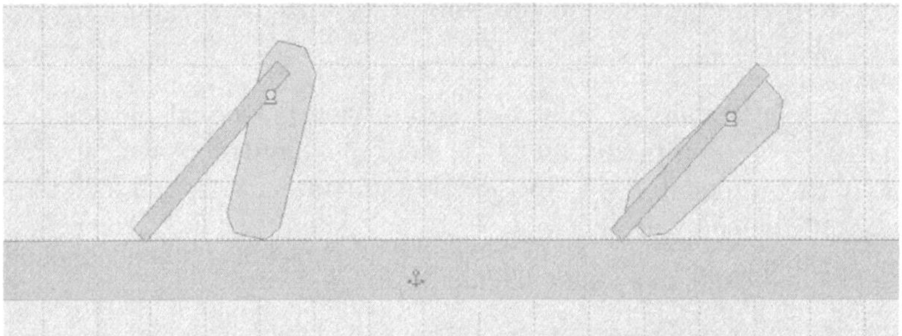

Figure 3. The change of shaft length on the flipping

In the real flying body, the mass distribution is not uniform; the main body chassis is much smaller than the top of the four rotating propellers. At this time, it is necessary to increase the length of the rotating rod. The total mass is 207g, diagonal distance between propellers is 27cm, and propeller radius is 7cm. After the experiment, the length of the rotate rod is chosen to be 20.6 cm.

3. Concept design

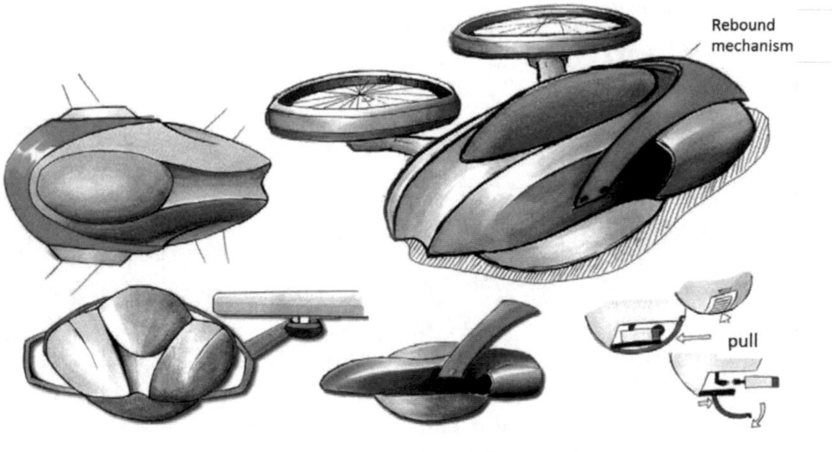

Figure 4. Idea schematic drawing

3.1. Physically flip the drone.

Naturally, the center of gravity and center of the object will be the lowest position. A dome frame protector was installed to replace circular protectors. The dome frame protector enlarges the volume of drone with small additional weight. We move the Lithium battery to lower position to keep the gravity center of drone lower. Computer simulations were successful predict the rollover. The result of prototype test deviated with the prediction. Consequently, covering structure in outer edge of the fan was adopted.

We try to lift the center of gravity to flip the drone. According to the results of simulation and parameters test, we explored several design concepts through brainstorming. One of the ideas shown in Figure 4. Outside the bracket, there is a circuit ring made of carbon fiber to enhance the structural strength and reduce the weight. Figure 5 proposes to add a covered structure (#12) on the outer edge of the fan's blade to prevent the possible damage [7] and surround matter into the blade rotation range. At the one end of the body chassis, there is a support shaft (#P1) with fulcrum near side of the body (#212) for turning over. The hoisting mechanism uses a remote signal (#S) to activate and control the servo motor (#22).

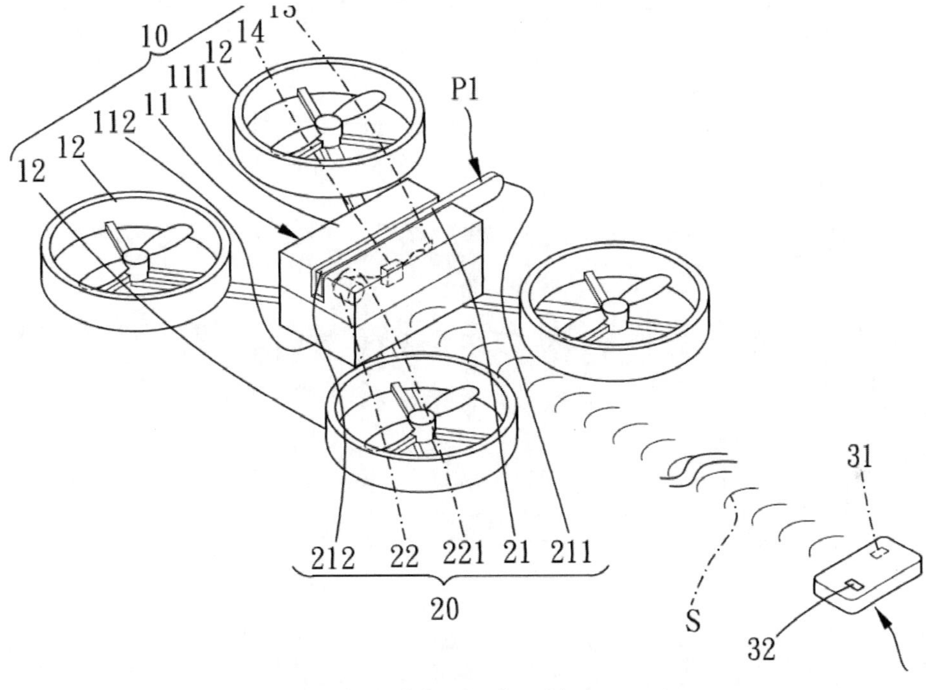

Figure 5. Outer edge of the fan is added covering structure

3.2. Prototyping

During prototyping step, designer need consider all the parts (chassis, motor, and controller) that obtained, and how to connect them to demonstrate major functions. The electronic control module that selected is Arduino-mini pro [8] with installation of Bluetooth module [9] and Servo motor (Figure 6). The user can start the rebound mechanism circuit by the signal from remote controller. There is a corresponding Bluetooth transmitter operate by 2.4G signal in remote controller. Limited by the size of Arduino module [10], motor and rebound mechanism makes the main chassis larger than a commercial product. However, during the mass production phase, the electronic control module can be customized to reduce the size of circuit board and chassis. For maintain the strength of overall structure, the main body was 3D printed with PLA material, but the four arms made of carbon pipes.

IOP Conf. Series: Materials Science and Engineering **280** (2017) 012043 doi:10.1088/1757-899X/280/1/012043

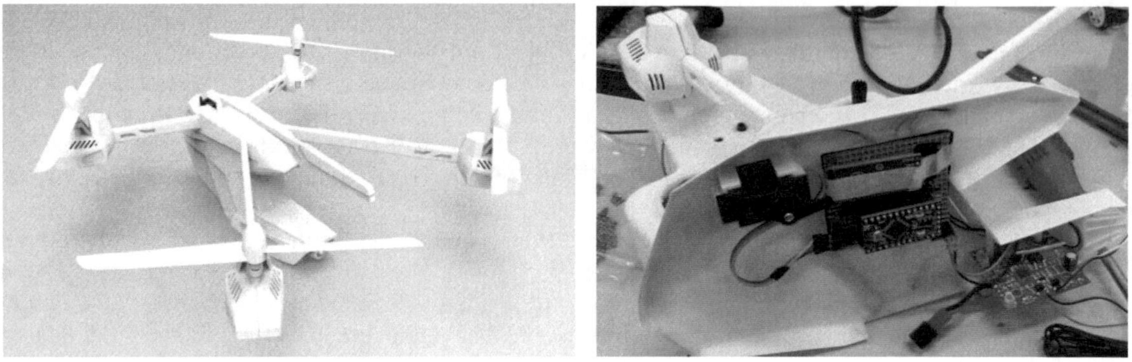

Figure 6. The servo-driven rebound lever: (left) structure, (right) Arduino board connections.

3.3. Evaluations

The servo-driven rebound lever turns the body back to the fly state. We verified the effect and possible pitfalls in the evaluation phase. The original drone (Figure 6) do not have protection frame around propeller blade. Sometimes the blade position causes it fails turnover correctly (Figure 7). Protection frames are added to each fan to fix this pitfall. The situation was significantly improved. We observed that almost every time the rebound angle is the same (Figure 8, 9). The protection frame around fan led to a drawback of additional weight. Lightweight materials and the frame structure which remove the unnecessary stuff need further investigation.

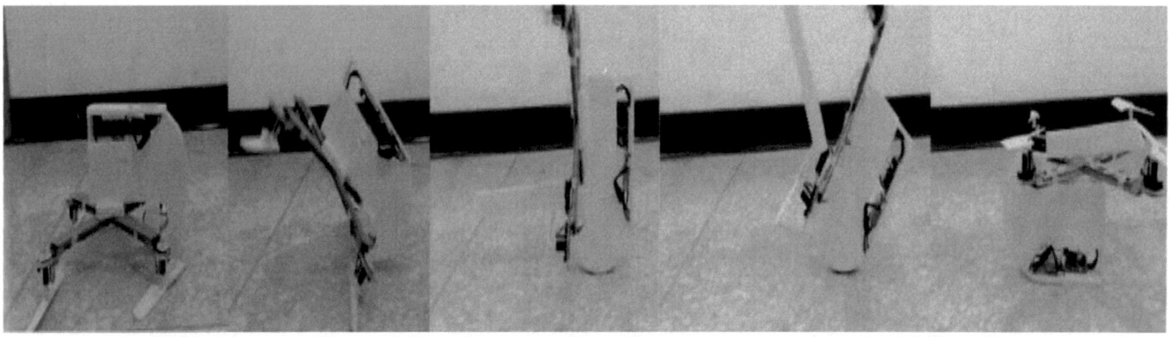

Figure 7. Propeller without protection ring: skew, cannot be successfully turned

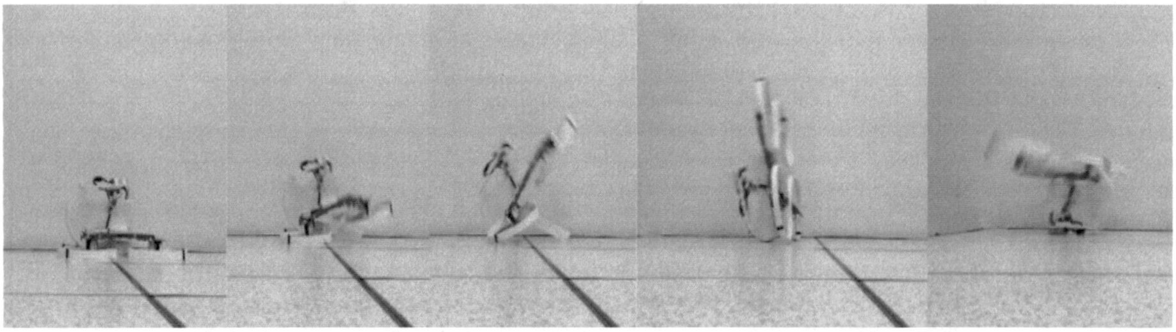

Figure 8. Propeller with protection ring: can successfully stand up

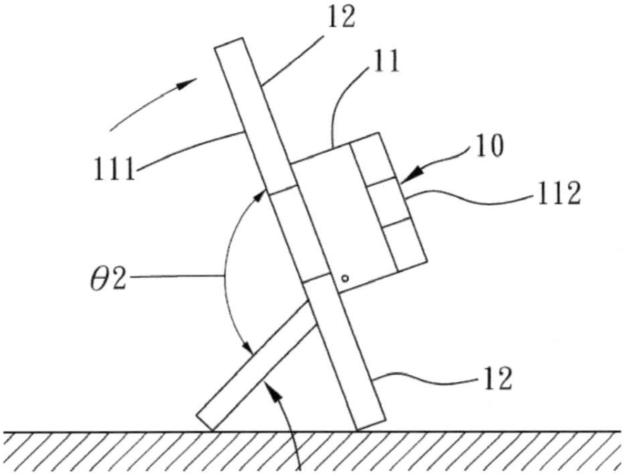

Figure 9. Schematic diagram of rotate rod [7], where Θ is rotation angle

4. Conclusions

The design process use prototype and simulation to verify design situations, designer can find the appropriate design parameters. The prototype constructed through integration of mechanical part and the remote control circuit through the Bluetooth linking channel. Protection frames were added around each propeller to improve the flip over and change the drone position to facilitate the take-off. Limited by current size of Arduino module, motor and rebound mechanism made the main chassis larger than a commercial product. However, we will improve this through built-in sensor and circuit miniaturization in future. Remote app interface enables user in knowing the control conditions. This feature reduces the number of attempts and increases the opportunities for successful practice.

References

[1] K. P. Valavanis, Ed. 2011 Special issue on unmanned aerial vehicles, *J. Intell. Robot. Syst.*, vol. 61, pp 1–585

[2] David Cabecinhas, Roberto Naldi, 2016 Robust Landing and Sliding Maneuver Hybrid Controller for a Quadrotor Vehicle, *IEEE Transactions On Control Systems Technology*, 24, no. 2, pp 400-408

[3] Taiwan patent, M-518111, http://twpat-simple.tipo.gov.tw/tipotwoc/tipotwkm?!!FR^M518111

[4] Taiwan patent, M-510456, http://twpat-simple.tipo.gov.tw/tipotwoc/tipotwkm?!!FR^M510456

[5] US patent, US3273835A, https://patents.google.com/patent/US3273835A

[6] http://news.berkeley.edu/2016/12/06/wall-jumping-robot-is-most-vertically-agile-ever-built/

[7] Taiwan patent, I586414, http://twpat-simple.tipo.gov.tw/tipotwoc/tipotwkm?!!FR^I586414

[8] A. D' Ausilio, 2011 Arduino: a low-cost multipurpose lab equipment, *Behaviour Research Methods*, vol. 44, Issue 2, pp 305-313

[9] Arduino shield list: http://shieldlist.org.

[10] A. Garrig, D. Marroqui, J. M. Blanes, R. Gutierrez, 2017 Designing Arduino electronic shields: experiences from secondary and university courses, *IEEE Global Engineering Education Conference (EDUCON)*, Athens, Greece, pp 935-937

IOP Conf. Series: Materials Science and Engineering **280** (2017) 012044 doi:10.1088/1757-899X/280/1/012044

Extend Course for Product Designer in Digital Mobile Era

Fang-Lin CHAO[1, *], Tzu-Heng LIU[1] and Xian-Chun HUANG[1]

[1]Department of Industrial Design, Chaoyang University of Technology, Taichung, 436, Taiwan, R.O.C.

*flin@cyut.edu.tw

Abstract: Product design refers to a system of processes from confirming a product's specifications to product's structure. Form, technology, and needs must be considered simultaneously to ensure qualities. In recent years, with the advancement of smartphone technology, many products are connected with apps. Designers cannot exclude themselves from this new wave of the trend. In this article, household hydroponic products design is used as an example, to show the close relationship between digital mobile technology and product design in the contemporary world. Regularly measure the amount of liquid to be added is difficult for a consumer who has no professional experience. To facilitate the introduction of small-scale aquaculture systems into the home, we proposed sensor hardware combined with App software, measured EC and pH value and transmitted to the phone. The app can calculate and display the amount of added and control the amount of inserted through a Bluetooth connection. The physical design needs to take into account the connection between the electronic parts and the circuit board, and interface operation. Thus, not only the model of the product but also the user interface has to be integrated to show the product's quality completely. Besides, authors made reflection upon the necessity for adjustments for interdisciplinary courses under the changing digital mobile era. Also, under the current curriculum structure, possible teaching approach is expressed for extending student's feasibility.

1. Introduction

Product design which refers to a series of work from the product specifications to determine the product structure is an essential part of product development. Product designer must take into account the form, technology, and needs, through the appropriate product form to indicate a specific differentiated value. In recent years, the progress of mobile communication technology, product, and portable app combination; a designer in the front, must not be outside the wave. In addition to the shape of the product, the interface must be integrated to complete the quality of design. In this paper, the design of home-use water farming products is provided to show the design process today. To reflect the digital mobile product changes, cross-domain curriculum and the teaching content adjustment also offer to extend the existing curriculum structure.

MIT App Inventor is an informal online learning environment [1]. Xe (2016) analyzes the user skill; they can build complex applications with existing modules to enhance development capabilities [2]. Therefore, the computer science as the dominant learning approach will adjust in future. Teacher [3] explores the effectiveness of the high school student programming camp. Beginners can also build programs by dragging and dropping existing modules. The mobile application created from the student shows the smoothness of the teaching process.

A small program can also be useful; one can immediately feel the response. Learning fast is a high motivation for both new students and programmers. Students were challenged to evaluate a previously developed app and provide a suggestion for improvement. The achievements are very positive.

Microcontroller open-platforms have gained attention recently by educators and creative designer. Wide availability enriches the utilization of modern electronics and enables new possibilities in education [4, 5]. Related hardware (Arduino shields) can be plugged on top of the main Arduino board and extend its capabilities [6]. By combining electronic hardware, the designer had an opportunity to create and realized their concepts. A low-cost multi-meter and a digital temperature monitor designed and demonstrated in [7]. Arduino platform could combine with software and hardware in open-platforms, which provide teaching for the under-graduates industrial designer. Students have to master skills [8] associated with the Arduino include connecting Arduino circuit, learning Arduino IDE, and applying a variety of sensors (Figure 1). 3D printer technology was then introduced to support prototyping of hardware and design. Students acquired knowledge (programming, mechanical design, and electronic circuit) add to one's possessions extended their design abilities [9].

Figure 1. Apply Arduino to course of Industrial Design, making Hi-Fi concept models [8]

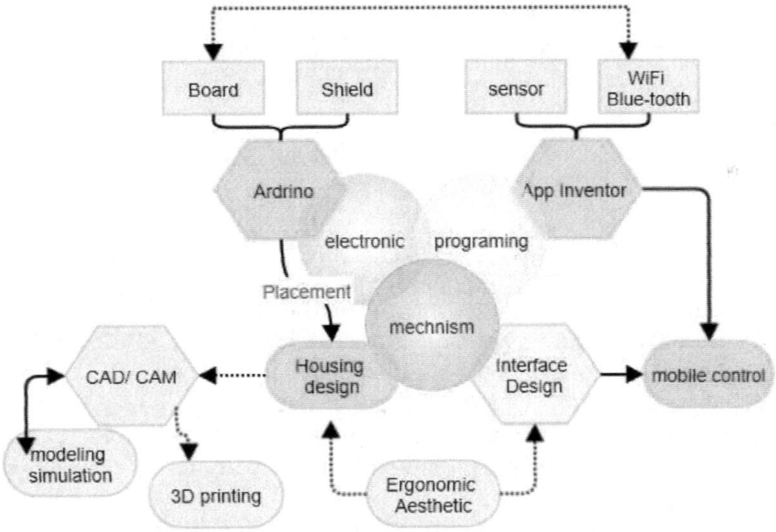

Figure 2. Digital mobile product and cross-domain associations

Refer to Figure 2; digital product design requires cross-link between structure (configuration), electronics (hardware), and information design (software and interface). In a digital mobile era, new generation designer need extend his knowledge. "cross-disciplinary" does not require the individual designer to complete the work alone, but understand system elements and their associations. App Inventor and Arduino (Figure 2) are building block based tools. The simplicity features enable designer in programming and electronic functional (mechanism) design. Designers can develop concepts firmly and realize ideas through cross-domain cooperation.

2. Case study: home-use hydroponics sensor module design

Growing plants in nutrient solutions without the use of soil is Hydroponics. The NFT system (Nutrient Film Technique) is quite popular with home hydroponic growers as well, because of its simplicity. NFT relies on a continuously running pump to deliver a low but continuous flow of nutrient over the plant roots. As with all re-circulating systems, the nutrient concentration will fall over time as the plants absorb the dissolved minerals [10] and transpiration from plant leaves. The electrical conductivity (EC) and pH of the solution need maintain by adding additional nutrients, and the whole solution will also change periodically.

Nutrient A and B contain different mineral ion. Nutrient A, B are required to be added separately according to the EC measured the concentration [10], and then look up table to add the appropriate amount of liquid. Evaporation also affects the pH balance. After a period of use, the user needs measured pH value to determine the amount of Phosphate to add to the target value.

2.1. Sensor module design: physical

Regularly estimate the amount of liquid to be added is difficult for the consumer who has no professional experience. People need a friendly procedure to facilitate the introduction of small-scale aquaculture systems into the home. The design proposal based on the sensor hardware combined with App software to transmit measured value to the phone. The sensor and the submersible motor placed in the tank with a bucket of pipe to control the amount of added (Figure 3).

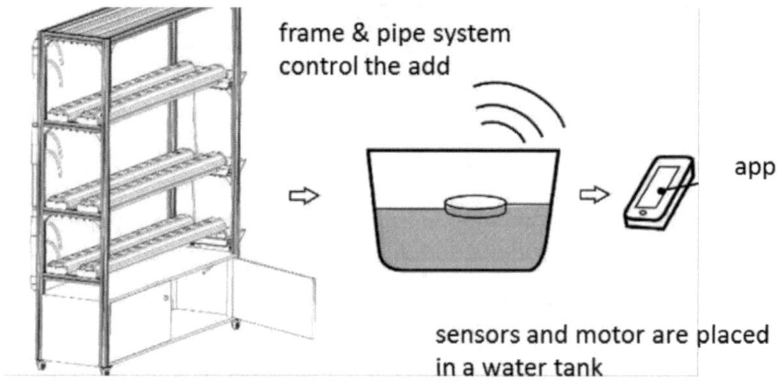

Figure 3. Proposal of the sensors and app.

In physical design, one take account the connection between the electronic parts and the circuit board. Figure 4 shows the relative position of the LCD, sensor and functional elements. Ergonomic and aesthetic concerns will make the operation more user-friendly. There are four circuit boards, including Arduino main board, EC, pH and temperature circuit. First, determine the internal circuit and parts configuration. Large interior space results in chassis volume increases which is not suitable for applications. Final circuit placement (Figure 4) laid parts in series for reducing the total wiring length.

We used 3D CATIA tool to construct three-dimensional appearance model tool, and the 1:1 parts model with 3D-printing equipment. The surface of each parts model is repaired to improve the surface quality (Figure 5). Then, parts and circuit board assembled within the main chassis for functional testing.

IOP Conf. Series: Materials Science and Engineering **280** (2017) 012044 doi:10.1088/1757-899X/280/1/012044

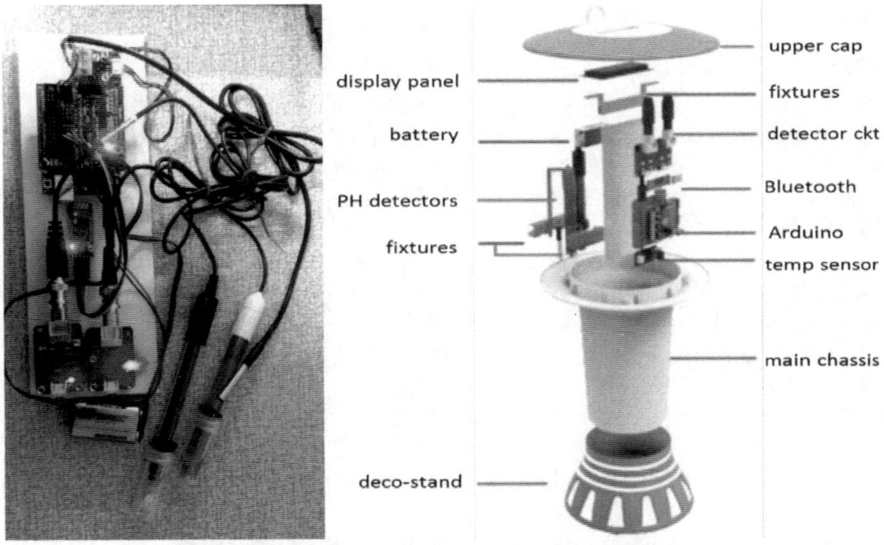

Figure 4. Internal component explosion diagram

Figure 5. Parts made by 3D-printing (before surface treatment)

2.2. *Sensor module mobile app design*

App design needs to proceed with user study, through the appropriate visual design and operation procedure; steps and configuration of three screens indicate in Figure 6. App Inventor programs blocks (Figure 7) represent sensing the measured value by Arduino through the Bluetooth connection. The interface layout was simplified to adapt the programming environment.

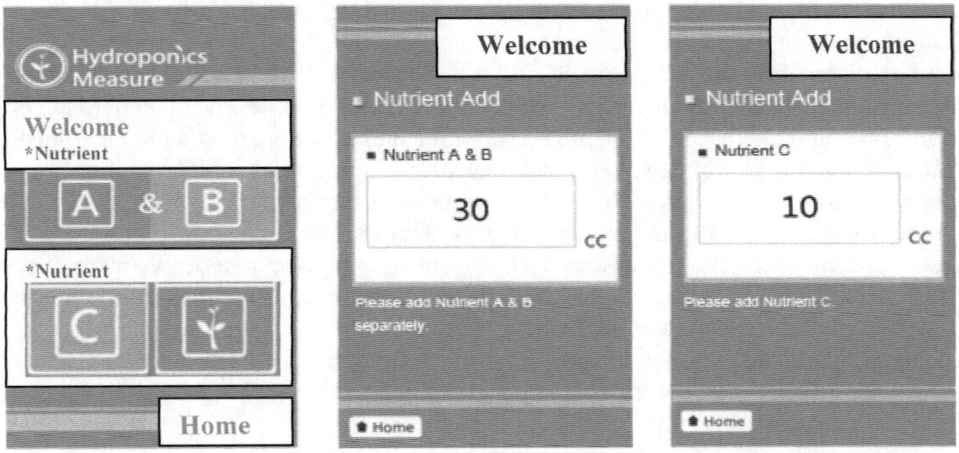

Figure 6. App interface design of measure, display, and nutrition setting steps

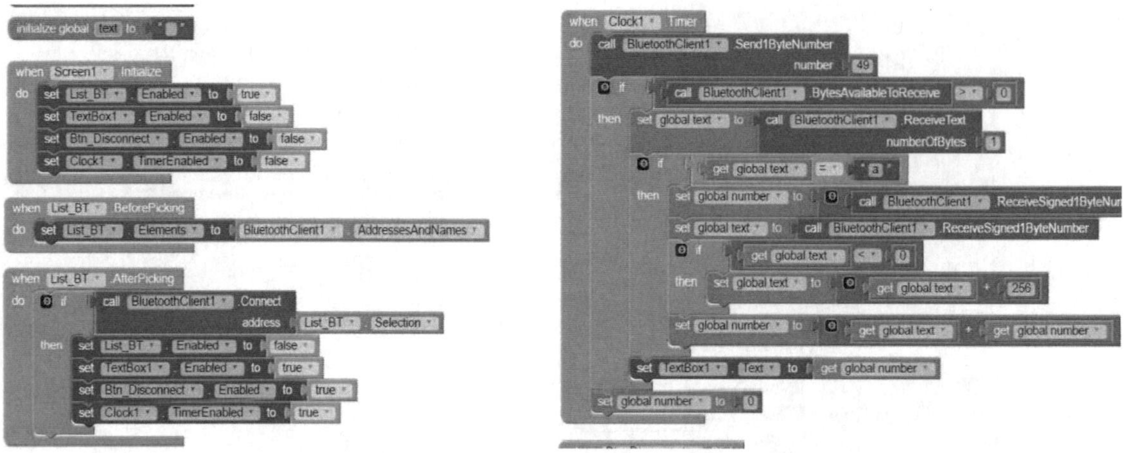

Figure 7. App Inventor program blocks of Bluetooth picking response (by YJ Wong)

3. Digital Mobile Product: Cross-Domain Curriculum and Experiences

Through the experience of hydroponics sensor module, we aware that digital mobile design needs close cooperation. To fulfill the lacking knowledge, we learn and communicate with programmer and circuit expert. While entering digital era, nevertheless, our design course had not been adopted. Understanding the relevance elements lead to inspiring through cross-domain collaboration. Therefore, designers require the extended cross-domain curriculum.

3.1. Current curriculum of Industrial Design

Industrial design needs digest different knowledge. There is separatet focusing between various departments. Some emphasize on design method and inspiration of ideas, and some on engineering and practical skills. Design professional developments mostly are studio based core courses such as basic design, product design, and thematic design [11]. Writing programs are beyond student's perspective at the moment. Programming learns systematic inference and teamwork, which used in all professions. The difficulty of learning leads low intention and willingness to provide such courses by the department.

3.2. Extend course for digital mobile product

Two extension course modules were proposed namely "introducing App Inventor" and "Arduino prototyping" for undergraduate junior level students each module spans 9-week period. Owing to the limited teaching hour available, For design students without the background in electronic and information technology, different materials and teaching methods are necessary. The teaching materials used in the courses are "Practice of App Inventor II" [12] and "Arduino prototyping projects." In a typical programming class, students have different levels of expertise. Progressive learning [13, 14] removes the detail of programming syntax, encourage students to transform ideas into animated stories and art right away. It is used accordingly to reduce initial barriers.

In second module "Arduino prototyping," the purpose is familiar with some essential projects. A student can utilize or extend it in the concept stage so that they wouldn't be limited by knowledge or experience to increase the integrity of creation. The instructor used nine experimental examples to help the students understand necessary technology. Project implementations include automatic light sensing, notification of water temperature in coffee brewers, and wet diaper alarm system…etc. Students learned circuit diagram, electronic symbols and practiced wiring of circuits. Secondly, the instructor covered a conceptual description of programming, and students should be able to use them repeatedly in their project.

3.3. Preliminary feedback on extending course:

Each module spans only 9-week period; it wasn't possible to allocate enough time in the implementation so that just some of the students finished all the programming blocks. Students will have to reduce their conceptual ideas and focus on the most critical elements and achieve them through programming. For example, when the number of screens increases, the "Tiny DB" command will need to exchange parameters between screens, and this often was troublesome for the beginner-level user. Below is an excerpt from the feedback from students which reveal positive responses.

This time I learned a lot of exciting stuff! In the beginning, I was very unfamiliar with it and thought that the connections between each coding block to be confusing. I didn't quite know what coding block to put in what. Afterwards, I slowly became familiar with the relationships between different coding blocks. At the very end, as I use my phone to scan the QR code and see the finished product, I thought, "That was cool!" If I have time, I'd like to explore more!

You could create different sparks in creativity, and give a product a modified form of expression or way of use. Although I have only learned a rudimentary bit, I still found much interesting application through it. I can start creating my own program, and this is fairly challenging.

4. Conclusions

For students without a background in information technology, cross-discipline learning will require different teaching materials and methods, and this still has to be structured out. By introducing progressive education, students' difficulties, in the beginning, can be reduced. As students use prototype tools in the concept phase, they sense the concepts that realizable. In the implementation phase, cross-discipline collaboration is needed to increase the integrity of the design. Designers should experience the inner workings of technology, to collaborate with others. Programming and Arduino prototyping offer a way to learn to think and do systematic inference, as well as team collaboration. These are new skills in which designers in a digital age can capture.

References

[1] Gestwicki, Paul, Khuloud Ahmad, 2011 App Inventor for Android with studio-based learning, *Journal of Computing Sciences in Colleges*, p 55-63.

[2] B. Xie and H. Abelson, 2016 Skill progression in MIT app inventor, *IEEE Symposium on Visual Languages and Human-Centric Computing*, Cambridge, p 213-217.

[3] Maloney, M. Resnick, N. Rusk, B. Silverman, The Scratch programming language and environment, 2010 *ACM Transactions on Computing Education*, **10**, Issue 4, no 16.

[4] Filomena Soares, 2017 A student-friendly way in teaching-learning, *25th Mediterranean Conference on Control and Automation (MED),* Valletta, Malta, p1295-1298.

[5] A. D' Ausilio, Arduino: a low-cost multipurpose lab equipment, 2011 *Behaviour Research Methods*, Vol. 44, Issue 2, p 305-313.

[6] Arduino shield list: http://shieldlist.org.

[7] A. Garrig, D. Marroqui, J. M. Blanes, R. Gutierrez, Designing Arduino electronic shields: experiences from secondary and university courses, 2017 *IEEE Global Engineering Education Conference (EDUCON)*, Athens, Greece, p 935-937.

[8] Fu Jiuqiang, Teaching method research based on Arduino platform, 2015 *Sixth International Conference on Intelligent Systems Design and Engineering Applications*, p 797-801.

[9] Anif Jamaluddin, Wireless Water Flow Monitoring Based On Android Smartphone, 2016 *2nd ICIMECE*, p 244-247.

[10] http://www.ecaa.ntu.edu.tw/weifang/Hort/Chap08.htm

[11] Yang Minying, Xiong Wanlai, Lin Shenghong, 2009 Study on the Students' Learning Status and Career-Related Issues in Industrial Design Department, *Journal of Design*, 8(3).

[12] Tsai Yi-Tan, 2015 ***App Inventor,*** (Flag Technology Co., Ltd)

[13] Wolber, David, et al., 2011 ***App Inventor,***(O'Reilly Media, Inc.)

[14] Gray J, Abelson H, Wolber D, et al., Teaching CS principles with app inventor, 2012 *Proceedings of the 50th Annual Southeast Regional Conference. ACM*, p 405-406.

Design of A Cyclone Separator Using Approximation Method

Bong-Su Sin[1], Ji-Won Choi[1] and Kwon-Hee Lee[1,*]

[1]Department of Mechanical Engineering, Dong-A University, 840 Handan-2-Dong, Saha-Gu, Busan 604-714, Korea

*Email: leekh@dau.ac.kr

Abstract. A Separator is a device installed in industrial applications to separate mixed objects. The separator of interest in this research is a cyclone type, which is used to separate a steam-brine mixture in a geothermal plant. The most important performance of the cyclone separator is the collection efficiency. The collection efficiency in this study is predicted by performing the CFD (Computational Fluid Dynamics) analysis. This research defines six shape design variables to maximize the collection efficiency. Thus, the collection efficiency is set up as the objective function in optimization process. Since the CFD analysis requires a lot of calculation time, it is impossible to obtain the optimal solution by linking the gradient-based optimization algorithm. Thus, two approximation methods are introduced to obtain an optimum design. In this process, an L_{18} orthogonal array is adopted as a DOE method, and kriging interpolation method is adopted to generate the metamodel for the collection efficiency. Based on the 18 analysis results, the relative importance of each variable to the collection efficiency is obtained through the ANOVA (analysis of variance). The final design is suggested considering the results obtained from two optimization methods. The fluid flow analysis of the cyclone separator is conducted by using the commercial CFD software, ANSYS-CFX.

1. Introduction

A cyclone separator is a device that makes mixture flow into the inlet, generates a cyclone, and separates particulates from a gas, air or liquid stream by the use of centrifugal force. Usually, the cyclone separator has a simple structure, thus it is easy to manufacture the product. The cyclone separator is widely used in various industries, such as those involved with removing dusts, collecting microparticles, cleaning equipment, biosensors, and air purification systems. The cyclone separator investigated in the present study is a device that removes brine particulates from the mixed brine and steam. This cyclone separator is used as a unit of the geothermal power plant [1].

The previous studies [2-7] have focused on the numerical analysis and design to improve the separator performance. Most researches have investigated the effect of the shape and process conditions such as the dimensions of each part of the cyclone, the shape of the inlet, flow rate, and/or temperature, on the performance of the cyclone separator. Starimand [2] has expressed the size of the cyclone separator, maximizing its efficiency. However, the research was based on the trial and error method. Robert [3] and Grane [4] tried to improve the collection efficiency of the cyclone by designing a multiport shaking device and parallel cyclone. Weddin [5] predicted the efficiency of the cyclone separator through theoretical model. Suh [6, 7] suggested the working formula for the channel design of a decanter-type centrifuge.

Most researches have relied on the empirical method or the simplified mathematical model to predict the performance of the cyclone separator. Thus, there are limits to finding an optimum design.

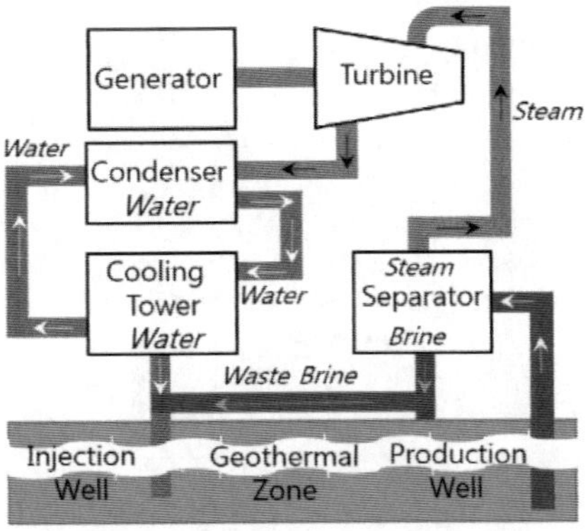

Figure 1. Geothermal power plant

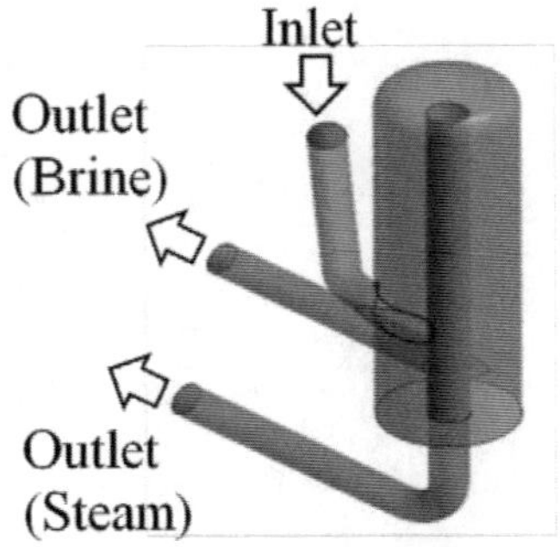

Figure 2. Initial CAD model of cyclone separator

In this study, six design variables related to its shape and the presence of filter are defined as the design variables to find an optimum design, maximizing the collection efficiency. The six design variables are selected by assuming that they are significant factors with respect to the collection efficiency. The collection efficiency in this cyclone separator is calculated by dividing a mass flow of the discharged brine at the outlet by the mass flow that entered at the inlet. The collection efficiency can be regarded as an index to measure the separation amount [1]. Thus, the collection efficiency is set up as the objective function in the optimization process.

This research adopts the DOE and the meatmodel based optimization technique to determine an optimum design of the cyclone separator. The L_{18} orthogonal array [8, 9] is introduced to perform the DOE. The effects of interactions in the orthogonal array are evenly distributed among the columns with exception of the relationship between columns 1 and 2 [9]. The effect on the collection efficiency by each design variable is evaluated, following 18 CFD analyses. The sensitivity information of each design variable to the collection efficiency is calculated quantitatively in an ANOVA table. After that, the optimum level that maximizes the collection efficiency of the cyclone separator is statistically predicted. However, the optimum is selected in the discrete values. Thus, the metamodel based optimization using the kriging is performed to overcome the difficulty. Finally, the optimum design considering the results obtained from two methods is suggested as the final design. All the CFD analyses in this research are performed by using ANSYS-CFX [10].

2. Initial Design and Numerical Analysis

2.1. Geothermal energy and initial design

In this study, we are interested in the types of geothermal power generation as shown in Figure 1. The schematic plot of the cyclone separator is shown as Figure 2. The initial design is determined on the basis of the shape which is widely and experientially used. When a steam-brine mixture flows into the main body through the inlet, a swirling motion down through the inner wall of the cyclone separator is formed. This swirling motion pushes the brine towards the wall by a centrifugal force, and the brine accumulates at the bottom of the system. In contrast, separated steam moves up from the bottom and discharges through the outlet resulting the brine to be separated from a steam-brine mixture.

An initial model is made by utilizing CATIA followed by a numerical analysis model using an ANSYS-workbench. The numerical model for CFD analysis is made of 27,003 grids and 131,169 meshes. To predict the collection efficiency of initial design, a commercial software ANSYS-CFX [10] is utilized. The velocity of fluid flow to the inlet is set at 33m/s and the mass ratio of the brine to steam is 1:9. The boundary condition at the outlet is set as an atmospheric pressure.

2.2 Analysis results of the initial design

The collection efficiency is defined as:

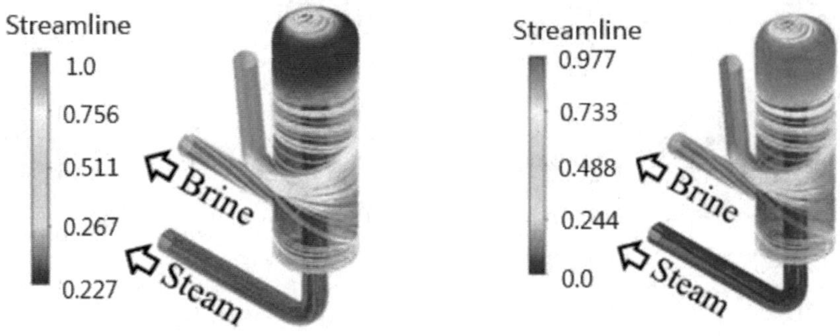

(a) Volume fraction of steam (b) Volume fraction of brine

Figure 3. Volume fraction result of initial design

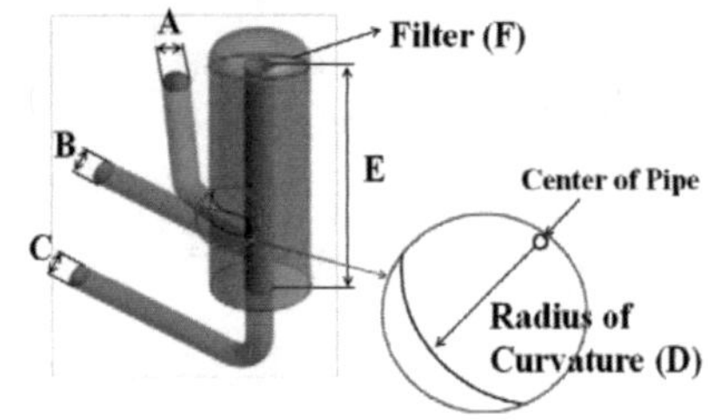

Figure 4. Six design variables

$$\eta = \frac{\dot{m}_{out}}{\dot{m}_{in}} \quad (1)$$

where $\dot{m}_{out}$ is the mass flow rate of the discharged brine at the outlet, and $\dot{m}_{in}$ is the mass flow rate that entered at the inlet. That is, it can be said that the larger the collection efficiency is, the better the performance of the cyclone separator is. The fluid flow inside the separator is investigated through CFD analysis.

The volume fractions and stream lines of steam and brine are represented in Figure 3. From Figure 3(b), it can be seen that the cyclone makes particulates of the brine gather on the top of cyclone body. Then, they are dropped to its bottom, and leave the cyclone body through the outlet. The collection efficiency of the initial design is calculated as 57.3% by the CFD analysis. The value is slightly larger than the usual collection efficiency of around 55.0% used in the plant.

3. Optimization Using DOE and Kriging Metamodel

3.1. Definition of design variables

The six design variables are defined to find a design that maximizes the collection efficiency. They are diameter of the inlet(A), diameter of the brine outlet(B), diameter of the steam outlet(C), radius of curvature of the nozzle which is inserted into the body from the inlet(D), length of the pipe extended to the steam outlet inside the body(E), and the presence of a filter helping the separation process inside the body(G). The design variables are represented in Figure 4. The objective in this study is to find an optimum value of each design variable for maximizing the collection efficiency.

3.2. Optimization using OA

The design of experiments using orthogonal array is adopted to select an optimum settings of the design variables to maximize the collection efficiency. For five design variables, a number of levels is set to three. The second level is set up as the initial value. The first and third levels are fixed by the lower and upper ones around the initial value, respectively. The levels of design variables for an orthogonal array are determined as shown in Table 1. Then, an appropriate orthogonal array is selected by considering the number of design variables and levels.

Table 1. Levels of design variables

level	*A*(mm)	*B*(mm)	*C*(mm)	*D*(mm)	*E*(mm)	*G*
1	637	637	637	2700	7758	O
2	708	708	708	3000	8167	X
3	778	778	778	3300	8575	

Table 2. Experiments using orthogonal array $L_{18}(2^1 \times 3^7)$

Exp. No	*G*	*A*	*B*	*C*	*D*	*E*	*error*	*error*	η
1	1	1	1	1	1	1	1	1	57.3
2	1	1	2	2	2	2	2	2	58.3
3	1	1	3	3	3	3	3	3	50.8
4	1	2	1	1	2	2	3	3	54.1
5	1	2	2	2	3	3	1	1	50.9
6	1	2	3	3	1	1	2	2	58.6
7	1	3	1	2	1	3	2	3	55.3
8	1	3	2	3	2	1	3	1	57.8
9	1	3	3	1	3	2	1	2	57.6
10	2	1	1	3	3	2	2	1	49.6
11	2	1	2	1	1	3	3	2	55.0
12	2	1	3	2	2	1	1	3	58.6
13	2	2	1	2	3	1	3	2	60.6
14	2	2	2	3	1	2	1	3	56.5
15	2	2	3	1	2	3	2	1	57.6
16	2	3	1	3	2	3	1	2	54.0
17	2	3	2	1	3	1	2	3	54.0
18	2	3	3	2	1	2	3	1	57.9

Table 3. ANOVA for collection efficiency

Factor	*S*	*Ø*	*V*	F_0
G	0.50	1	0.50	0.06
A	7.09	2	3.54	0.44
B	9.93	2	4.96	0.62
	17.41	2	8.71	1.09
D	32.28	2	16.14	2.01
E	45.13	2	22.57	2.82
error	48.10	6	8.02	

Table 4. Pooled ANOVA for collection efficiency

Factor	*S*	*Ø*	*V*	F_0	*F*(0.05)	*F*(0.1)
C	17.41	2	8.71	1.99	3.98	2.86
D	32.28	2	16.14	3.69	3.98	2.86
E	45.13	2	22.57	5.16	3.98	2.86
error	48.10	11	4.37			

The $L_{18}(2^1 \times 3^7)$ orthogonal array is selected for the DOE. The orthogonal array has an advantage that the interaction are distributed evenly across from third column to eighth column except first two columns [8, 9]. The variables, *A*, *B*, *C*, *D* and *E* are assigned to the second-sixth columns. The last two columns are filled with errors. The collection efficiencies obtained from each experiment are shown in Table 2. Based on the analyses results, the relative importance of each variable to the collection efficiency can be obtained through an ANOVA. Tables 3 and 4 show ANOVA results at 0.05 and 0.1 significance levels. For 0.05 significance level, it can be seen from the Tables 3 and 4 that the most sensitive variable is *E*, and the variables, *G, A, B* and *C* are insignificant. On the contrary, for 0.1 significance level, the variables, *E* and *D* are significant. From the ANOVA tables, the effect on the collection efficiency of each design variable can be found. Based on the ANOM (analysis of mean), the optimum design is selected as $A_2B_3C_2D_1E_1G_2$. The predicted collection efficiency at the optimum design is calculated as:

$$\eta_{pred} = \mu_f + a_i + b_j + c_k + d_l + e_m + g_n \tag{2}$$

where η_{pred} is a predicted value of the collection efficiency, μ_f is the overall mean, a_i, b_j, c_k, d_l, e_m and g_n are the main effects of *A, B, C, D, E* and *G* at *i, j, k, l, m* and *n* level, respectively. The predicted collection efficiency 61.7% is obtained by Eq. (2), which is the point estimator. The collection efficiency of 55.1~68.2% is obtained with the confidence interval 95%. The true value of the collection efficiency at the optimum levels determined from the ANSYS-CFX is 60.4%.

3.3. Optimization using Kriging model

Kriging model is one of metamodel utilized in complex design problem. Kriging is an interpolation method named after a South African mining engineer named D. G. Krige. In general, the response function $\eta(\mathbf{x})$ is represented as[11-14]

$$\eta(\mathbf{x}) = \beta + z(\mathbf{x}), \tag{3}$$

where $\mathbf{x}=[x_1\ x_2\ x_3\ x_4\ x_5]=[A\ B\ C\ D\ E]$, β is a constant, and $z(\mathbf{x})$ has a zero mean and variance σ^2 following the Gaussian distribution. For the presence and absence of filter(design variable G), each kriging model of $\eta(\mathbf{x})$ is built.

Let $\hat{\eta}(\mathbf{x})$ be an approximation model of the collection efficiency. When the mean squared error between $\eta(\mathbf{x})$ and $\hat{\eta}(\mathbf{x})$ is minimized, $\hat{\eta}(\mathbf{x})$ becomes:

$$\hat{\eta}(\mathbf{x}) = \hat{\beta} + \mathbf{r}^T(\mathbf{x})\mathbf{R}^{-1}(\boldsymbol{\eta} - \hat{\beta}\mathbf{i}). \tag{4}$$

where **R** is the correlation matrix, **r** is the correlation vector, $\boldsymbol{\eta}$ is the collection efficiency vector obtained from Table 2, and **i** is the unit vector.

Correlation matrix and correlation vector are defined as

$$R(\mathbf{x}^j, \mathbf{x}^k) = Exp\left[-\sum_{i=1}^{5}\theta_i\left|x_i^j - x_i^k\right|^2\right], \quad (j = 1,\ldots,18,\ k = 1,\ldots,18) \tag{5}$$

$$\mathbf{r}(\mathbf{x}) = [R(\mathbf{x},\mathbf{x}^{(1)}),\ R(\mathbf{x},\mathbf{x}^{(2)}),\ \ldots,\ R(\mathbf{x},\mathbf{x}^{(18)})]^T \tag{6}$$

where superscript of **x** is the row number in Table 2.

The unknown parameters θ_1, θ_2,…, θ_n are obtained from the following equation.

$$maximize \quad -\frac{[18ln(\hat{\sigma}^2) + ln|\mathbf{R}|]}{2}, \tag{7}$$

where θ_i (i=1,2,3,4,5) > 0.

The optimization process suggested in this section is summarized in Figure 5. At the first step, the sample points are defined, which are the same as Table 2. From 18 analysis results, the kriging metamodel of the collection efficiency represented as Eq. (4) is built. As the final step, the sequential

IOP Conf. Series: Materials Science and Engineering **280** (2017) 012045 doi:10.1088/1757-899X/280/1/012045

quadratic programming built in VisualDOC [15] is adopted to calculate an optimum design. For the presence case of filter, the predicted collection efficiency is calculated as 58.6%. On the contrary, for the absence case of the filter, the predicted collection efficiency is calculated as 60.6%, which is the 13th response of 18 responses in Table 2. Because the Kinging is an interpolation method, a prediction at the *i*-th sample point, the estimator gives the *i*-th response. The optimum parameters of the kriging model and the optimum design variables are summarized in Table 5. The volume fraction at the optimum design are represented as Figure 6.

4. Conclusions

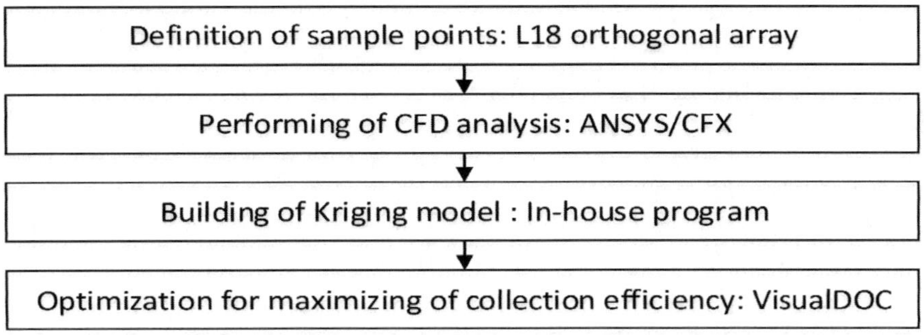

Figure 5. Design process using metamodel

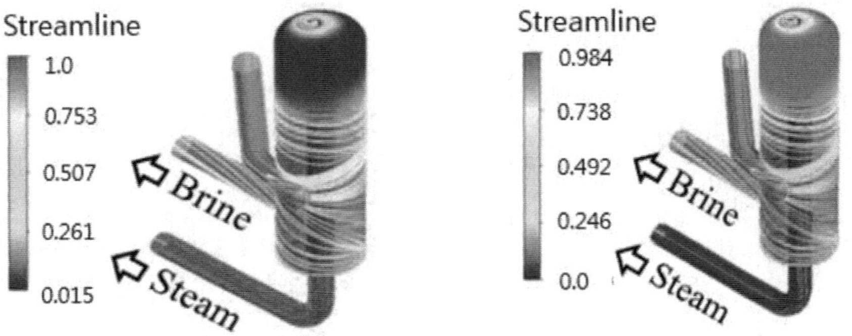

(a) Volume fraction of steam (b) Volume fraction of brine

Figure 6. Volume fraction result of optimum design

Table 5. Optimum parameters and optimum design values (absence of filter)

	A	*B*	*C*	*D*	*E*
Optimum parameters(θ_i in Eq. (5))	32.8	6.7	18.1	0.1	8.2
Optimum design variables(mm)	708	637	708	3300	7758

In this study, the collection efficiency of the cyclone separator for the plant is calculated through the numerical analysis. Furthermore, the optimum design for maximizing the collection efficiency is suggested by defining the design variables and introducing the DOE scheme and the Kriging interpolation method. Based on the ANOVA, the relative importance of each variable to the collection efficiency is investigated. From the ANOVA, it can be seen that the most sensitive variables are *E* and *D*, and the variables, *A, B, C* and *G* are relatively insensitive. Finally, the optimum design is determined based on the results obtained from two optimization methods. The collection efficiency of the suggested optimal design is increased up to 60.6%, comparing with the 57.3% of the initial design.

IOP Conf. Series: Materials Science and Engineering **280** (2017) 012045 doi:10.1088/1757-899X/280/1/012045

Acknowledgements
This research was financially supported by Valve Center from the Regional Innovation Center (RIC), Program of Ministry of Trade, Industry and Energy (MOTIE).

References
[1] Sin BS, Praveen K S and Lee KH 2013 Numerical analysis of the cyclone separator for geothermal plant *Proc. of the Korea Society of Mechanical Engineers(Busan)* 23~24 May pp 188-9
[2] Stairmand CJ 1951 The design and performance of cyclone separators Trans. Instn. Chem. Engrs. **29**, 356-83
[3] Robert ED Jr 1990 A model for the prediction of the collection efficiency characteristics of a small, cylindrical aerosol sampling cyclone *Aerosol Sci. Technol.* **12**(4), 1055-66
[4] Grane RL, Barbaris LN and Behrouzi P 1992 Particulate behavior in cyclone separators with secondary gas extraction *J. Aerosol Sci.* **23**, 765-8
[5] Wedding JB, Weigand MA and Carney TA 1982 A 10μm cut point inlet for the dichotomous sampler *Environ. Sci. Technol.* **16**, 602-6
[6] Suh YK 2003 Channel design of decanter-type centrifuge(Ⅰ)–particles' suspension and the channel size *J. Korean Society Prec. Eng.* **20**(10), 148-55
[7] Suh YK 2003 Channel design of decanter-type centrifuge(Ⅱ)–particles' sediment and bowl length *J. Korean Society Prec. Eng.* **20**(11), 108-16
[8] Roy RK 2001 *Design of Experiments Using the Taguchi Approach* (New York: Wiley & Sons)
[9] Peace GS 1992 *Taguchi Methods-A Hands-On Approach* (Massachusetts: Addison-Wesley Publishing Company)
[10] ANSYS Inc. *ANSYS CFX 11.0 User`s Manual*
[11] Guinta A and Watson L. 1998 A comparison of approximation modeling techniques: polynomial versus interpolating models *Proc 7th AIAA/USAF/NASA/ISSMO Symp on Multidisciplinary Analysis and Optimization* **2**, 392-440(AIAA-98-4758)
[12] Fang KT, Li R and Sudjianto A. 2006 *Design and Modeling for Computer Experiments* (Chapman & Hall/CRC)
[13] Kim GY, Han SH and Lee KH 2014 Structural optimization of a knuckle with consideration of stiffness and durability requirements *Sci. World J.* Vol. 2014 Article ID 763692
[14] Jang BH and Lee KH 2014 Analysis and design of a ball joint, considering manufacturing process *J. Mech. Eng. Sci.* **228**(1), 146-51
[15] VR & D, Inc. 2004 *VisualDOC User Manual Ver. 5.0*, (Colorado Springs: VR & D)

Compound Half-Backed Weave Design For Digital Jacquard Fabric

Meng Zhang[1] and Jiu Zhou[1,*]

[1]Zhejiang Sci-Tech University(Engineering Research Center of Clothing of Zhejiang Province), Hangzhou, Zhejiang, 310018, China

*E-mail: zhoujiu34@126.com

Abstract: Based on layered-combination design mode and compound structure, this paper presents a design method, named compound half-backed weave in order to achieve innovating weave structure and surface effect of fabric. This design method includes primary weaves chosen, half-backed technical points set up and half-backed weave databases established. The fabric produced using compound half-backed weave designed by this method can exhibit a unique half-backed effect that only half of the threads on the fabric surface remain in a state of being covered by adjacent wefts. Compound half-backed weave can not only meets the design need of jacquard fabric with different digital images and effectively improves the efficiency of structural design, but also puts forward new theory and method for innovative design of digital jacquard fabric.

1. Introduction

Jacquard woven fabric is made from dyed yarns and different weaves [1, 2]. It exhibits complex pictorial patterning effects on the face of fabrics and is one of the most value-added textile products [3]. The design process of traditional jacquard fabric is divided into three steps. First, a colourful pattern is drawn freehand. Then, the weave structure of jacquard fabric is designed to copy the effect of original pattern. Lastly, a weaving pattern is exported in jacquard CAD to control a jacquard loom and the jacquard fabric with the freehand pattern would be produced [4, 5]. Since consumers selected fabric firstly depending on their visual images on the face of fabric [6-8], it is important for designers to produce jacquard fabric with exquisite effect and abundant colours. However, the design cycle must last long. The design of layered-combination design mode and compound structure make digital images can be used directly in jacquard fabric design and shorten the pattern and technical designing time [9, 10].

Previous research works of layered-combination design mode, of which the key technology is the innovation of structure design, indicated that the compound structures can be divided into three basic types (figure1): backed, unbacked and half-backed [11]. The backed structure is a kind of compound structure being employed normally on design of conventional weft-backed fabric. The weft with longer float length cover the other one and only the colour information of the weft with longer float length is on the fabric surface [12, 13]. The unbacked compound structure, which can only be used on layered-combination design mode, is capable of expressing mixture colours accurately. There is no relationship among adjacent two wefts and the colour information of all wefts is on the fabric surface [4]. The theoretical and practical researches of unbacked compound structure, so called full-colour compound structure, have been introduced in the previous studies [14]. Several innovative design

IOP Conf. Series: Materials Science and Engineering **280** (2017) 012046 doi:10.1088/1757-899X/280/1/012046

works of digital jacquard with millions of mixed colours and various pattern motifs have been produced using full-colour effect compound structure, including simulation effect, superimposition effect, double-face effect, shot-effect, etc [14]. However, the half-backed compound structure is that each weft could be covered only by either the upper or the lower adjacent weft. An innovative method of compound structure, here called the compound half-backed weave design, based on the structural characteristics of the compound half-backed weave, is presented in this paper.

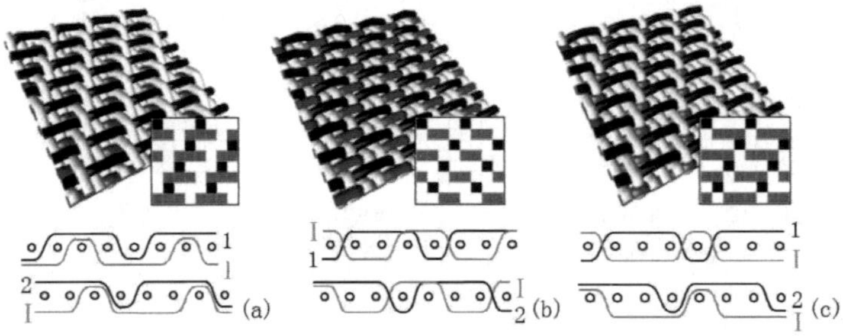

Figure 1. Compound structures design with two wefts: (a) Backed, (b) Unbacked, (c) Half-backed.

2. Design Principle and Method of compound half-backed weave

2.1. Design Principle of compound half-backed weave

Similar to the design method of compound full-colour weave, according to the feature of compound structures, the single weaves used for combination are called primary weaves [4]. Design principle and process of compound half-backed weave are as follows: in the wake of primary weaves chosen; half-backed technical points are set up for each of them; two series of shaded gamut weaves, named basic weaves and joint weaves, are established by enhancing weave points on two primary weaves without destroying half-backed technical points (figure 2). The combination of any single weave from basic weaves and another one from joint weaves has the characteristic of compound half-backed weave that do well out of the half-backed technical points. The half-backed technical points, resemble a kind of weave, are assembling of the weave points that make the combination of single weaves meet the requirements of half-backed effect. The fabric produced by this compound structure could express unique half-backed effect in which only half of the threads were covered by each other.

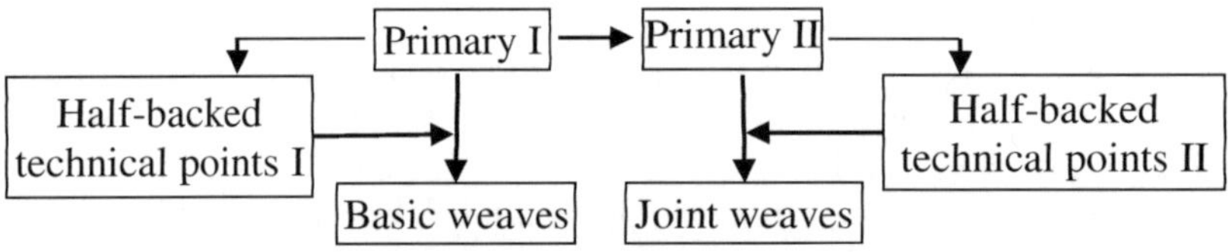

Figure 2. Design process of compound half-backed weave.

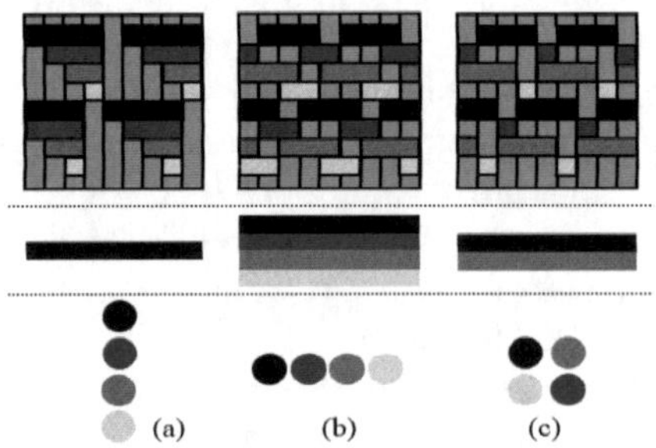

Figure 3. Mixed colouring with four wefts compound structure: (a) Backed, (b)Unbacked, (c) Half-backed.

Colouring effect on fabric surface is colour mixing of limited opaque colours. People only can get the mixing effect but the single colour from a normal distance [4]. The colour mixing is realized by interweaving warp and weft, and the colour proportion of yarn is decided by float length, which is to say, the colour proportion of yarn is decided by weave structure. Since compound half-backed weave is consist of at least two single weaves, even wefts are needed in half-backed effect jacquard design. With two wefts, the surface effect of backed structure is the same with the one of compound half-backed weave, so four wefts are the least in the application of compound half-backed weave. The colour mixing principle with four wefts is shown in figure 3. Weave structures, surface visual effect and covered relationship among wefts are shown from top to bottom.

2.2. Design Method of compound half-backed weave

According to the design process (figure 2), compound half-backed weave design method can be divided into four steps: choosing primary weaves, designing half-backed technical points and establishing compound half-backed weave.

Choosing primary weave I is similar to the compound full-colour weave design method, and twill and sateen both can be primary weaves. But primary weave II must be fixed and immovable when primary weave I is selected in the method presented here. 8-thread sateen, taken as an example, figure 4 shows the selection of primary weave II: primary weave I themselves, the derivative weaves of the primary weaves I, and the combinations of primary weave and derivative weaves are listed from top to bottom. Primary weave I is 8-thread weft-faced sateen, which provides seven weave variations through moving the starting point of the primary weave I. The eight derivative weaves (including the primary weave I) can generate eight combinations with the primary weave I in the form of 1:1 along weft direction. Only two combinations can satisfy the half-backed structural characteristics that each weft in a compound weave should in a covered relationship with at least one juxtaposed weft. When primary weave I is selected, primary weave II must be a weave shifted up one weft yarn from primary weave I or be a same weave with primary weave I. To control an appropriate length of floats on fabric surface, the best repeat range of primary weave should be between 5 × 5 and 48 × 48.

In compound half-backed weave design part, 16-thread sateen served as an example. As shown in figure 5, two primary weaves and their half-backed technical points were established. Primary weaves were selected using the same weave repeat of warp and weft. Primary weave II is the one shifted up one weft yarn from primary weave I. Then reverse the interlacing points on primary weaves I. The two interlacing points were strengthened upwards and downwards simultaneously, and half-backed technical points I were set by reversing the previous weave points again. Half-backed technical points II were set using the same method.

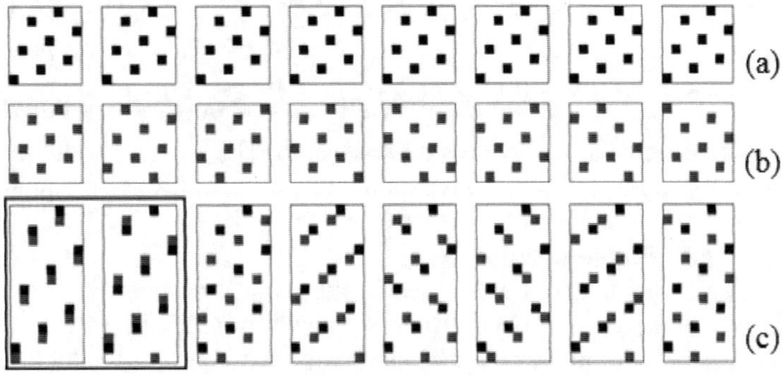

Figure 4. Selection of primary weave II: (a) Primary weave I; (b) Derivative weaves; (c) Combinations.

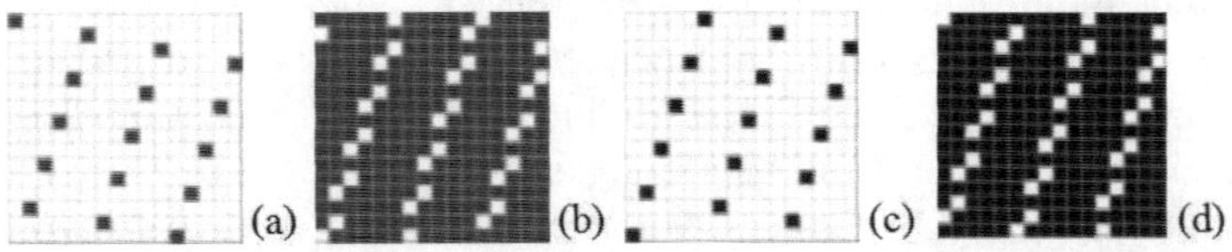

Figure 5. Primary weaves and their half-backed technical points: (a) Primary weave I, (b) Half-backed

technical points I, (c) Primary weave II, (d) Half-backed technical points II.

After confirming two primary weaves and their half-backed technical points, a series of gamut weaves, called basic weaves, were designed by reinforcing interlacing points step by step without destroying the half-backed technical points I upon the primary weave I. Another set of gamut weaves, called joint weaves, were designed in the same way upon the primary weave II. Figure 6 shows the impound half-backed weave when the enhancement number of shaded interlacing points is 16.

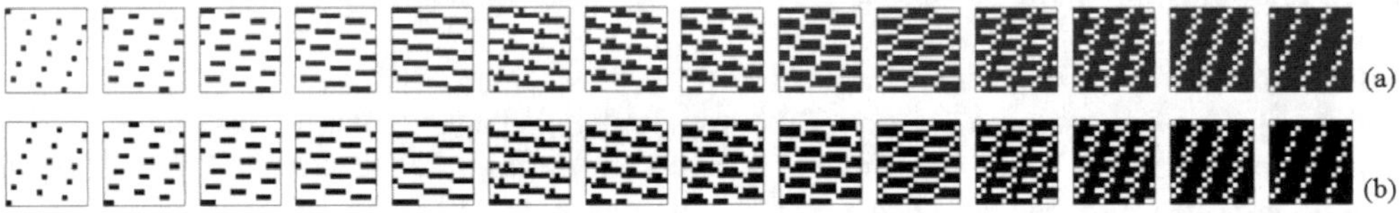

Figure 6. Compound half-backed weave: (a) Basic weaves, (b) Joint weaves.

These two series of weave can be used to design digital jacquard fabric with two wefts directly, and they also can be used to design fabric with four wefts by combination and separation (figure 7): basic weave and joint weave are combined in the form of 1:1 along weft direction, then disassemble the combination in form of 1:1:1:1 along weft direction into four weave databases used for jacquard with four wefts design. To paraphrase this method, Basic weaves can be turned into two weave databases, here called basic weave database I and basic weave database II designed by disassembling the odd wefts and even wefts from basic weaves. Similarly, joint weave database I and joint weave database II also can be established via the same way from joint weaves (figure 8).

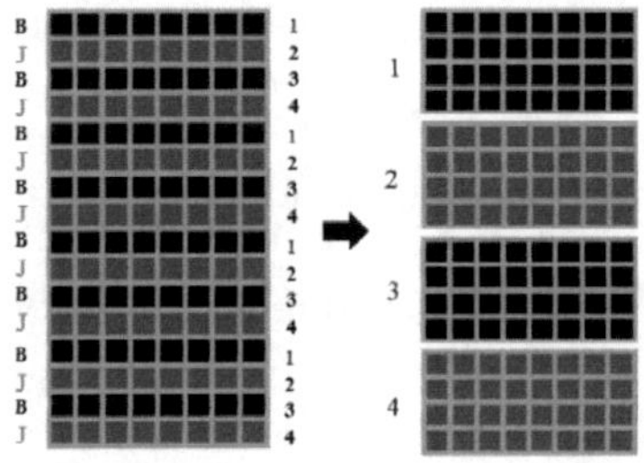

Figure 7. Combination and separation of weave databases.

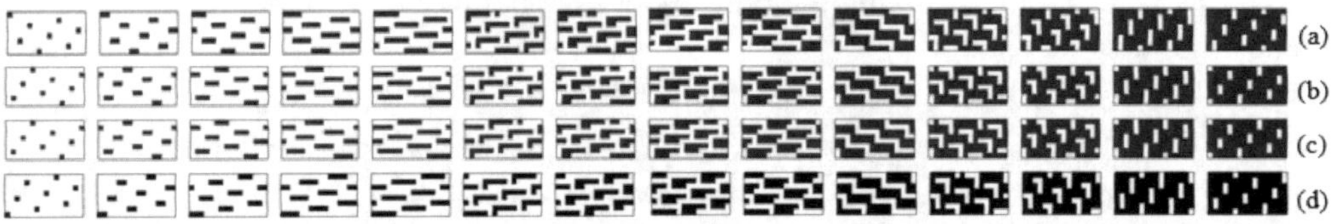

Figure 8. Compound half-backed weave databases: (a) Basic weave database I, (b) Basic weave-database II, (a) Joint weave database I, (b) Joint weave database II.

3. Application of compound half-backed weave

The key technical parameters of fabric are detailed in the technical specification table (table 1). A set of 22.2/24.4dtex × 2 warp silk yarns in white and four sets of 22.2/24.4dtex × 5 weft silk yarns in earthy yellow, grass green, black, and orange respectively are used to produce half-backed jacquard fabric in this paper.

In the design process of half-backed jacquard fabric, firstly pattern design need to be separated into four colour layers. Since the brightness of gray scale serves as the basic parameter for correspondence of weaves13 and one to one corresponding principle is indispensable for jacquard fabric, gray scale processing is necessary to taken in Photoshop to reduce the colour gradations less than or equal to the weave numbers of database. The first single-layer structure can be formed by the one-to-one correspondence between the first layer's gray gradations and basic weave-database I in jacquard CAD system. The second, third and fourth single-layer structures can be formed by joint weave-database I, basic weave-database II and joint weave-database II, respectively. Then, the compound fabric structure is combined by these four single-layer structures in 1:1:1:1 order along weft direction. Figure 9 illustrates that the one with four layers.

Table 1. Technical specification of half-backed fabric with four-layer fabric structure

	Parameters	
	Warp	**Weft**
Materials	22.2/24.4dtex × 2 (white)	22.2/24.4dtex × 5 (cyan/magenta/yellow/black)
Density	114 threads/cm	100 threads/cm
Composition	Silk 100 %	
Weave structure	16-thread gamut weaves, 14 gray grades	
Design repeat	2400 needles × 2624 fillings	
Pattern repeat	21 cm (width) × 26.4 cm (length)	

Figure 9. Digital image with four layers.

Figure 10. Compound fabric structure and half-backed jacquard fabric: (a) Compound fabric pattern; (b) Fabric effect; (c) Details of fabric.

The jacquard fabric with half-backed effect can be produced directly on a loom controlled by the compound fabric structure. Figure 10 presents the design effect of compound fabric structure, real fabric and detailed smooth shading effect. Only half of the threads are covered at random on the fabric surface due to the existence of half-backed technical points. Each weave database has 14 weaves means every single layer has 14 colour, so the fabric with four layers can express 14 × 14 × 14 × 14 = 38416 colour mixing effects on fabric surface that only can be realized in the layered-combination design mode. The other designers cannot copy the pattern and colour effect of the fabric accurately only on the basis of the surface features of final fabric. Therefore, such method proposed in this paper can help designers keep off the piracy in the technical level, and promote the value-added of digital jacquard fabric greatly.

4. Conclusion

This study of compound half-backed weave is a kind of design researches in the field of digital structure innovation of jacquard textile under the layered-combination design mode. The design of compound half-backed weave should be approached on the base of gamut weaves and its half-backed technical points. Once the weave databases were set up, it is timesaving for designers do not need to design weave-databases every times and it can be used in every design procedure. The fabric designed with compound half-backed weave design is capable of expressing random half-backed colour mixture effect and can present million-level mixed colours by only four sets of weft. It is much easier than the traditional method that using eight sets of wefts. And the fabric that produced in this way only can be realized under the layered-combination design mode, so it is effective for protecting the designers' copyright. It is envisaged that the fabrics produced by compound half-backed weave meet well the technical requirement of balanced interlacement and pose no problem in mass production.

Acknowledgement
Authors are thankful for providing the funds by key laboratory project of the ministry of culture of China (2017016).

References
[1] Mathur K, Hinks D, Seyam A F M and Donaldson R A 2009 Towards automation of colour/weave selection in jacquard design: Model verification Colour Research & Application p 225
[2] Osaki K 2003 High quality colour reproduction on jacquard silk textile from digital colour images AUTEX Research Journal p 173
[3] Kim S and Lee J S 2011 New production method for a plain weave figured fabric Fibers and Polymers p 137
[4] ZHOU J and Frankie Ng 2007 All-colouring compound construction of digital jacquard fabric Journal of Textile Research p 59
[5] Weinsdorfer H 2004 50 Years of weaving technology International Textile Bulletin p 54
[6] Deng Z and Wang L 2010 Enhanced visualization of simulated woven fabrics Fibers and Polymers p 531
[7] Na Y 2009 Fashion design styles recommended by consumers' sensibility and emotion Human Factors and Ergonomics in Manufacturing & Service Industries p 158
[8] Han Y, Zheng D, Baciu G, Feng X and Li M 2012 Fuzzy region competition-based auto-colour-theme design for textile images Textile Research Journal p 638
[9] ZHOU J and Frankie NG 2006 Innovative principle and method of design of colourful digital jacquard fabrics Journal of Textile Research p 6
[10] ZHOU J and ZHANG M 2015 Design of double-face jacquard fabric based on full-colour structure Journal of Textile Research p 39
[11] ZHOU J and JIANG Y J 2009 Innovative design of jacquard fabrics based on digital technology Journal of Textile Research p 53
[12] LUO B 2009 Development and production of jacquard-decorating fabric with overlap weft Shanghai Textile Science & Technology p 38
[13] JIANG X 2015 Design of double face fabric with pattern effect based on double weft Progress in Textile p 43
[14] Frankie Ng and ZHOU J 2009 A Study on Figured Double-face Jacquard Fabric with Full-colour Effect Textile Research Journal p 930

IOP Conf. Series: Materials Science and Engineering **280** (2017) 012047 doi:10.1088/1757-899X/280/1/012047

Research on the Design of Visually Impaired Interactive Accessibility in Large Urban Public Transport System

Weiru Zhang[1,*]

[1]Wuhan University of Technoogy No. 122 Luoshi Road, Hongshan district, Wuhan, Hubei Province

*Wangn_rur@yeah.net

Abstract. In medieval times, due to people's reliance on belief, public space of Christianity came into being. With the rise of secularization, religion gradually turned into private belief, and accordingly public space returned to private space. In the 21st century, due to people's reliance on intelligent devices, information-interactive public space emerges, and as information interaction is constantly constraining the visually impaired, public space regressed to the exclusive space of limited people[1]. Modernity is marked by technical rationality, but an ensuing basic problem lies in the separation between human action, ethics and public space. When technology fails to overcome obstacles for a particular group, the gap between the burgeoning intelligent phenomena and the increasing ratio of visually impaired is also expanding, ultimately resulting in a growing number of "blind spots" in information-interactive space. Technological innovation not only promotes the development of the information industry, but also promotes the rapid development of the transportation industry. Traffic patterns are diversifying and diverging nowadays, but it's a fatal blow for people with visually disabilities, Because they still can only experience the most traditional mode of transportation, sometimes even not go out. How to guarantee their interactive accessibility in large urban public transport system right, currently, is a very important research direction.

1. The Necessity of Concern for Information Interaction of the Blind

Interactive design has undergone only a few decades of development, which, however, is still increasing at an astounding pace. Science and technology have boosted the development of design, and promoted interactive design's further innovation towards information interaction direction. In 2008, IBM's CEO Peng Mingsheng first proposed the idea of "Smarter Planet", which means to adopt the new generation of information technology in combining the Internet with the Internet of things for the sake of realizing the integration of human society and physical system. Numerous large cities are also scrambling to join the ranks of smart city: "Data drives the world, software defines the world, automation is taking over the world, and the construction of smarter cities will become a vital carrier of the next wave to fuel IT world.[2]" The times are changing, whereas design is changing to change people's life, but there are actually a large number of people who haven't felt excited about such designs. In such a smart city, such urban public space, or even such public transportation system space, the visually impaired in ever-increasing numbers, do not have access to the humane merits brought by information interaction. The interaction of information, to a large extent, has left them out of account. Especially in a disturbing environment beyond hearing aid's availability, visual inability of the blind tends to intensify the circumstance that they cannot smoothly experience the conveniences brought by information interaction.

1.1. From the Perspective of Data

There is a depiction in the novel Blindness as follows: "As the amber light came on at a busy crossing, one of the cars stopped, and the panic-stricken driver desperately waved his hands, screaming, "I am blind! I am blind! I am blind!" Later, blindness spread rapidly, and the entire city plunged into an unprecedented catastrophe"[3]. Although our real life is not as dramatic as a novel, blindness does exist in the real world. A large number of references have cited the following data: "Dr. Thylefore, planning director for WHO prevention of deafness and blindness in Geneva, states that, China is a country with the most blind people in the world at 5 million or so, accounting for 18% of the world's blind population, and around 450,000 people are born blind every year in China." Such a huge number was virtually the statistic ten years ago. The World Health Organization updated the data five years later: In 2010, there are 8.288 million blind people, 67.24 million low-vision people, and 755.12 million people with visual impairment in China, which is the sum of the above two (see Figure 1)[4]. According to data graphs provided by ICEVI (see Figure 2), there are more than 6100 blind groups, 44,300 low-vision groups and 55,400 visually impaired groups out of every one million people in China. To express it with more intuitive data: there is one blind person, five low-vision persons and six visually impaired persons in every one hundred people.

Table 3. **Number of people visually impaired and corresponding percentage of the global impairment by WHO Region and country, 2010**

		Blindness	Low vision	Visual Impairment
WHO Region	Total population (millions)	No. in millions (percentage)	No. in millions (percentage)	No. in millions (percentage)
Afr	804.9 (11.9)	5.888 (15)	20.407 (8.3)	26.295 (9.2)
Amr	915.4 (13.6)	3.211(8)	23.401 (9.5)	26.612 (9.3)
Emr	580.2 (8.6)	4.918 (12.5)	18.581 (7.6)	23.499 (8.2)
Eur	889.2 (13.2)	2.713 (7)	25.502 (10.4)	28.215 (9.9)
Sear (India excluded)	579.1 (8.6)	3.974 (10.1)	23.938 (9.7)	27.913 (9.8)
Wpr (China excluded)	442.3 (6.6)	2.338 (6)	12.386 (5)	14.724 (5.2)
India	1181.4 (17.5)	8.075 (20.5)	54.544 (22.2)	62.619 (21.9)
China	1344.9 (20)	8.248 (20.9)	67.264 (27.3)	75.512 (26.5)
World	6737.5 (100)	39.365 (100)	246.024 (100)	285.389 (100)

Figure1. Visually impaired percentage by WHO

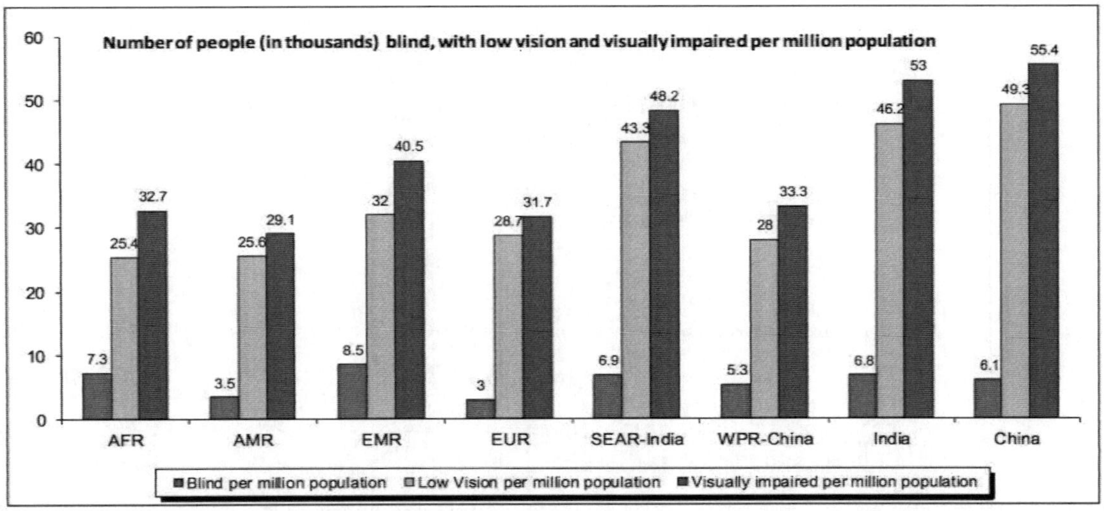

Figure 2. Numbers of people visually impaired per million populations

Meanwhile, Dr. Thylefors stressed once again in 2010's report, it is estimated that there are 450,000 people going blind each year in China, which indicates that there is a chance of blindness almost every minute in China[5]. Although WHO does not offer the latest data, it is thus estimated that the number of blind people in China will increase fourfold in 2020 at more than 50 million. Despite the ever-developing modern technology, blindness has also been "spreading" as depicted in the novel. The blind just can't wait to count on designers to gradually discover their needs, but most stimulate people's pressing concern about the visually impaired in proportion to the increasing number. In the era of information, the traditional way of interaction is gradually being replaced. A wide variety of intelligent devices have unfolded a colorful world of interactive experience to sighted people, but still leave endless darkness and helplessness to the visually impaired.

1.2 From the Perspective of Accessibility

According to Zygmunt Bauman's analysis, the reason for contemporary design's difficulty in drawing public attention to the disadvantaged despite their freedom of choice, the disadvantaged in the present context no longer possess that self-consciousness and sense of historical mission of the traditional working class, hence unattractive to the intellectuals, who, though take pity on these people, have always learned to restrain themselves and are reluctant to associate with the group and their misery[6]. Therefore, when an absolute interactive process hangs high over lives of the visually impaired, interaction here will lose its bite. Homogeneity gradually erases the diversity of design, which is then replaced by the philosophy of homogeneous universal design, and a convergence towards of information which has blurred the public space of the large number of the tendency of blind. For another simple example, public space of public transportation system in the 1980s was equipped with staff specializing in selling tickets, and such pure human-human interactions, to some extent, can better enhance the accessibility of interactive feedbacks from the visually impaired.

2. Interactive Space in Metropolis Public Transportation System under the Development of Information Technology

The public traffic space discussed in this paper refers to the specific space provided by traffic management, social development, technological progress and emergence of various traffic means for these social activities after the formation of transportation system. Currently, the composition of urban traffic means is divided into the following aspects (see Figure 3). Based on existing research on the blind in the city, excluding subsidiary and special traffic means such as high-speed rail, ferry and airplane, this paper has laid emphasis upon public transportation system space, metro system space and individual traffic space (urban pedestrian space).

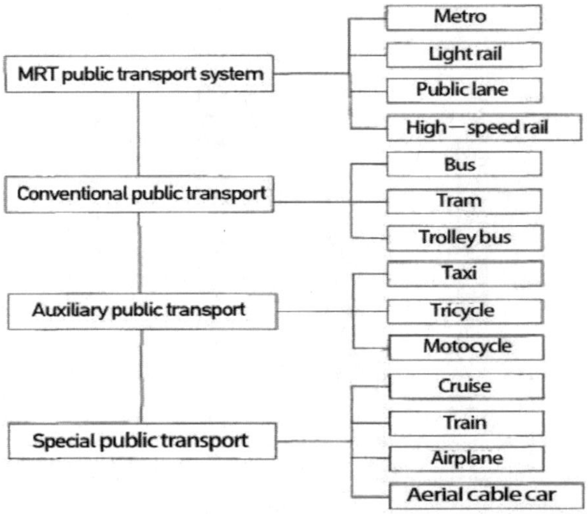

Figure 3. Currently composition of urban traffic

The interaction in public traffic space can be defined in a broad or narrow sense: in the broad sense, it refers to study on people's psychology for the purpose of constructing a stable, operable general environment and achieving the abstract nervous systematization of information exchanges and communications for the blind. On the other hand, interactive space refers to narrow interactive space of a variety of subsidiary functional zones, like the reception and waiting zone, which arises out of the continuous development of information technology and personalized management on the basis of a transportation system. The paper divides the representations of narrow public interactive space into the following three types: ticketing space (including online ticket), waiting zone, and traffic means in itself.

3. Classifications and Analysis of Interaction Modes of the Blind in Urban Public Transportation System

Public space develops from simple human-human interaction to complex human-computer interaction. When it comes to public space, interactive behaviors are no longer the aggregation of numerous individuals but a hybrid space made up of all sorts of visual and auditory touch points. The birth of the Internet in 1969 has launched the world's first deeply virtual public interactive space, and successfully blocked the blind out. The emergence of intelligence has altered the definition of public space for the blind and their cohesive relations with urban space. Catherine Grout describes modern public space in her Pour de I' art dans notre quotidian: "Public space today is so instable and is always floating in an unprecedented persistent state in absence of any sense of place." The "lost space" arisen in modern metropolises may slowly disappear with the isolation and rupture of interaction and sharing of public space, and the same is true for traffic space [7]. With the concerted efforts of the urban wave of information and the rise of data science, China is now vigorously advocating constructing smarter cities, but if the essence of intelligence is not built upon bringing benefit to everyone, the overall concept may seem increasingly implausible.

3.1 Interaction Accessibility in Individual Traffic Space

In the physical barrier of the visually impaired, the blind road becomes one of the most necessary conditions for the blind. In 2015, one hot post on Weibo attracts people to pay great attention to blind "Beijing blind road, why becomes "death road" (Figure 4). The normal way of interaction between the blind and the blind road is terminated by all kinds of damages. In fact, China has the world's longest, most widely distributed blind path in the world. Over the past 30 years, Beijing has more than 1,600 kilometers of blind facilities, But the visible or intangible obstacles still lie between the blind and the outside world, there is no smooth interactive space for them at all. Through oral interviews with 50 normal vision groups, 90 percent said that they had never seen the blind walk on the road.

Figure 4. "Death road for the blind"

3.2 Interaction Accessibility in Public Traffic Space

App Smart Bus has changed the traditional operation mode of bus via utilizing advanced GPS positioning technology, 3G communications technology, and GIS (Geographic Information System) technology, and can intelligently arrange the operating vehicles by combining with the operation characteristics of buses. People can track bus arrival at any time through the mobile App. As intelligence has replaced manual work, the stop-reporting service of traditional ticket staff is beingreplaced smart display screens at bus stops, but the visually impaired can neither send feedbacks to the stop-reporting system on vehicles via listening in such a noisy environment nor know the exact information of arriving buses via visually reading the smart screen, and Smart Bus on mobile also lacks the function of voice stop reporting. It is not until May 2015 that the smart bus blind-aid system was first promoted in Shenzhen, Guangdong Province, which includes blind-aid control devices, bus stop information, cloud bus information databases, cloud voice information databases and cloud mapping data. This system can offer practical, effective blind-aid service for the visually impaired via voice plus button interaction as well as cloud bus information in the system, but this system has hitherto been popularized in only a few cities.

3.3 Interaction Accessibility in Metro Traffic Space

Metro station is a quite typical public transportation space with a large flow of people and a relatively special space. First, unlike open space of bus stop, its space is rather confined; second, geographically restricted by spatial distinctiveness, the special space and intelligent devices there have hindered interactive behaviors of the visually impaired in taking metro. Barrier-free design for the visually impaired in Japanese metro has effectively protected their interactive interests. At most metro exits, due to the low-lying terrain, all the upper and lower staircase handrails are made in wavy shape (see Figure 5), and each wave on the handrail matches up with a step, and there are corresponding braille tips on each handrail at the turning point (see Figure 6).

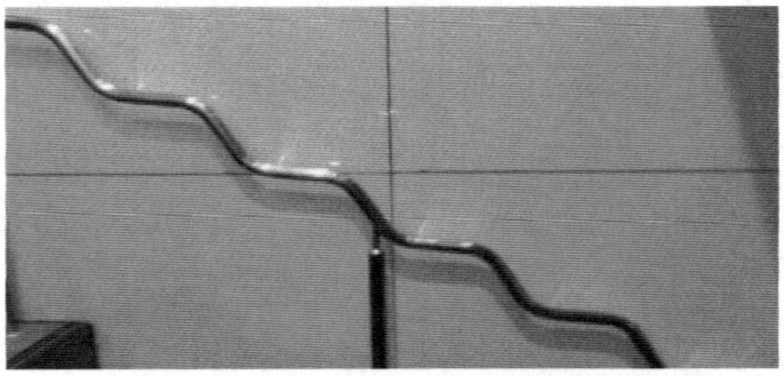

Figure 5. Wavy shape of Handrail

Figure 6. Braille tips on Handrail

Since the blind can recognize the whole metro platform space, in October 2014, several students of Tongji University designed a tactile map for the blind: they adopted 3D tangible dot-line-plane textured symbols to compose the map, in which blind tracks are represented by discontinuous dashed lines, fare gate by two horizontal lines, and passengers' direction by oblique arrow. It can realize the intangible accessibility in the interaction process, and help the visually impaired grasp in-station layout quickly.

The British charity Royal London Society for the Blind (RLSB) and the digital product design company Ustwo jointly developed a smart mobile application named Wayfindr specially for the blind. They have set a number of signal transponders in metro stations, which can connect with signals on matching smart device of the visually impaired, and then remind them of corresponding interior metro structure, barriers and staircases via bone-conduction earphones. Nevertheless, this system requires tremendous investment in manpower and material resources to maintain its normal operation, while the visually impaired also have to bear relevant equipment funds. The launch of this design is far less smooth as planned, but its original design intention of "technology changes lives of the visually impaired" in the premise of technological development is in itself a process of "converting the invisible into the visible", which can help the visually impaired experience the great power of technology and design despite their inability to see the well-developed information world clearly.

4. Create More "Unseen but Known" Designs

The "known and seen" iconic culture puts forward in Gombrich's theory of "iconography" is perfectly in line with the theory of symbolic interaction in information interaction, which indicates that people can transform the symbolic information they see into operation information they know for carrying out subsequent interactive motions, but apparently, this theory cannot be applied to interaction fields of the blind. Both the above-mentioned wavy staircase handrails and the recently developed App Wayfindr have evaded the seen but adopted other senses well known to the blind for design and research. Serious designers are men who never blindly follow the general trend of new technology, for they are experts in recognizing changes. Technology art in this smart era emphasizes human's co-evolution with the technological ecology. Realizing the interaction accessibility is nowadays of great urgency, so designers should create more excellent "unseen but known" designs to bring light to the darkness sensed by the blind in this unfamiliar, smart society.

References

[1] Lei Sun 2013 Action, Ethic and Public Space: Research on Hannah Arendt's Communicative Political Philosophy (Bei Jing: Beijing Normal University Publishing House) p 15-18
[2] Bing Long, Mingzi Wang J 2015 Technical Routes of Urban Village Reconstruction Planning in the Big-Data Era. Architecture and culture, (4) 3
[3] José Saramago 2014 Ensaio sobre a Cegueira (Hai Nan:Nan Hai Publish Co.) p 2-6
[4] Correspondence to: Silvio P. Mario, World Health Organization, 20 Avenue Appia, 1211 Geneva 27, Switzerland
[5] Press Release WHO/46 6 September 1999
[6] Bowman 2011 Modernity and the Holocaust (Jiang Su:Yi Lin Press) p 1-5
[7] Catherine Grew 2005 Art involved in the space. (Guang Xi: Guangxi Normal University Press)

Institute of Physics
Dirac House, Temple Back
Bristol BS1 6BE UK

ISSN: 1757-8981
ISBN 978-1-5108-5498-7
9 781510 854987

CURRAN ASSOCIATES INC.
proceedings.com